Birkhäuser

Advanced Courses in Mathematics
CRM Barcelona

Centre de Recerca Matemàtica

Managing Editor:
Carles Casacuberta

More information about this series at http://www.springer.com/series/5038

Gebhard Böckle • David Burns • David Goss
Dinesh Thakur • Fabien Trihan • Douglas Ulmer

Arithmetic Geometry over Global Function Fields

Editors for this volume:
Francesc Bars (Universitat Autònoma de Barcelona)
Ignazio Longhi (Xi'an Jiaotong-Liverpool University)
Fabien Trihan (Sophia University, Tokyo)

 Birkhäuser

Gebhard Böckle
Interdisciplinary Center for Scientific Computing
Universität Heidelberg
Heidelberg, Germany

David Burns
Department of Mathematics
King's College London
London, UK

David Goss
Department of Mathematics
The Ohio State University
Columbus, OH, USA

Dinesh Thakur
Department of Mathematics
University of Rochester
Rochester, NY, USA

Fabien Trihan
Department of Information
   and Communication Sciences
Sophia University
Tokyo, Japan

Douglas Ulmer
School of Mathematics
Georgie Institute of Technology
Atlanta, GA, USA

ISSN 2297-0304   ISSN 2297-0312 (electronic)
ISBN 978-3-0348-0852-1   ISBN 978-3-0348-0853-8 (eBook)
DOI 10.1007/978-3-0348-0853-8
Springer Basel Heidelberg New York Dordrecht London

Library of Congress Control Number: 2014955449

Mathematics Subject Classification (2010): Primary: 11R58; Secondary: 11B65, 11G05, 11G09, 11G10, 11G40, 11J93, 11R23, 11R65, 11R70, 14F05, 14F43, 14G10, 33E50

Printed on acid-free paper

Springer Basel is part of Springer Science+Business Media (www.birkhauser-science.com)

Contents

Cohomological Theory of Crystals over Function Fields and Applications

Gebhard Böckle

On Geometric Iwasawa Theory and Special Values of Zeta Functions

David Burns and Fabien Trihan
with an appendix by Francesc Bars

The Ongoing Binomial Revolution

David Goss

Arithmetic of Gamma, Zeta and Multizeta Values for Function Fields

Dinesh S. Thakur

Curves and Jacobians over Function Fields

Douglas Ulmer

Foreword

This book collects the notes from five advanced courses presented at the Centre de Recerca Matemàtica (CRM), February 22nd to March 5th and April 6th to 16th 2010, as part of the year-long Research Programme in Arithmetic Geometry. The January to April term of the thematic year was devoted to (global) function fields of characteristic $p > 0$, with special emphasis on topics related to the arithmetic of L-functions, and this is the general subject of the texts collected in this volume.

In the function field context, L-functions appear in many different forms: on the one hand, there are the classical $\mathbb{C}$-valued functions first investigated by Artin and Weil, which inspired Iwasawa theory and for which one can find p-adic avatars; on the other hand, there is a genuine characteristic p theory, which was already pioneered by Carlitz in the 1930s, but really took momentum from the 1970s, following the works of Drinfeld, Hayes and Goss, and which by now includes L- and Γ-functions, modular forms, Bernoulli numbers, motivic and Hodge-theoretic approaches and many results on transcendence of special values. The main focuses of the CRM programme, namely Iwasawa theory and characteristic p L-functions, touched both sides, and both of them are represented in this book, that will hopefully provide a source of stimulation to further explore the beautiful connections and analogies between these aspects of arithmetic.

For characteristic-0-valued L-functions, we have the notes by Ulmer and Burns–Trihan, both dealing with topics close to the theory of abelian varieties over number fields: they give respectively a thorough introduction to the arithmetic of Jacobians over function fields and a state-of-the-art survey of Geometric Iwasawa Theory. The other three courses, by Böckle, Thakur and Goss, examine some of the most important recent ideas in the positive characteristic theory: they cover respectively crystals over function fields (with a number of applications to L-functions of t-motives), gamma and zeta functions in characteristic p, and the binomial theorem. All courses offer the reader a number of open problems, pointing the way to future research.

Let us describe the contents of each contribution in more detail. After a review of basics on Jacobians and the relevant version of the Mordell–Weil theorem (seen under both the Lang–Néron and the Shioda–Tate approach), Ulmer presents the web of conjectures summarizable under the names of Tate, Artin–Tate, and Birch and Swinnerton-Dyer: implications and known results are carefully discussed. Finally, he briefly surveys his own and Berger's work on the construction of elliptic curves and Jacobians with large Mordell–Weil rank in the Kummer tower.

The study of the Birch and Swinnerton-Dyer conjecture in towers and p-adic interpolation of L-functions naturally leads into Iwasawa theory. In their paper, Burns and Trihan explain the proofs of various versions of the Iwasawa Main Conjecture over compact p-adic Lie extensions of global function fields for direct images of flat, smooth $\mathbb{Z}_p$-sheaves on separated schemes and for abelian varieties. They also discuss several concrete consequences concerning the leading terms and values of equivariant zeta functions. An appendix by Francesc Bars introduces some ideas by which one can deal with (abelian) extensions when the Iwasawa algebras are non-noetherian.

The cohomological techniques which lie at the heart of the results in characteristic 0 inspired Böckle and Pink to develop an analogous theory in order to give an algebraic proof of the rationality of Goss' characteristic p L-functions. Böckle's lectures provide an introduction to this theory and a number of applications, such as Anderson's trace formula, meromorphy of L-functions, special values at negative integers, a new proof of the Goss–Sinnott theorem and a cohomological description of Drinfeld modular forms allowing one to attach Galois representations to them.

Thakur's contribution yields a wealth of information on the many analogues of the gamma and zeta functions in positive characteristic. In particular, sundry versions of gamma are introduced: for each, as for the zeta functions, the author discusses basic properties, interpolations, functional equations, and special values. The algebraic relations among the latter are studied by various methods and connections with periods of Drinfeld modules and t-motives are indicated. The author also briefly surveys Taelman's class number formula and his own work on multizeta values. Plenty of open questions make these lectures an exciting and stimulating reading.

After sketching the history of the binomial theorem, Goss' short paper describes applications to non-archimedean and characteristic p analysis. Moreover this apparently elementary result yields new insight into automorphisms of $\mathbb{Z}_p$ and possibly in symmetries of characteristic p L-series.

Next to the advanced courses collected in this volume, the Function Field term of the CRM Research Programme hosted a number of activities: we mention here two workshops (April 6–10 and 12–16) and various series of seminars.[1] We would like to thank all participants for their contributions, and in particular U. Hartl and A. Pál for having graciously offered a series of lectures each.

The Research Programme and the sub-programme on Function Fields were supported by the Spanish project Ingenio Mathematica (i-MATH) and by the Centre de Recerca Matemàtica. We particularly want to express our gratitude to the Director and all the staff of the Centre de Recerca Matemàtica for their excellent job. We would also like to thank the NSF for grant DMS-0968709 to Douglas Ulmer which helped students and postdocs to attend the different activities.

Francesc Bars, Ignazio Longhi and Fabien Trihan

---

[1]Function field arithmetic has seen a number of dramatic advances in recent years: for an introduction to some aspects which were also touched upon in the Barcelona activities, but are absent from this book, the reader might look at the proceedings (U. Hartl, ed.; in preparation) of the 2009 workshop *t-motives: Hodge structures, transcendence and other motivic aspects* held at the Banff International Research Station (BIRS).

Cohomological Theory of Crystals over Function Fields and Applications

Gebhard Böckle

Cohomological Theory of Crystals over Function Fields and Applications

Gebhard Böckle

Introduction

This lecture series introduces in the first part a cohomological theory for varieties in positive characteristic with finitely generated rings of this characteristic as coefficients developed jointly with Richard Pink. In the second part various applications are given.

The joint work with Pink was carried out in order to give an algebraic proof of the rationality of some L-functions defined by D. Goss. Prior to our work an analytic proof using methods à la Dwork had been given by Y. Taguchi and D. Wan in [45]. Moreover by an approach dual to ours and closer in spirit to D-modules in characteristic zero, one should also be able to obtain an algebraic proof from the work [14] of M. Emerton and M. Kisin.

Our expectation that such a cohomological theory should exist came from the cohomological theory of ℓ-adic étale sheaves developed by Grothendieck and his coworkers to prove the rationality of the ζ-functions introduced and studied by Weil, Hasse et al. In this case there had first been an analytic proof by Dwork. However it was only the cohomological method which in the hands of Deligne eventually led to a full proof of the Weil conjecture.

Let me be more concrete. For a variety X of finite type over $\operatorname{Spec}\mathbb{Z}$ one considers, following Weil, the ζ-function

$$\zeta_X(s) := \prod_{x \in |X|} \left(\left(1 - T^{d_x} \right)^{-1} \right) \Big|_{T = p_x^{-s}}.$$

Here $|X|$ is the set of closed points x of X; for such an x the residue field k_x is a finite field; p_x denotes its characteristic and d_x the degree of k_x over $\mathbb{F}_{p_x}$; s is a complex number. This infinite Euler product converges absolutely for $\Re(s) > \dim X$. It is conjectured that ζ_X has a meromorphic continuation to $\mathbb{C}$. If X is irreducible one furthermore conjectures that ζ_X has at most a simple pole at $s =$

$\dim X$. Except for a few cases, this conjecture is wide open. If X is irreducible and its generic points are of characteristic zero, then $s \mapsto \zeta_X(s)/\zeta_{\mathrm{Spec}\,\mathbb{Z}}(s - \dim X + 1)$ has a holomorphic continuation to $\Re(s) > \dim X - 1/2$; see Exercise 0.1 and [37].

For $X = \mathrm{Spec}\,\mathbb{Z}$, the Euler product ζ_X is simply the Riemann ζ-function. For X the ring of integers of a number field it is the Dedekind ζ-function. For these X it is known that ζ_X has a meromorphic continuation to $\mathbb{C}$ with a simple pole at $s = 1$. Its residue is of arithmetic significance.

We can rearrange the above product as follows. For any prime p let $X_p := X \times_{\mathrm{Spec}\,\mathbb{Z}} \mathrm{Spec}\,\mathbb{F}_p$ be the fiber of X above p. Then the closed points of X_p are precisely the closed points of X with $p_x = p$. Define

$$Z(X_p, T) := \prod_{x \in |X_p|} \left(1 - T^{d_x}\right)^{-1} \in 1 + T\mathbb{Z}[[T]]. \tag{1}$$

This is the Weil ζ-function of X_p. It can also be defined in an entirely different way by counting closed points of X_p over the fields $\mathbb{F}_{p^n}$, $n \to \infty$, namely $Z(X_p, T) = \exp\left(\sum_{r \geq 1} N_r t^r / r\right)$ with $N_r = \#X_p(\mathbb{F}_{p^r})$.

Using $Z(X_p, T)$ it is easy to verify that

$$\zeta_X(s) = \prod_p Z\left(X_p, p^{-s}\right).$$

The Weil conjecture, whose proof has been completed by Deligne [10], makes predictions about $Z(X_p, T)$.

(a) It asserts that $Z(X_p, T)$ is a rational function in T. If X_p is a smooth projective variety over $\mathbb{F}_p$ of dimension n then by Grothendieck et al. more refined assertions are true:

$$Z(X_p, T) = \prod_{j=0}^{2\dim X_p} \det\left(1 - T\,\mathrm{Frob}_p^{-1} \mid H_{\mathrm{et}}^j(X_p, \mathbb{Q}_\ell)\right)^{(-1)^{j+1}}; \tag{2}$$

(b) $Z(X_p, T)$ satisfies the functional equation $Z(X_p, \frac{1}{p^n T}) = \pm p^{En/2} T^E Z(X_p, T)$ where $E = \sum_{j=0}^{2n} (-1)^i B_j$ with $B_j = \dim H^j$, $-$ or, equivalently, E denotes the self intersection number of the diagonal $\Delta \subset X \times X$;

(c) the eigenvalues of Frob_p acting on $H_{\mathrm{et}}^j(X_p, \mathbb{Q}_\ell)$ are algebraic integers all of whose complex absolute values are of size $p^{j/2}$ (they are Weil numbers of weight j).

The equality of the right-hand sides of (1) and (2) is derived from a *Lefschetz trace formula*. It is a key assertion of the cohomological approach toward proving the Weil conjecture. Recommended references are [11, 15, 32, 38].

Exercise 0.1. This exercise may require additional reading on the assertions of the Weil conjecture.

(a) Suppose X_p is irreducible and of dimension n (but not necessarily smooth). Then $N_r - (p^n)^r = O\left(p^{(n-\frac{1}{2})r}\right)$.

(b) Suppose that X is irreducible and generically of characteristic zero. Then $s \mapsto \zeta_X(s)/\zeta_{\mathrm{Spec}\,\mathbb{Z}}(s - \dim X + 1)$ has a holomorphic continuation to $\Re(s) > \dim X - 1/2$.

To explain the situation we will be interested in, we consider a slightly different setting. Let X be as above and let $\mathcal{A} \to X$ be an abelian scheme over X, e.g., an elliptic curve over X. (This means that the morphism is smooth projective and flat, it carries a section – the 0-section of the abelian scheme – and all fibers $\mathcal{A}_x$ are abelian varieties.) For simplicity we assume that $X = X_p$ for some fixed prime p. For the abelian variety $\mathcal{A}_x$ it is known that $H^1_{\mathrm{et}}(\mathcal{A}_x, \mathbb{Q}_\ell)$ is the dual of the ℓ-adic Tate module of $\mathcal{A}_x$, tensored with $\mathbb{Q}_\ell$ over $\mathbb{Z}_\ell$:

$$H^1_{\mathrm{et}}(\mathcal{A}_x, \mathbb{Q}_\ell)^\vee \cong \mathrm{Tate}_\ell(\mathcal{A}_x) \otimes_{\mathbb{Z}_\ell} \mathbb{Q}_\ell.$$

Moreover $H^j_{\mathrm{et}}(\mathcal{A}_x, \mathbb{Q}_\ell) = \Lambda^j H^1_{\mathrm{et}}(\mathcal{A}_x, \mathbb{Q}_\ell)$, the jth exterior power of H^1. An often studied L-function in this context is

$$\prod_{x \in |X|} \det\left(1 - T^{d_x} \mathrm{Frob}_x^{-1} \mid H^1_{\mathrm{et}}(\mathcal{A}_x, \mathbb{Q}_\ell)\right)^{-1}$$
$$= \prod_{x \in |X|} \det\left(1 - T^{d_x} \mathrm{Frob}_x^{-1} \mid \mathrm{Tate}_\ell(\mathcal{A}_x)\right)^{-1} \in 1 + T\mathbb{Z}[[T]].$$

By Grothendieck's more general formulation of the conjectures of Weil, this is also a rational function in T.

An analog of the right-hand side can be defined within the framework of function field arithmetic. Consider a smooth projective curve over a finite field. Let A be the coordinate ring of the affine curve obtained from the projective curve by removing a single closed point. For such A one has the notion of Drinfeld A-module (or more generally A-motive). For every place $\mathfrak{p}$ of A, and any Drinfeld A-module φ (or more generally any A-motive M) one can associate the $\mathfrak{p}$-adic Tate module $\mathrm{Tate}_\mathfrak{p}(\varphi)$. If φ is defined over a finite field k_x, then the corresponding Frobenius endomorphism acts on the Tate module and one obtains

$$\det\left(1 - T^{d_x} \mathrm{Frob}_x^{-1} \mid \mathrm{Tate}_\ell(\varphi)\right) \in 1 + T^{d_x} A[[T^{d_x}]].$$

Suppose now that φ is a Drinfeld A-module over a scheme X of finite type over $\mathbb{F}_p$ – one should think of φ as a family of Drinfeld A-modules over X. For such φ Goss conjectured that

$$L(\varphi, X, T) := \prod_{x \in |X|} \det\left(1 - T^{d_x} \mathrm{Frob}_x^{-1} \mid \mathrm{Tate}_\ell(\varphi_x)\right)^{-1} \in 1 + TA[[T]]$$

is in fact a *rational function over* A. This was first proved by Taguchi and Wan and later by R. Pink and the author.

Goss also defined an analog of the global ζ-function considered above. For this, observe that every Drinfeld A-module φ_x has a characteristic. This yields a morphism of schemes $X \to \operatorname{Spec} A$ (and the same for A-motives). As before, one considers the various fibers $X_{\mathfrak{p}} := X \times_{\operatorname{Spec} A} \operatorname{Spec} A/\mathfrak{p}$ and defines

$$L^{\mathrm{glob}}(\varphi, X, s) := \prod_{\mathfrak{p} \in \mathbf{Max}(A)} L\big(\varphi\big|_{X_{\mathfrak{p}}}, X_{\mathfrak{p}}, T\big)\big|_{T=\mathfrak{p}^{-s}};$$

here $s \in S_\infty := \mathbb{Z}_p \times \mathbb{C}_\infty^*$, which can be regarded as an analog of the complex plane; we skip the definition of $\mathfrak{p}^{-s}$ but note that the product only converges in an analog of a right half-plane. Goss defines what it means for a function

$$S_\infty \to \mathbb{C}_\infty$$

to be *meromorphic* and *essentially algebraic*. It is shown for $A = \mathbb{F}_q[t]$ in the work [45] of Taguchi and Wan and for general A in [3] that the global L-functions of Goss possess these two properties. More will be explained in the upcoming lectures.

The aims of the present lecture series are the following:

- Introduce the cohomological theory of Pink and myself, which is applicable to families Anderson's A-motives (and more generally).
- Prove a Lefschetz trace formula within this theory following an argument by Anderson and obtain an algebraic proof of Goss' conjectures on L-functions of families of t-motives.
- Discuss the following topics related to the above theory:
  (a) A cohomological formula for special values of Goss' global L-function at negative integers.
  (b) Goss' conjectures about the meromorphy of global L-functions as well as results and conjectures on the distribution of their zeros.
  (c) The link between the theory of Pink and myself to the étale theory of sheaves of $\mathbb{F}_p$-vector spaces on varieties X of characteristic p.
  (d) An alternative proof of a theorem of Goss and Sinnott on the relation between components on the class groups of torsion fields of Drinfeld modules and the divisibility of L-values.
  (e) The association of Galois representations (or more general A-motive like objects) to Drinfeld modular forms.

References

A detailed account of the cohomological theory treated in this course is given in the monograph [8]. The results on meromorphy of global L-functions are from [3]. The cohomological treatment of Drinfeld modular forms stems from [4]. A very important article regarding a trace formula for Goss' L-function is Anderson's [2]. Much background on Drinfeld modules and t-motives can be found in [24]. Another rich source is [49]. Further references are given throughout the text. Some of the results we present have not yet appeared in print or preprint form.

Acknowledgment

I would like to thank the CRM at Barcelona for the invitation to present this lecture series during an advanced course on function field arithmetic from February 22 to March 5, 2010 and for the pleasant stay at CRM in the spring of 2010 during which a preliminary version of these lecture notes were written. I also thank the NCTS in Hsinchu, Taiwan, and in particular Chieh-Yu Chang and Winnie Li for inviting me in September 2010 for giving another lecture series on the above results. It much helped with the revisions of the original notes. I thank D. Thakur for his many remarks on the Goss L-function of the Carlitz module. For help with the correction of a preliminary version, I thank A. Karumbidza and I. Longhi. I acknowledge financial support by the Deutsche Forschungsgemeinschaft through the SFB/TR 45.

Notation

Let p be a prime number and q a power of p. We fix a finite field k with q elements. All schemes X, Y, Z, U etc. are assumed to be noetherian and separated over k. All morphisms, fiber products, tensor products of modules and algebras are taken over k unless specified otherwise. By a (quasi-)coherent sheaf on a scheme X we will always mean a (quasi-)coherent sheaf of $\mathcal{O}_X$-modules. Any homomorphism of such sheaves is assumed to be $\mathcal{O}_X$-linear, and any tensor product of such sheaves is taken over $\mathcal{O}_X$. The Frobenius morphism on X over k, which acts on functions by $x \mapsto x^q$, is denoted $\sigma \colon X \to X$.

Throughout most of the lectures we fix a scheme C which is assumed to be a localization of a scheme of finite type over k. The notation is intended to reflect the role of C as a <u>C</u>oefficient system. To guarantee the existence of sufficiently many functions we assume that C is affine; thus $C = \operatorname{Spec} A$, where A is a localization of a finitely generated k-algebra. Interesting special cases of such A are the coordinate ring of any smooth affine curve over k, any field which is finitely generated over k, and any finite Artin ring over k.

The assumptions on C imply that $X \times C$ is noetherian for every noetherian scheme X over k. This is useful in dealing with coherent sheaves on $X \times C$. As a general rule, sheaves on X will be distinguished from those on $X \times C$ by an index (__)$_0$. Throughout we let $\operatorname{pr}_1 \colon X \times C \to X$ denote the projection to the first factor. For any coherent sheaf of ideals $\mathcal{I}_0 \subset \mathcal{O}_X$ we abbreviate $\mathcal{I}_0 \mathcal{F} := (\operatorname{pr}_1^{-1} \mathcal{I}_0) \mathcal{F}$.

In the special case where C is an irreducible smooth affine curve over k whose smooth compactification is obtained by adjoining precisely one closed point ∞, we define: K as the fraction field of A, K_∞ as the completion of K at ∞ and $\mathbb{C}_\infty$ as the completion of the algebraic closure of K_∞. Similarly, for any place v of K we denote by K_v the completion of K at v and by $\mathcal{O}_v$ the ring of integers of K_v and by k_v the residue field of K_v.

Lecture 1

First Basic Objects

In this lecture we shall introduce τ-sheaves. These are the first building blocks in the theory developed with Pink. We shall see how they arise from (families of) Drinfeld A-modules and A-motives.

1.1 τ-sheaves

Definition 1.1. A τ-*sheaf over* A *on* X is a pair $\underline{\mathcal{F}} := (\mathcal{F}, \tau_{\mathcal{F}})$ where $\mathcal{F}$ is a quasi-coherent sheaf on $X \times C$ and $\tau_{\mathcal{F}}$ is an $\mathcal{O}_{X \times C}$-linear homomorphism

$$(\sigma \times \mathrm{id})^* \mathcal{F} \xrightarrow{\tau_{\mathcal{F}}} \mathcal{F}.$$

As A remains fixed for the most part, we usually speak of τ-*sheaves on* X.

A *homomorphism* of τ-sheaves $\underline{\mathcal{F}} \to \underline{\mathcal{G}}$ on X is a homomorphism of the underlying sheaves $\varphi \colon \mathcal{F} \to \mathcal{G}$ such that

$$
\begin{array}{ccc}
(\sigma \times \mathrm{id})^* \mathcal{F} & \xrightarrow{\tau_{\mathcal{F}}} & \mathcal{F} \\
{\scriptstyle (\sigma \times \mathrm{id})^* \varphi} \downarrow & & \downarrow {\scriptstyle \varphi} \\
(\sigma \times \mathrm{id})^* \mathcal{G} & \xrightarrow{\tau_{\mathcal{G}}} & \mathcal{G}
\end{array}
$$

commutes.

The sheaf underlying a τ-sheaf $\underline{\mathcal{F}}$ will always be denoted $\mathcal{F}$. We will mostly abbreviate $\tau = \tau_{\mathcal{F}}$ (if the underlying sheaf is clear from the context).

Exercise 1.2. Let $f \colon Y \to X$ be a morphism of schemes and let $\mathcal{F}$ and $\mathcal{G}$ be sheaves of $\mathcal{O}_X$- and $\mathcal{O}_Y$-modules, respectively. Prove that there is a natural isomorphism

$$\mathrm{Hom}_{\mathcal{O}_Y}(f^* \mathcal{F}, \mathcal{G}) = \mathrm{Hom}_{\mathcal{O}_X}(\mathcal{F}, f_* \mathcal{G})$$

of abelian groups, called *adjunction*. In the case where X and Y are affine schemes, reformulate adjunction in terms of modules.

Exercise 1.3.

(i) Suppose that N is an S-module and $\tilde{\sigma} \colon S \to S$ is a ring homomorphism. Denote by $N_{\tilde{\sigma}}$ the same underlying abelian group however with S acting via $\tilde{\sigma}$, i.e., $s \cdot_{\tilde{\sigma}} n := \tilde{\sigma}(s) \cdot n$. Call a morphism $\alpha \colon N \to N$ $\tilde{\sigma}$-linear if $\alpha(sn) = \tilde{\sigma}(s)\alpha(n)$. Denote by $S[\tilde{\sigma}]$ the not necessarily commutative polynomial ring in $\tilde{\sigma}$ over S with the commutation rule $\tilde{\sigma}s = \tilde{\sigma}(s)\tilde{\sigma}$ for $s \in S$. Then the following are equivalent for a map $\alpha \colon N \to N$.

   (a) α is $\tilde{\sigma}$-linear.

   (b) $\alpha \colon N \to N_{\tilde{\sigma}},\ n \mapsto \alpha(n)$ is linear.

   (c) $\alpha^{\mathrm{lin}} \colon S^{\tilde{\sigma}} \otimes_S N \to N,\ s \otimes n \mapsto s\alpha(n)$ is S-linear where the map $S \to S$ used in the tensor product is $\tilde{\sigma}$.

   (d) N is a left $S[\tilde{\sigma}]$-module via $\sum s_i \tilde{\sigma}^i \cdot n = \sum_i s_i \alpha^i(n)$, i.e., via the unique action extending that of S on N so that $\alpha(n) = \tilde{\sigma} \cdot n$.

(ii) On any affine chart $\operatorname{Spec} R \subset X$ a τ-sheaf over A corresponds to an $(R \otimes A)$-module M together with a $\sigma \otimes \mathrm{id}$-linear homomorphism $\tau \colon M \to M$. In other words, it corresponds to a left module over the non-commutative polynomial ring $(R \otimes A)[\tau]$, defined by the commutation rule $\tau(u \otimes a) := (u^q \otimes a)\tau$ for all $u \in R$ and $a \in A$.

Definition 1.4. The category formed by all τ-sheaves over A on X and with the above homomorphisms is denoted $\mathbf{QCoh}_\tau(X, A)$. The full subcategory of all *co-herent* τ-sheaves (those $\underline{\mathcal{F}}$ for which $\mathcal{F}$ is coherent) is denoted $\mathbf{Coh}_\tau(X, A)$.

The above two categories are abelian A-linear categories, and all constructions like kernel, cokernel, etc. are the usual ones on the underlying quasi-coherent sheaves, with the respective τ added by functoriality. In particular, the formation of kernel, cokernel, image and coimage is preserved under the inclusions $\mathbf{Coh}_\tau(X, A) \subset \mathbf{QCoh}_\tau(X, A)$.

Exercise 1.5. Find an example of a non-zero τ-sheaf which contains no coherent τ-subsheaf except for 0. Show that any quasi-coherent sheaf (without τ) is the direct limit of its coherent subsheaves.

Because of the above example and for various technical reasons, in [8] we also introduce the category of ind-coherent τ-sheaves, i.e., τ-sheaves which are the filtered direct limit of their coherent τ-subsheaves.

Proposition 1.6.

   (a) $\mathbf{Coh}_\tau(X, A) \subset \mathbf{QCoh}_\tau(X, A)$ *is a Serre subcategory* (*see Definition* 2.9).

   (b) $\mathbf{QCoh}_\tau(X, A)$ *is a Grothendieck category, i.e., it is closed under exact filtered direct limits and it possesses a generator: an element* $\underline{\mathcal{U}}$ *such that every element of* $\mathbf{QCoh}_\tau(X, A)$ *is a quotient of* $\oplus_I \underline{\mathcal{U}}$ *for some index set* I.

The proof of the first part is obvious; that of the second can be found in [8, Theorem 3.2.7]

1.2 (Algebraic) Drinfeld A-modules

Throughout this section, we assume that $C = \operatorname{Spec} A$ is an irreducible smooth curve over k whose smooth compactification $\bar{C}$ is obtained by adjoining precisely one closed point called ∞. Our prime example will be $A = k[t]$. For any non-zero element $a \in A$, we set $\deg(a) := \log_q \#(A/Aa)$.

By a line bundle L on X we mean a group scheme over X which is Zariski locally isomorphic to the additive group scheme $\mathbb{G}_a \times X$. Its endomorphism ring $\operatorname{End}_k(L)$ consists of all k-linear endomorphisms as a group scheme over X. If $X = \operatorname{Spec} R$ is affine and L is trivial (and thus $L = \operatorname{Spec} R[x]$), one can identify $\operatorname{End}_k(L)$ with the non-commutative polynomial ring $R[\tau]$ defined by the commutation rule $\tau u := u^q \tau$ for all $u \in R$. Here τ acts on a polynomial $f = \sum r_i x^i \in R[x]$ as $\tau f = \sum a_i x^{qi}$ and thus on $r \in R = \mathbb{G}_a(\operatorname{Spec} R)$ as $\tau(r) = r^q$. This means that τ is simply the Frobenius endomorphism on the sections $\mathbb{G}_a(\operatorname{Spec} R) = R$.

For arbitrary X, let $\mathcal{L}$ denote the invertible sheaf of sections of L over X. Since Frobenius is exponentiation to the power q on sections, it defines a homomorphism $\mathcal{L} \to \mathcal{L}^{\otimes q}$, and thus one obtains a q-linear homomorphism only if composed with a linear homomorphism $\mathcal{L}^{\otimes q} \to \mathcal{L}$. From this one deduces that $\operatorname{End}_k(L)$ is isomorphic to the module of global sections of $\oplus_{n \geq 0} \mathcal{L}^{\otimes(q^n-1)}$.

Definition 1.7. A *Drinfeld A-module of rank $r > 0$ on X* consists of a line bundle L on X and a ring homomorphism $\varphi \colon A \to \operatorname{End}_k(L)$, $a \mapsto \varphi_a$, such that for all points $x \in X$ with residue field k_x the induced map

$$\varphi_x \colon A \longrightarrow \operatorname{End}_k(L|x) \cong k_x[\tau], \quad a \longmapsto \sum_{i=0}^{\infty} u_i(a)\tau^i$$

has coefficients $u_i(a) = 0$ for $i > r\deg(a)$ and $u_{r\deg(a)}(a) \in k_x^*$.

A *homomorphism* $(L, \varphi) \to (L', \varphi')$ of Drinfeld A-modules over X is a homomorphism of line bundles $L \to L'$ that is equivariant with respect to the actions φ and φ'. It is called an *isogeny* if it is non-zero on any connected component of X. The latter implies that its kernel is a finite subgroup scheme of L.

The *characteristic of* (L, φ) is the morphism of schemes $\operatorname{Char}_\varphi \colon X \to C$ corresponding to the ring homomorphism

$$d\varphi \colon A \to \operatorname{End}_{\mathcal{O}_X}(\operatorname{Lie}(L)) \cong \Gamma(X, \mathcal{O}_X).$$

A Drinfeld module over a field is of *generic characteristic* if $d\varphi_a$ is invertible for all $a \in A \smallsetminus \{0\}$; else $\operatorname{Ker}(d\varphi)$ is a non-zero prime ideal of A, and the Drinfeld module is called of *special characteristic* $\operatorname{Ker}(d\varphi)$. For arbitrary X, an element $a \in A$ is prime to the characteristic of φ if $d\varphi_a$ is a unit in $\Gamma(X, \mathcal{O}_X)$.

Exercise 1.8. Verify that the above definitions agree with the "usual ones" in the case $X = \operatorname{Spec} F$ for any field F of characteristic p.

Example 1.9. In the special case where $X = \operatorname{Spec} R$ is affine and L is trivial over X, any Drinfeld A-module is isomorphic to one in the standard form

$$\varphi \colon A \longrightarrow R[\tau], \quad a \longmapsto \sum_{i=0}^{r \deg(a)} u_i(a)\tau^i,$$

where $u_{r \deg(a)}(a)$ is a unit in R for all $a \in A \smallsetminus \{0\}$. The characteristic of φ is the morphism corresponding to the ring homomorphism $A \to R$, $a \mapsto u_0(a)$.

An isogeny $\varphi \to \varphi'$ between Drinfeld modules in standard form over $\operatorname{Spec} R$ is given by some $\psi \in R[\tau]$ with leading coefficient a unit in R such that $\varphi'_a\psi = \psi\varphi_a$ for all $a \in A$. Such a ψ defines an isomorphism if its degree is zero.

Further results on Drinfeld modules such as their analytic definition via lattices, a discussion of their torsion points and the existence of isogenies and on Drinfeld–Hayes modules can be found in Appendix 10.7.

1.3   *A*-motives

The following construction due to Drinfeld attaches a coherent τ-sheaf to any Drinfeld A-module (L, φ) of rank r on X. The functor

$$U \longmapsto \operatorname{Hom}_k\big(L|U, \ \mathbb{G}_a \times U\big)$$

defines a quasi-coherent sheaf of $\mathcal{O}_X$-modules on X. Letting each $a \in A$ act via right composition with φ_a defines on it the structure of a sheaf of $(\mathcal{O}_X \otimes A)$-modules. Let $\mathcal{M}(\varphi)$ be the corresponding quasi-coherent sheaf of $(\mathcal{O}_{X \times C})$-modules on $X \times C$.

Example 1.10. In Example 1.9 the module underlying $\mathcal{M}(\varphi)$ is $M(\varphi) := R[\tau]$. Here R and τ act by left multiplication, and $a \in A$ by right multiplication with φ_a. It is easy to see that $M(\varphi)$ is finitely generated over $R \otimes A$. In the special case $A = k[t]$ it is free over $R \otimes A \cong R[t]$ with basis $\{1, \tau, \tau^2, \ldots, \tau^{r-1}\}$. If $\varphi_t = \theta + \alpha_1 + \cdots + \alpha_r\tau^r$ with $\alpha_r \in R^*$, then the matrix representing τ is given by

$$\tau = \begin{vmatrix} 0 & 0 & \cdots & 0 & \frac{t-\theta}{\alpha_r} \\ 1 & 0 & \cdots & 0 & \frac{-\alpha_1}{\alpha_r} \\ 0 & 1 & \cdots & 0 & \frac{-\alpha_2}{\alpha_r} \\ \vdots & \vdots & \ddots & \vdots & \\ 0 & 0 & \cdots & 1 & \frac{-\alpha_{r-1}}{\alpha_r} \end{vmatrix} \big(\sigma_R \times \operatorname{id}_t\big).$$

Exercise 1.11. For any Drinfeld A-module (L, φ) of rank r on X, the sheaf $\mathcal{M}(\varphi)$ is a locally free sheaf on $X \times C$ of rank r. (*Hint:* Reduce to X affine, φ of standard form; treat first $A = \mathbb{F}_q[t]$; reduce the general case to this.)

Let $\sigma \in \text{End}_k(\mathbb{G}_a \times X)$ denote the Frobenius endomorphism relative to X. Left composition with σ defines an $\mathcal{O}_{X \times C}$-linear homomorphism $\mathcal{M}(\varphi) \to (\sigma \times \text{id})_* \mathcal{M}(\varphi)$, and thus via adjunction an $\mathcal{O}_{X \times C}$-linear homomorphism

$$\tau \colon (\sigma \times \text{id})^* \mathcal{M}(\varphi) \longrightarrow \mathcal{M}(\varphi).$$

The resulting τ-sheaf is denoted $\underline{M}(\varphi)$. This construction is functorial in (L, φ). Moreover $\text{Coker}(\tau)$ is supported on the graph of Char_φ and locally free of rank 1 over X.

The following definition is essentially due to Anderson. We fix a morphism $\text{Char} \colon X \to C$.

Definition 1.12. A *family of A-motives over* X, or short an *A-motive on* X, of *rank* r *and of characteristic* Char is a coherent τ-sheaf $\underline{M}$ on X such that

(a) the underlying sheaf $\mathcal{M}$ is locally free of rank r over $\mathcal{O}_{X \times C}$, and

(b) as subsets of $X \times C$ the support of $\text{Coker}(\tau)$ is a subset of the support of the graph of Char.

When X is the spectrum of a field F, $A = \mathbb{F}_q[t]$ and the module corresponding to $\underline{M}$ is finitely generated over $F[\tau]$, then $\underline{M}$ is a t-motive in the sense of Anderson [1].

Exercise 1.13.

(a) Denote by $\mathcal{O}_X[\tau]$ the sheaf of rings on X defined on affine charts $\text{Spec}\, R \subset X$ by $R[\tau]$ and with the obvious gluing morphisms. Given a Drinfeld module (L, φ) on X we defined $\underline{M}(\varphi)$ as $\mathcal{H}om_{k/X}(L, \mathbb{G}_{a,X})$ with the induced τ and $\mathcal{O}_{X \times C}$-actions. Show that one may recover (L, φ) as follows:

Define a functor on X-schemes $\pi \colon Y \to X$ by assigning to π the value $\text{Hom}_{\mathcal{O}_Y[\tau]}(\pi^* \mathcal{M}(\varphi), \mathcal{O}_Y)$ where $\mathcal{O}_Y$ is considered as a sheaf on Y with the obvious action by $\mathcal{O}_Y[\tau]$. Show that this functor is represented by L on X and that one can recover φ from the A-action on $\underline{M}(\varphi)$. (*Hints:* Exercise 1.3; use affine covers.)

(b) The assignment $(L, \varphi) \to \underline{M}(\varphi)$ defines a contravariant functor which is fully faithful. For $X = \text{Spec}\, F$, the image is the set of those A-motives which are free over $F[\tau]$ of rank 1.

(c) Call a map $\underline{M} \to \underline{N}$ between A-motives on X an *isogeny* if its kernel and cokernel are finite over X. Show that any isogeny has trivial kernel and that $(L, \varphi) \to \underline{M}(\varphi)$ maps isogenies of Drinfeld A-modules to isogenies of A-motives.

(d) Show that for $X = \text{Spec}\, F$ any non-zero subobject of $\underline{M}(\varphi)$ is isogenous to $\underline{M}(\varphi)$, i.e., that it is an object which is irreducible up to isogeny. (*Hint:* use that $\text{End}(\varphi)$ is an order in a division algebra.)

The category of Drinfeld A-modules does not permit the formation of direct sums or tensor products or related operations from linear algebra. The passage to Anderson's t-motives, and more generally to A-motives, adds this missing flexibility.

We will see that A-crystals are even more flexible in that they form an abelian category with tensor product, which possesses a cohomology theory with compact support with many of the usual properties.

Lecture 2

A-crystals

In the previous lecture we introduced the first basic objects. Their definition was natural in light of the definition of families of Drinfeld modules and A-motives. In Section 2.1 we shall revisit the motivation given at the beginning of this lecture series. This will indicate that τ-sheaves are not in all respects suitable for the sought-for cohomological theory. Namely it suggests that we should find a new category built out of τ-sheaves in which those homomorphisms of τ-sheaves whose kernel and cokernel have nilpotent endomorphism τ into isomorphisms. The formal procedure to obtain this category is localization. We briefly recall this in Sections 2.2 and 2.3 and refer to [8, § 1.2] for further details and further references. In Section 2.4 we introduce the important notions of nilpotent τ-sheaf and of nil-isomorphism. Their understanding is a prerequisite to Section 2.5 where we introduce the category of A-crystals. This is the category for which we shall in the following lectures investigate the cohomological formalism introduced in [8].

From this section on, until the end of Lecture 7, the symbol A will again denote the localization of a finitely generated k-algebra.

2.1 Motivation II

The objects of $\mathbf{Coh}_\tau(X, A)$ are pairs of a coherent $\mathcal{O}_{X \times C}$-module and an endomorphism. For such pairs, the definition of inverse image, $\otimes$ and direct image can be defined in an obvious way, and we will do this later. One problem with direct image is that coherence is not preserved. But this does not come unexpectedly: already direct image between categories of quasi-coherent sheaves does not preserve coherence.

What is lacking at this point?

- For a trace formula, we need a functor $Rf_!$, direct image with proper support, for any morphism f of finite type. The standard construction in the setting of schemes is $Rf_! = R\bar{f}_* \circ j_!$ where $f = \bar{f} \circ j$ with $\bar{f}$ proper and j an open immersion. (Such a factorization is called a (relative) compactification.) It remains to define $j_!$ for j an open immersion.

Note that $j_!$ from quasi-coherent sheaves is not a useful functor here, since $j_!$ of a coherent sheaf is not necessarily quasi-coherent. On the other hand, in the present setting we would like $j_!$ to preserve coherence.

- If we are mainly interested in L-functions, we should regard pairs $(\mathcal{F}, 0)$ with the zero morphism as zero. More generally we should regard pairs $(\mathcal{F}, \tau)$ with τ nilpotent as zero!
- Expanding on the previous example, we might like to regard a homomorphism $\underline{\mathcal{F}} \to \underline{\mathcal{G}}$ as an isomorphism if its kernel and cokernel have nilpotent τ.
- We would like to have a simple categorical characterization of objects $\underline{\mathcal{F}}$ to which we can attach an L-function, i.e., to which we can attach a pointwise L-factor at all closed points.

For this further motivation, suppose that X is of finite type over $\mathbb{F}_q$ and that $\mathcal{F}$ is the pullback of a coherent sheaf $\mathcal{F}_0$ on X. To any such $\underline{\mathcal{F}}$ one can assign an L-function as a product of pointwise L-factors as follows.

For any $x \in |X|$ let k_x denote its residue field and d_x its degree over k. Then the pullback $\mathcal{F}_x$ of $\mathcal{F}$ to $x \times C$ is equal to the pullback of $\mathcal{F}_0$ from X to x, pulled back under $x \times C \to x$; hence it corresponds to a free $(k_x \otimes A)$-module of finite rank M_x. The induced homomorphism $\tau_x \colon (\sigma_x \times \mathrm{id})^* \mathcal{F}_x \to \mathcal{F}_x$ corresponds to a $\sigma_x \otimes \mathrm{id}_A$-linear endomorphism $\tau_x \colon M_x \to M_x$. The iterate $\tau_x^{d_x}$ of the latter is $k_x \otimes A$-linear, and one can prove that

$$\det{}_{k_x \otimes A}\left(\mathrm{id} - t^{d_x}\tau_x^{d_x} \mid M_x\right) = \det{}_A\left(\mathrm{id} - t\tau_x \mid M_x\right).$$

This is therefore a polynomial in $1 + t^{d_x} A[t^{d_x}]$. Since there are at most finitely many $x \in |X|$ with fixed d_x, the following product makes sense:

Definition 2.1. The *naive L-function* of $\underline{\mathcal{F}}$ is

$$L^{\mathrm{naive}}(X, \underline{\mathcal{F}}, T) := \prod_{x \in |X|} \det{}_A\left(\mathrm{id} - T\tau_x \mid M_x\right)^{-1} \in 1 + TA[[T]].$$

For the trace formula suppose first that X is proper over k. Then for every integer i the coherent cohomology group $H^i(X, \mathcal{F}_0)$ is a finite-dimensional vector space over k. Moreover, the equality $\mathcal{F} = \mathrm{pr}_1^* \mathcal{F}_0$ yields a natural isomorphism $H^i(X \times C, \mathcal{F}) \cong H^i(X, \mathcal{F}_0) \otimes A$. This is therefore a free A-module of finite rank. It also carries a natural endomorphism induced by τ; hence we can consider it as a coherent τ-sheaf on $\mathrm{Spec}\, k$, denoted by $H^i(X, \underline{\mathcal{F}})$. The first instance of the trace formula for L-functions then states:

Theorem 2.2. *Suppose A is a domain that is the localization of a finitely generated k-algebra. Then*

$$L^{\mathrm{naive}}(X, \underline{\mathcal{F}}, T) = \prod_{i \in \mathbb{Z}} L^{\mathrm{naive}}\left(\mathrm{Spec}\, k, H^i(X, \underline{\mathcal{F}}), T\right)^{(-1)^i}.$$

A standard procedure to extend this formula to non-proper X is via cohomology with compact support. For this we fix a dense open embedding $j \colon X \hookrightarrow \overline{X}$ into a proper scheme of finite type over k. (The existence of such a compactification is a result due to Nagata; [39, 40] or [36].) We want to extend the given $\underline{\mathcal{F}}$ on X to a coherent τ-sheaf $\widetilde{\underline{\mathcal{F}}}$ on $\overline{X}$ *without changing the L-function*. Any extension whose τ_z on $\widetilde{\mathcal{F}}_z$ is zero for all $z \in |\overline{X} \smallsetminus X|$ has that property, and it is not hard to construct such an extension. In fact, any coherent sheaf on $\overline{X}$ extending $\mathcal{F}$, multiplied by a sufficiently high power of the ideal sheaf of $\overline{X} \smallsetminus X$, does the job. However, there are many choices for this $\widetilde{\underline{\mathcal{F}}}$, and none is functorial. Thus there is none that we can consider a natural extension by zero "$j_!\underline{\mathcal{F}}$" in the sense of τ-sheaves. For the purpose of L-functions, however, any such is a reasonable choice since it satisfies

$$L^{\mathrm{naive}}\left(X, \underline{\mathcal{F}}, T\right) \;=\; L^{\mathrm{naive}}\left(\overline{X}, \widetilde{\underline{\mathcal{F}}}, T\right).$$

One can in fact show the following: Given any two extensions $\widetilde{\underline{\mathcal{F}}}_1$, $\widetilde{\underline{\mathcal{F}}}_2$ of $\underline{\mathcal{F}}$ to X, there exists a third one $\widetilde{\underline{\mathcal{F}}}_3$ and injective homomorphisms of τ-sheaves $\varphi_i \colon \widetilde{\underline{\mathcal{F}}}_3 \to \widetilde{\underline{\mathcal{F}}}_i$, $i = 1, 2$, whose cokernels have nilpotent τ. In particular one would like to regard all such $\underline{\mathcal{F}}_i$ as isomorphic.

Ignoring the ambiguity in the definition of $j_!\underline{\mathcal{F}}$ for the moment, let us nevertheless provisionally regard $H^i(\overline{X}, \widetilde{\underline{\mathcal{F}}})$ as the cohomology with compact support $H^i_c(X, \underline{\mathcal{F}})$. Then from Theorem 2.2 we obtain a more general trace formula:

Theorem 2.3. *Suppose A is a domain that is the localization of a finitely generated k-algebra. Then*

$$L^{\mathrm{naive}}(X, \underline{\mathcal{F}}, T) \;=\; \prod_{i \in \mathbb{Z}} L^{\mathrm{naive}}\left(\mathrm{Spec}\, k, H^i_c(X, \underline{\mathcal{F}}), T\right)^{(-1)^i}.$$

Since the factors on the right-hand side are polynomials in $1 + tA[t]$ or inverses of such polynomials, the *rationality* of $L^{\mathrm{naive}}(X, \underline{\mathcal{F}}, T)$ is an immediate consequence.

The order of presentation of the above theorems is for expository purposes only. In [8, Ch. 8], Theorem 2.3 is proved first when X is regular and affine over k and then generalized to arbitrary X by dévissage. The proof in the affine case is based on a trace formula by Anderson from [2]. While Anderson formulated it only for $A = k$, Taguchi and Wan [46] already noted that it holds whenever A is a field, and [8, Ch.s 8 and 9] extends it further. Also, the formula in [2] is interpreted in [8] as the Serre dual of the one in Theorem 2.3. This explains the absence of cohomology in the trace formula given in [2].

2.2   Localization

Let $\mathfrak{C}$ be a category and let $\mathcal{S}$ denote a collection of morphisms in $\mathfrak{C}$. Morphisms in $\mathcal{S}$ are drawn as double arrows $\Longrightarrow$ to distinguish them from arbitrary morphisms $\longrightarrow$ in $\mathfrak{C}$.

Definition 2.4. The collection $\mathcal{S}$ is a *multiplicative system* if it satisfies the following three axioms:

(a) $\mathcal{S}$ is closed under composition and contains the identity morphism for every object of $\mathfrak{C}$.

(b) For any $t\colon N' \Rightarrow N$ in $\mathcal{S}$ and any $f\colon M \to N$ in $\mathfrak{C}$, there exist $s\colon M' \Rightarrow M$ in $\mathcal{S}$ and $f'\colon M' \to N'$ in $\mathfrak{C}$ such that the following diagram commutes:

$$
\begin{array}{ccc}
M' & \xrightarrow{\ f'\ } & N' \\
{\scriptstyle s}\big\Downarrow & & \big\Downarrow{\scriptstyle t} \\
M & \xrightarrow{\ f\ } & N.
\end{array}
$$

The same statement with all arrows reversed is also required.

(c) For any pair of morphisms $f, g\colon M \to N$ the following are equivalent:

    (i) There exists $s \in \mathcal{S}$ such that $sf = sg$.

    (ii) There exists $t \in \mathcal{S}$ such that $ft = gt$.

Suppose $\mathcal{S}$ is a multiplicative system. Then one constructs a new category $\mathcal{S}^{-1}\mathfrak{C}$ as follows:

The objects are those of $\mathfrak{C}$, i.e., $\mathrm{Ob}(\mathcal{S}^{-1}\mathfrak{C}) := \mathrm{Ob}(\mathfrak{C})$.

Morphisms will be equivalence classes of certain diagrams. A *right fraction from M to N* is a diagram

$$
M \xrightarrow{\ f\ } L \overset{s}{\Longleftarrow} N
$$

in $\mathfrak{C}$ with $s \in \mathcal{S}$. Two right fractions $M \xrightarrow{f_i} L_i \overset{s_i}{\Longleftarrow} N$ are called *equivalent* if there is a commutative diagram

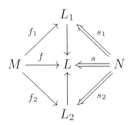

in $\mathfrak{C}$ with $s \in \mathcal{S}$.

Exercise 2.5. Using the axioms 2.4, show that this defines an equivalence relation on the class of all right fractions from M to N.

One defines
$$\mathrm{Hom}_{\mathcal{S}^{-1}\mathcal{C}}\left(M, N\right)$$
as the set of equivalence classes of right fractions, provided the following set the-oretic condition is satisfied:

Definition 2.6. $\mathcal{S}$ is called *essentially locally small* if for all M and N the class of right fractions from M to N possesses a set of representatives.

Due to the symmetry in the definition of multiplicative systems one can also work with left fractions $M \Leftarrow L \to N$. The axioms 2.4 imply that every equivalence class of left fractions corresponds to a unique equivalence class of right fractions, and vice versa, so that one obtains the same result. Using axiom 2.4 (b) one defines the composition of right fractions. Thus under the condition 2.6 one obtains a well-defined category $\mathcal{S}^{-1}\mathcal{C}$, called the *localization of $\mathcal{C}$ by $\mathcal{S}$*.

There is a natural *localization functor* $q \colon \mathcal{C} \to \mathcal{S}^{-1}\mathcal{C}$ mapping any object to itself and any morphism $M \xrightarrow{f} N$ to the equivalence class of the right fraction $M \xrightarrow{f} N \xLeftarrow{\mathrm{id}} N$. To distinguish morphisms in $\mathcal{S}^{-1}\mathcal{C}$ from those in $\mathcal{C}$ we often denote them by dotted arrows $\dashrightarrow$. A morphism of the form $q(f)$ is, by abuse of notation, also denoted by a solid arrow. We will often abbreviate $\overline{\mathcal{C}} := \mathcal{S}^{-1}\mathcal{C}$ when $\mathcal{S}$ is clear from the context.

Exercise 2.7. Let $\mathcal{S}$ be an essentially locally small multiplicative system $\mathcal{S}$ in a category $\mathcal{C}$.

(a) For every $s \in \mathcal{S}$ the morphism $q(s)$ is an isomorphism in $\overline{\mathcal{C}}$.

(b) For any category $\mathcal{D}$ and any functor $F \colon \mathcal{C} \to \mathcal{D}$ such that $F(s)$ is an iso-morphism for all $s \in \mathcal{S}$, there exists a unique functor $\bar{F} \colon \overline{\mathcal{C}} \to \mathcal{D}$ such that $F = \bar{F} q$.

(c) If $\mathcal{C}$ is an additive category, then so is $\overline{\mathcal{C}}$ and q is an additive functor. In this case, if the functor F in (b) is additive, then so is $\bar{F}$.

(d) Suppose $\mathcal{S}$ and $\mathcal{S}'$ are multiplicative systems of categories $\mathcal{C}$ and $\mathcal{C}'$, respec-tively. If a functor $F \colon \mathcal{C}' \to \mathcal{C}$ satisfies $F(\mathcal{S}) \subset \mathcal{S}'$, then there is an induced functor $\bar{F} \colon \overline{\mathcal{C}'} \to \overline{\mathcal{C}}$.

Remark: The properties 2.7 (a)–(b) characterize $\overline{\mathcal{C}}$ and q up to equivalence of categories.

Definition 2.8. A multiplicative system $\mathcal{S}$ is *saturated* if, in addition to 2.4 (a)–(c), it also satisfies the following condition:

(d) For any morphism $f \colon M \to N$ in $\mathcal{C}$, if there exist morphisms $g \colon N \to N'$ and $h \colon M' \to M$ such that gf and fh are in $\mathcal{S}$, then f is in $\mathcal{S}$.

If $\mathcal{S}$ if saturated, one easily shows that for any morphisms $L \xrightarrow{f} M \xrightarrow{g} N$, if two of f, g, and gf are in $\mathcal{S}$, then so is the third. This property is useful in simplifying arguments.

2.3   Localization for abelian categories

Definition 2.9. A full subcategory $\mathfrak{B}$ of an abelian category $\mathfrak{A}$ which is closed under taking subobjects, quotients, extensions, and isomorphisms, is called a *Serre subcategory*.

Exercise 2.10. ([52, Ex. 10.3.2]) For any abelian category $\mathfrak{A}$ there is a bijection between the class of saturated multiplicative systems $\mathcal{S}$ and the class of Serre subcategories $\mathfrak{B}$. Explicitly, given $\mathfrak{B}$ one defines $\mathcal{S}$ as the class of those morphisms whose kernel and cokernel are in $\mathfrak{B}$. Conversely, given $\mathcal{S}$ one defines $\mathfrak{B}$ as the full subcategory consisting of those objects M of $\mathfrak{A}$ such that $0 \to M$ is in $\mathcal{S}$.

Definition 2.11. An abelian category is called *locally small* if for every object the equivalence classes of subobjects form a set.

Exercise 2.12. Suppose $\mathfrak{A}$ is locally small and let $\mathcal{S}$ be the multiplicative system associated to any Serre subcategory $\mathfrak{B}$. Let $f\colon M \to N$ denote a morphism in $\mathfrak{A}$. Then:

(a) (See [52, Ex. 10.3.2]) $\mathcal{S}$ is essentially locally small, the localized category $\bar{\mathfrak{A}} := \mathcal{S}^{-1}\mathfrak{A}$ is abelian and the functor $q\colon \mathfrak{A} \to \bar{\mathfrak{A}}$ is exact.

(b) (i) The object $q(M) \in \bar{\mathfrak{A}}$ is zero if and only if $M \in \mathfrak{B}$.

   (ii) The morphism $q(f)$ is zero if and only if $\operatorname{Im} f \in \mathfrak{B}$.

   (iii) The morphism $q(f)$ is a monomorphism if and only if $\operatorname{Ker} f \in \mathfrak{B}$.

   (iv) The morphism $q(f)$ is an epimorphism if and only if $\operatorname{Coker} f \in \mathfrak{B}$.

   (v) The morphism $q(f)$ is an isomorphism if and only if both $\operatorname{Ker} f$ and $\operatorname{Coker} f$ are in $\mathfrak{B}$.

(c) If $q(f)$ is an isomorphism, then f can be factored as $f = gh$ where h is an epimorphism with kernel in $\mathfrak{B}$ and h is a monomorphism with cokernel in $\mathfrak{B}$.

(d) Every short exact sequence in $\bar{\mathfrak{A}}$ is isomorphic to the image of a short exact sequence in $\mathfrak{A}$.

(e) Every complex in $\bar{\mathfrak{A}}$ is isomorphic to the image of a complex in $\mathfrak{A}$.

   Next recall that an object $M \in \mathfrak{A}$ is *noetherian* if every increasing sequence of subobjects becomes stationary.

Exercise 2.13. If $M \in \mathfrak{A}$ is noetherian, then $q(M) \in \bar{\mathfrak{A}}$ is noetherian.

Exercise 2.14. Is the category of A-motives on a scheme X abelian?

   Show that the category which is the localization of the category of A-motives at the set of isogenies is an F-linear abelian tensor category and that any morphism is given by a diagram

$$\underline{M} \Longleftarrow \underline{H} \longrightarrow \underline{N}.$$

2.4 Nilpotence

For a τ-sheaf $\underline{F}$ one defines the iterates $\tau_{\underline{F}}^n$ of $\tau_{\underline{F}}$ by setting inductively

$$\tau_{\underline{F}}^0 := \mathrm{id} \quad \text{and} \quad \tau_{\underline{F}}^{n+1} := \tau_{\underline{F}} \circ (\sigma \times \mathrm{id})^* \tau_{\underline{F}}^n.$$

Thus

- $\tau^n : (\sigma^n \times \mathrm{id})^* \mathcal{F} \to \mathcal{F}$ is an $\mathcal{O}_{X \times C}$-linear homomorphism.
- Each $(\sigma^n \times \mathrm{id})^* \underline{F} := \big((\sigma^n \times \mathrm{id})^* \mathcal{F}, (\sigma^n \times \mathrm{id})^* \tau_{\mathcal{F}} \big)$ is a τ-sheaf.
- $\tau_{\underline{F}}^n : (\sigma^n \times \mathrm{id})^* \underline{F} \to \underline{F}$ is a homomorphism of τ-sheaves.

Definition 2.15. (a) A τ-sheaf $\underline{F}$ is called *nilpotent* if $\tau_{\underline{F}}^n$ vanishes for some, or equivalently all, $n \gg 0$.

(b) A τ-sheaf $\underline{F}$ is called *locally nilpotent* if it is a union of nilpotent τ-subsheaves.

The full subcategories of $\mathbf{QCoh}_\tau(X, A)$ formed by all nilpotent *and* coherent, respectively locally nilpotent, τ-sheaves are denoted $\mathbf{Nil}_\tau(X, A) \subset \mathbf{LNil}_\tau(X, A)$.

We have the following inclusions of categories:

$$\begin{array}{ccc}
\mathbf{Nil}_\tau(X, A) & \lhook\joinrel\longrightarrow & \mathbf{LNil}_\tau(X, A) \\
\big\uparrow & & \big\uparrow \\
\mathbf{Coh}_\tau(X, A) & \lhook\joinrel\longrightarrow & \mathbf{QCoh}_\tau(X, A),
\end{array} \qquad (2.1)$$

where furthermore $\mathbf{Nil}_\tau(X, A) = \mathbf{Coh}_\tau(X, A) \cap \mathbf{LNil}_\tau(X, A)$ (essentially by definition).

Remark 2.16. We observe the following obvious fact. Suppose that X is a scheme of finite type over k and $\underline{F}$ is a locally free τ-sheaf. Then, if $\underline{F}$ is nilpotent, its L-function is trivial, i.e., $L(X, \underline{F}, T) = 1$.

Proposition 2.17. *The categories in* (2.1) *are Serre subcategories of* $\mathbf{QCoh}_\tau(X, A)$.

Proof. The non-trivial part is to prove invariance under extensions for the category $\mathbf{LNil}_\tau(X, A)$. For this, consider a short exact sequence $0 \to \underline{F}' \xrightarrow{\alpha} \underline{F} \xrightarrow{\beta} \underline{F}'' \to 0$ in $\mathbf{QCoh}_\tau(X, A)$ with $\underline{F}'$ and $\underline{F}''$ in $\mathbf{LNil}_\tau(X, A)$. We claim that every coherent subsheaf $\mathcal{G}$ is contained in a nilpotent coherent τ-subsheaf of $\underline{F}$. By hypothesis on $\underline{F}''$, the image $\beta(\mathcal{G})$ is contained in a sheaf underlying a nilpotent coherent τ-subsheaf $\underline{G}''$ of $\underline{F}''$. In particular, there exists n such that $\tau_{\mathcal{G}''}^n = 0$.

Hence $\tau^n((\sigma^n \times \mathrm{id})^* \mathcal{G}) \subset \mathcal{F}'$. Now we apply our hypothesis on $\underline{F}'$. It yields a nilpotent coherent τ-subsheaf containing $\tau^n((\sigma^n \times \mathrm{id})^* \mathcal{G})$. In particular there exists n' such that $\tau^{n'}(\tau^n(\mathcal{G})) = 0$. One easily deduces that $\sum_{i=0}^{n+n'} \tau^i((\sigma^i \times \mathrm{id})^* \mathcal{G})$ is a nilpotent coherent τ-subsheaf of $\underline{F}$ which contains $\mathcal{G}$. $\qquad \square$

By Proposition 2.10, the Serre subcategory $\mathbf{LNil}_\tau(X, A)$ defines a corresponding multiplicative system:

Definition 2.18. A homomorphism of τ-sheaves is called a *nil-isomorphism* if both its kernel and its cokernel are locally nilpotent.

Note that by diagram (2.1) a homomorphism of *coherent* τ-sheaves is a nil-isomorphism if and only if its kernel and cokernel are *nilpotent*.

The following characterization of nil-isomorphisms will be useful. Note that inverse image by $\sigma^n \times \mathrm{id}$ always preserves coherence.

Proposition 2.19. *A homomorphism of τ-sheaves $\varphi \colon \underline{F} \to \underline{G}$ is a nil-isomorphism if there exist $n \geq 0$ and a homomorphism of τ-sheaves α making the following diagram commute:*

$$
\begin{array}{ccc}
(\sigma^n \times \mathrm{id})^*\underline{F} & \xrightarrow{\;\tau^n\;} & \underline{F} \\
{\scriptstyle (\sigma^n \times \mathrm{id})^*\varphi}\Big\downarrow & {\scriptstyle \alpha}\nearrow & \Big\downarrow{\scriptstyle \varphi} \\
(\sigma^n \times \mathrm{id})^*\underline{G} & \xrightarrow{\;\tau^n\;} & \underline{G}.
\end{array}
\qquad (2.2)
$$

If $\underline{F}$ and $\underline{G}$ are coherent, the converse is also true.

Proof. For this proof we abbreviate $\underline{H}_{(n)} := (\sigma^n \times \mathrm{id})^*\underline{H}$ for any τ-sheaf $\underline{H}$. We only give the proof of the first assertion. The reader is advised to try to prove the converse by herself. Let us suppose that α exists, so that we have the commutative diagram

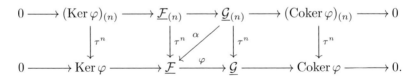

Here the bottom row is exact and, since inverse image on quasi-coherent sheaves is a right exact functor, the top row is a complex whose right half is exact. It is straightforward to deduce that the outer vertical homomorphisms vanish. This shows that $\mathrm{Ker}\,\varphi$ and $\mathrm{Coker}\,\varphi$ are nilpotent; hence φ is a nil-isomorphism, as desired. $\square$

Applying Proposition 2.19 to $\varphi = \tau_{\underline{F}}^n$ and $\alpha = \mathrm{id}_{\underline{F}}$ yields:

Corollary 2.20. *For every τ-sheaf and every $n \geq 0$ the homomorphism $\tau_{\underline{F}}^n \colon (\sigma^n \times \mathrm{id})^*\underline{F} \to \underline{F}$ is a nil-isomorphism.*

Remark 2.21. From 2.16 we deduce the following rather trivial observation. Suppose X is a scheme of finite type over k and $\underline{F} \to \underline{G}$ a homomorphism of locally free τ-sheaves whose kernel and cokernel is locally free as well. Then $L(X, \underline{F}, T) = L(X, \underline{G}, T)$.

2.5 *A*-crystals

By Proposition 2.17 the categories in the upper row of (2.1) are Serre subcategories of the categories in the lower row. Proposition 2.10 identifies the corresponding saturated multiplicative systems with the respective classes of nil-isomorphisms. In this section we introduce the associated localized categories.

Regarding existence, it is shown in [8, Proposition 2.6.1 and Theorem 3.2.7] that $\mathbf{QCoh}_\tau(X, A)$ is a Grothendieck category and so in particular it is locally small. Thus, by Exercise 2.12, localization at the corresponding multiplicative system exists for any Serre subcategory.

Definition 2.22. In the following commutative diagram the lower row is obtained from the upper row by localization with respect to nil-isomorphisms, the vertical arrows are the respective localization functors, and the lower horizontal arrows are obtained from the upper horizontal arrows by the universal property of localization:

$$\begin{array}{ccc} \mathbf{Coh}_\tau(X, A) & \hookrightarrow & \mathbf{QCoh}_\tau(X, A) \\ {\scriptstyle q}\downarrow & & {\scriptstyle q}\downarrow \\ \mathbf{Crys}(X, A) & \longrightarrow & \mathbf{QCrys}(X, A). \end{array}$$

We refer to the objects of $\mathbf{Crys}(X, A)$ as *A-crystals on X*, and to the objects of $\mathbf{QCrys}(X, A)$ as *A-quasi-crystals on X*. As A usually remains fixed, we mostly speak only of *(quasi-)crystals on X*.

Both $\mathbf{Crys}(X, A)$ and $\mathbf{QCrys}(X, A)$ are A-linear abelian categories and the horizontal functor in the bottom row of the diagram is fully faithful.

Remark 2.23. Regarding the notation for arrows, we retain the conventions described above Exercise 2.7. For instance we shall often write $\Longrightarrow$ to denote nil-isomorphisms in $\mathbf{QCoh}_\tau(X, A)$.

There is a standard way to represent homomorphisms of crystals which is derived from Proposition 2.19 for nil-isomorphisms between coherent τ-sheaves.

Proposition 2.24. *Any homomorphism $\varphi \colon \underline{\mathcal{F}} \dashrightarrow \underline{\mathcal{G}}$ in $\mathbf{Crys}(X, A)$ can be represented for suitable n by a diagram*

$$\underline{\mathcal{F}} \overset{\tau^n}{\Longleftarrow} (\sigma^n \times \mathrm{id})^* \underline{\mathcal{F}} \longrightarrow \underline{\mathcal{G}}.$$

Proof. The rather straightforward proof, building on Proposition 2.19, is left to the reader. □

Based on this one can give the following alternative description of crystals. The category $\mathbf{Crys}(X, A)$ has the same objects as the category $\mathbf{Coh}_\tau(X, A)$. Given coherent τ-sheaves $\underline{\mathcal{F}}$ and $\underline{\mathcal{G}}$, the set of morphisms from $\underline{\mathcal{F}}$ to $\underline{\mathcal{G}}$ in $\mathbf{Crys}(X, A)$ is defined as

$$\mathrm{Hom}_{\mathrm{crys}}(\underline{\mathcal{F}}, \underline{\mathcal{G}}) := \left(\bigcup_{n \in \mathbb{N}} \mathrm{Hom}_\tau \left((\sigma^n \times \mathrm{id})^* \underline{\mathcal{F}}, \underline{\mathcal{G}} \right) \right) / \sim,$$

where the equivalence relation $\sim$ is defined as follows: $\varphi \colon (\sigma^n \times \mathrm{id})^* \underline{\mathcal{F}} \to \underline{\mathcal{G}}$ and $\psi \colon (\sigma^m \times \mathrm{id})^* \underline{\mathcal{F}} \to \underline{\mathcal{G}}$ are equivalent if there exists $\ell \geq \max\{m, n\}$ such that

$$\varphi \circ (\sigma^n \times \mathrm{id})^* (\tau^{\ell-n}) = \psi \circ (\sigma^m \times \mathrm{id})^* (\tau^{\ell-m}).$$

Composition of morphisms in $\mathbf{Crys}$ is defined in the obvious way, i.e., the composite of $\varphi \colon (\sigma^n \times \mathrm{id})^* \underline{\mathcal{F}} \to \underline{\mathcal{G}}$ and $\psi \colon (\sigma^m \times \mathrm{id})^* \underline{\mathcal{G}} \to \underline{\mathcal{H}}$ is defined as

$$\psi \circ (\sigma^m \times \mathrm{id})^* \varphi \colon (\sigma^{m+n} \times \mathrm{id})^* \underline{\mathcal{F}} \longrightarrow \underline{\mathcal{H}}.$$

Lecture 3

Functors on τ-sheaves and A-crystals

We indicate the basic construction of all functors on τ-sheaves and A-crystals from [8]. For τ-sheaves these are inverse and direct image, tensor product and change of coefficients. For crystals we have in addition an exact functor extension by zero. Since our approach follows closely the well-known constructions for coherent sheaves we mostly omit details. In Section 3.1 or 3.3, respectively, inverse image and extension by zero on crystals are discussed in greater detail. The inverse image functor has properties different from those known for coherent sheaves. Extension by zero is not derived from a functor on coherent sheaves.

3.1 Inverse image

We fix a morphism $f\colon Y \to X$.

Definition 3.1. For any τ-sheaf $\underline{\mathcal{F}}$ on X we let $f^*\underline{\mathcal{F}}$ denote the τ-sheaf on Y consisting of $(f \times \mathrm{id})^*\mathcal{F}$ and the composite homomorphism

$$
\begin{array}{ccc}
(\sigma \times \mathrm{id})^*(f \times \mathrm{id})^*\mathcal{F} & \xrightarrow{\;\;\tau_{f^*\mathcal{F}}\;\;} & (f \times \mathrm{id})^*\mathcal{F}. \\
\| & \nearrow {\scriptstyle (f\times\mathrm{id})^*\tau_{\mathcal{F}}} & \\
(f \times \mathrm{id})^*(\sigma \times \mathrm{id})^*\mathcal{F} & &
\end{array}
$$

For any homomorphism $\varphi\colon \underline{\mathcal{F}} \to \underline{\mathcal{F}}'$ we abbreviate $f^*\varphi := (f \times \mathrm{id})^*\varphi$.

This defines an A-linear functor

$$
f^* \colon \mathbf{QCoh}_\tau(X, A) \longrightarrow \mathbf{QCoh}_\tau(Y, A)
$$

which is clearly right exact. When f is flat, it is exact. In general, its exactness properties are governed by associated Tor-objects.

Proposition 3.2.

(a) *If φ is a nil-isomorphism, then so is $f^*\varphi$.*

(b) *The functor f^* induces a functor*

$$f^*\colon \mathbf{QCrys}(X, A) \longrightarrow \mathbf{QCrys}(Y, A)$$

which preserves coherence, i.e., $f^*(\mathbf{Crys}(X, A)) \subset \mathbf{Crys}(Y, A)$.

Proof. Note that f^* is in general not exact. Thus (a) is not entirely trivial. However it can be easily reduced to the case of nil-isomorphisms where either kernel or cokernel are zero. These cases are easier to treat. For coherent τ-sheaves, a direct proof is obtained by applying Proposition 2.19. Once part (a) is proved, part (b) is immediate.   □

The following result, whose proof we omit, shows that quasi-crystals behave like sheaves.

Proposition 3.3. *Let $X = \bigcup_i U_i$ be an open covering with embeddings $j_i\colon U_i \hookrightarrow X$.*

(a) *A quasi-crystal $\underline{\mathcal{F}}$ on X is zero if and only if $j_i^*\underline{\mathcal{F}}$ is zero in $\mathbf{QCrys}(U_i, A)$ for all i.*

(b) *A homomorphism φ in $\mathbf{QCrys}(X, A)$ is a monomorphism, an epimorphism, an isomorphism, respectively zero, if and only if its inverse image $j_i^*\varphi$ has that property for all i.*

Next consider any point $x \in X$. Let k_x denote its residue field and $i_x\colon x \cong \operatorname{Spec} k_x \hookrightarrow X$ its natural embedding. The object $i_x^*\underline{\mathcal{F}}$ can be viewed as the *stalk of $\underline{\mathcal{F}}$ at x* in the category of crystals! This is justified by the following result.

Theorem 3.4. *The following assertions hold in* **Crys***:*

(a) *A crystal $\underline{\mathcal{F}} \in \mathbf{Crys}(X, A)$ is zero if and only if the crystals $i_x^*\underline{\mathcal{F}}$ are zero for all $x \in X$.*

(b) *The functors i_x^* are exact on $\mathbf{Crys}(X, A)$ for all $x \in X$.*

Proof. The 'only if' part of (a) follows from the well-definedness of i_x^* on **Crys**. For the 'if' part denote by $\underline{\mathcal{F}}$ also a τ-sheaf representing the crystal $\underline{\mathcal{F}}$. The images of $\tau_{\mathcal{F}}^n$ form a decreasing sequence of coherent subsheaves of $\mathcal{F}$. Thus their supports form a decreasing sequence of closed subschemes $Z_n \subset X \times C$. As $X \times C$ is noetherian, we deduce that $Z_\infty := Z_n$ is independent of n whenever $n \gg 0$. By replacing $\underline{\mathcal{F}}$ by the image of $\tau_{\mathcal{F}}^n$ for $n \gg 0$ we may assume $\operatorname{Supp}(\operatorname{Im}(\tau_{\mathcal{F}}^n)) = Z_\infty$ for all $n \geq 0$. If $Z_\infty = \varnothing$ we are done. Otherwise let η be a generic point of Z_∞ and set $x := \operatorname{pr}_1(\eta)$. We shall deduce a contradiction.

For this we replace X by its localization at x, after which $X = \operatorname{Spec} R$ for a noetherian local ring R, and x corresponds to the maximal ideal $\mathfrak{m} \subset R$. Let M be the $(R \otimes A)$-module corresponding to $\mathcal{F}$. By construction the support of M lies in $x \times C$. Since M is of finite type over the noetherian ring $R \otimes A$ it follows that $\mathfrak{m}^r M = 0$ for any $r \gg 0$, say for $r \geq r_0$.

If $i_x^* \mathcal{F}$ is nilpotent, we have $\tau^s M \subset \mathfrak{m} M$ for some $s \geq 1$. For every $i \geq 0$ this implies

$$\tau^{i+s} M \subset \tau^i(\mathfrak{m} M) \subset \mathfrak{m}^{q^i} M,$$

and for $i \gg 0$, so that $q^i \geq r_0$, it follows that $\tau^{i+s} M = 0$. This contradicts our assumption on Z_∞ and thus proves (a).

To prove (b) we may again replace X by its localization at x, so that $X = \operatorname{Spec} R$ for a noetherian local ring with maximal ideal $\mathfrak{m} \subset R$. We consider an arbitrary short exact sequence of $(R \otimes A)[\tau]$-modules which are finitely generated over $R \otimes A$:

$$0 \longrightarrow M' \longrightarrow M \longrightarrow M'' \longrightarrow 0.$$

The long exact Tor-sequence induces the exact sequence

$$\cdots \longrightarrow \operatorname{Tor}_1^{R \otimes A}\left(M'', (R/\mathfrak{m}) \otimes A\right) \longrightarrow M'/\mathfrak{m} M' \longrightarrow M/\mathfrak{m} M \longrightarrow M''/\mathfrak{m} M'' \longrightarrow 0.$$

Thus it suffices to show that the left-hand term is nilpotent for any M''. Using Lemma 3.6 below, we may write M'' as the quotient of an $(R \otimes A)[\tau]$-module P which is free of finite type over $R \otimes A$. From the resulting short exact sequence

$$0 \longrightarrow K \longrightarrow P \longrightarrow M'' \longrightarrow 0$$

and the long exact Tor-sequence one now obtains the exact sequence

$$0 \longrightarrow \operatorname{Tor}_1^{R \otimes A}\left(M'', (R/\mathfrak{m}) \otimes A\right) \longrightarrow K/\mathfrak{m} K \longrightarrow P/\mathfrak{m} P \longrightarrow M''/\mathfrak{m} M'' \longrightarrow 0.$$

It yields $\operatorname{Tor}_1^{R \otimes A}\left(M'', (R/\mathfrak{m}) \otimes A\right) \cong (K \cap \mathfrak{m} P)/\mathfrak{m} K$. By the Artin–Rees lemma there exists j_0 such that for all $j \geq j_0$ we have

$$K \cap \mathfrak{m}^j P = \mathfrak{m}^{j-j_0}\left(K \cap \mathfrak{m}^{j_0} P\right).$$

For any ℓ with $q^\ell > j_0$ we then have

$$\tau^\ell\left(K \cap \mathfrak{m} P\right) \subset K \cap \mathfrak{m}^{q^\ell} P = \mathfrak{m}^{q^\ell - j_0}\left(K \cap \mathfrak{m}^{j_0} P\right) \subset \mathfrak{m} K.$$

Thus the endomorphism of $(K \cap \mathfrak{m} P)/\mathfrak{m} K$ induced by τ is nilpotent, as desired. $\qquad \square$

Remark 3.5. The analogous statement for quasi-crystals is false. An example is given in [8, Remark 4.1.8].

Lemma 3.6. *If X is affine, every coherent τ-sheaf on X is the quotient of a coherent τ-sheaf whose underlying coherent sheaf is free.*

Proof. Suppose that $X = \operatorname{Spec} R$ and let M be the $(R \otimes A)[\tau]$-module correspond-ing to a coherent τ-sheaf on X. As M is of finite type over $R \otimes A$, we may write it as a quotient of a free module of finite type $N := (R \otimes A)^r$. Since N is free, the semi-linear endomorphism τ of M can be lifted to a semi-linear endomorphism of N. $\qquad\square$

Using Theorem 3.4, the following assertion can be reduced to the case of fields, where it is rather obvious, see [8, Thm. 4.1.12]:

Theorem 3.7. *The functor $f^* \colon \mathbf{Crys}(X, A) \longrightarrow \mathbf{Crys}(Y, A)$ is exact.*

One also has the following more difficult result, see [8, Sec. 4.6], regarding stalks:

Theorem 3.8. *Suppose that X is of finite type over k. Then the following hold:*

(a) *For any $\underline{F} \in \mathbf{Crys}(X, A)$, its crystalline support $\operatorname{Crys-Supp}(\underline{F}) := \{x \in X \mid i_x^* \underline{F} \neq 0 \text{ in } \mathbf{Crys}\}$ is a constructible subset of X.*

(b) *A crystal $\underline{F} \in \mathbf{Crys}(X, A)$ is zero if and only if the crystals $i_x^* \underline{F}$ are zero for all closed points $x \in X$.*

3.2 Further functors deduced from functors on quasi-coherent sheaves

Tensor product: The assignment $(\underline{F}, \underline{G}) \mapsto (\mathcal{F} \otimes_{\mathcal{O}_{X \otimes C}} \mathcal{G}, \tau_{\mathcal{F}} \otimes \tau_{\mathcal{G}})$ with the usual tensor product of homomorphisms defines an A-bilinear bi-functor

$$\otimes \colon \mathbf{QCoh}_\tau(X, A) \times \mathbf{QCoh}_\tau(X, A) \longrightarrow \mathbf{QCoh}_\tau(X, A)$$

which is right exact in both variables. Its exactness properties are governed by associated Tor-objects.

Coefficient change: For any homomorphism $h \colon A \to A'$, the assignment $\underline{F} \mapsto (\mathcal{F} \otimes_A A', \tau_{\mathcal{F}} \otimes_A \operatorname{id}_{A'})$ with the usual change of coefficients of homomorphisms defines an A-bilinear functor

$$\otimes_A A' \colon \mathbf{QCoh}_\tau(X, A) \longrightarrow \mathbf{QCoh}_\tau(X, A')$$

which is right exact. Its exactness properties are again governed by associated Tor-objects.

Direct image: Consider a morphism $f \colon Y \to X$ and $\underline{F} \in \mathbf{QCoh}_\tau(Y, A)$. Using $\sigma^* f_* \to f_* \sigma^*$ deduced from adjunction of inverse and direct image, one obtains a functorial assignment

$$\underline{F} \longmapsto \left((f \times \operatorname{id})_* \mathcal{F}, \tau \text{ induced from} (f \times \operatorname{id})_* \tau_{\mathcal{F}} \right).$$

With the usual direct image of homomorphisms this defines an A-linear functor

$$f_* \colon \mathbf{QCoh}_\tau(Y, A) \longrightarrow \mathbf{QCoh}_\tau(X, A')$$

which is left exact. Its exactness properties are governed by associated higher derived images.

The above three functors all preserve nil-isomorphisms and thus pass to functors on crystals.

One has the following remarkable property which also explains the term *crystal*, describing something which does not deform (note that the canonical morphism $X_{\mathrm{red}} \to X$ is finite radicial and surjective).

Theorem 3.9. *If f is finite radicial and surjective, the adjunction homomorphism* $\mathrm{id} \to f_* f^*$ *is an isomorphism in* $\mathbf{QCrys}(X, A)$ *and the functors*

$$\mathbf{QCrys}(X, A) \xrightleftharpoons[f_*]{f^*} \mathbf{QCrys}(Y, A)$$

are mutually quasi-inverse equivalences of categories.

3.3 Extension by zero

Let $j\colon U \hookrightarrow X$ be an open immersion and $i\colon Z \hookrightarrow X$ be a closed complement with ideal sheaf $\mathcal{I}_0$.

The following is the main result:

Theorem 3.10.

(a) *For every crystal $\underline{\mathcal{F}}$ on U there exists a crystal $\underline{\widetilde{\mathcal{F}}}$ on X such that $j^*\underline{\widetilde{\mathcal{F}}} \cong \underline{\mathcal{F}}$ and $i^*\underline{\widetilde{\mathcal{F}}} = 0$ in $\mathbf{Crys}(Z, A)$.*

(b) *The pair in (a) consisting of $\underline{\widetilde{\mathcal{F}}}$ and the isomorphism $j^*\underline{\widetilde{\mathcal{F}}} \cong \underline{\mathcal{F}}$ is unique up to a unique isomorphism; it depends functorially on $\underline{\mathcal{F}}$.*

(c) *For any $\underline{\mathcal{F}}$ and $\underline{\widetilde{\mathcal{F}}}$ as in (a) and any quasi-crystal $\underline{\widetilde{\mathcal{G}}}$ on X, inverse image under j induces a bijection*

$$j^*\colon \mathrm{Hom}_{\mathbf{QCrys}}\left(\underline{\widetilde{\mathcal{F}}}, \underline{\widetilde{\mathcal{G}}}\right) \longrightarrow \mathrm{Hom}_{\mathbf{QCrys}}\left(\underline{\mathcal{F}}, j^*\underline{\widetilde{\mathcal{G}}}\right).$$

(d) *The assignment $\underline{\mathcal{F}} \mapsto \underline{\widetilde{\mathcal{F}}}$ with $\underline{\widetilde{\mathcal{F}}}$ from (a) defines an A-linear extension by zero functor*

$$j_!\colon \mathbf{Crys}\,(U, A) \longrightarrow \mathbf{Crys}\,(X, A).$$

One should be aware that $j_!$ is *not* induced from any functor of coherent τ-sheaves, because in general a homomorphism $j_!\underline{\mathcal{F}} \dashrightarrow j_!\underline{\mathcal{G}}$ in $\mathbf{Crys}(X, A)$ lifts to a homomorphism in $\mathbf{Coh}_\tau(X, A)$ only after $\underline{\mathcal{F}}$ or $\underline{\mathcal{G}}$ is replaced by a nil-isomorphic τ-sheaf.

Remarks 3.11. Property (c) is the expected universal property of extension by zero. Property (a) is technically a very simple characterization of $j_!\underline{\mathcal{F}}$.

From (a) one easily deduces that $L(U, \underline{\mathcal{F}}, T) = L(X, j_!\underline{\mathcal{F}}, T)$ whenever the left-hand side is defined.

Proof of Theorem 3.10 (a). Let $\underline{\mathcal{F}}$ be a coherent τ-sheaf representing the same-named crystal. The first step is to construct a coherent extension of $\mathcal{F}$ to $X \times C$. This is standard, e.g., [9, no. 1, Cor. 2] or [28, Ch. II, Ex. 5.15]. For the convenience of the reader, we repeat the short argument.

Observe that $(j \times \mathrm{id})_* \mathcal{F}$ is a quasi-coherent extension of $\mathcal{F}$ to $X \times C$. Thus we can write $(j \times \mathrm{id})_* \mathcal{F}$ as a filtered direct limit $\varinjlim_{i \in I} \mathcal{F}_i$ over its coherent subsheaves $\mathcal{F}_i$ (with no τ). It follows that $\mathcal{F} \cong \varinjlim_{i \in I} j^* \mathcal{F}_i$. As $\mathcal{F}$ is coherent and the $j^* \mathcal{F}_i$ are still filtered, there exists an i such that $\mathcal{F} = j^* \mathcal{F}_i$. Thus $\widetilde{\mathcal{F}} := \mathcal{F}_i$ is a coherent extension of $\mathcal{F}$.

Next we wish to extend τ. As $\widetilde{\mathcal{F}} \subset (j \times \mathrm{id})_* \mathcal{F}$, the homomorphism $\tau_{\mathcal{F}}$ yields a homomorphism
$$\tau \colon (\sigma \times \mathrm{id})^* \widetilde{\mathcal{F}} \longrightarrow (j \times \mathrm{id})_* \mathcal{F}.$$

We would like the morphism to factor via $\widetilde{\mathcal{F}}$. Consider the image of $(\sigma \times \mathrm{id})^* \widetilde{\mathcal{F}}$ in the (quasi-coherent) quotient sheaf $(j \times \mathrm{id})_* \mathcal{F} / \widetilde{\mathcal{F}}$. Being the image of a coherent sheaf, it is itself coherent. Since it also vanishes on $U \times C$, it is annihilated by $\mathcal{I}_0^n$ for some integer $n \geq 0$. In other words, we have
$$\mathcal{I}_0^n \tau \left((\sigma \times \mathrm{id})^* \widetilde{\mathcal{F}} \right) \subset \widetilde{\mathcal{F}}.$$

Select an integer m with $(q - 1)m > n$. Since $\sigma^* \mathcal{I}_0 \subset \mathcal{I}_0^q$, we can calculate
$$\tau \left((\sigma \times \mathrm{id})^* (\mathcal{I}_0^m \widetilde{\mathcal{F}}) \right) \subset \mathcal{I}_0^{qm} \tau \left((\sigma \times \mathrm{id})^* \widetilde{\mathcal{F}} \right)$$
$$\subset \mathcal{I}_0^{qm-n} \widetilde{\mathcal{F}}$$
$$\subset \mathcal{I}_0 (\mathcal{I}_0^m \widetilde{\mathcal{F}}).$$

Thus after replacing $\widetilde{\mathcal{F}}$ by $\mathcal{I}_0^m \widetilde{\mathcal{F}}$ the homomorphism $\tau_{\mathcal{F}}$ extends to a homomorphism $(\sigma \times \mathrm{id})^* \widetilde{\mathcal{F}} \to \mathcal{I}_0 \widetilde{\mathcal{F}}$. Let $\underline{\widetilde{\mathcal{F}}}$ be the corresponding crystal on X. Then the first condition of (a) holds by construction, and the second follows from the fact that $\tau_{i_* \widetilde{\mathcal{F}}}$ vanishes.

For the proof of the remaining assertions we refer to [8]. $\qquad\square$

Example 3.12. Let $X = \mathbb{A}^1 = \operatorname{Spec} k[\theta]$ and $C = \mathbb{A}^1 = \operatorname{Spec} k[t]$. Let $\underline{\mathcal{C}}$ denote the Carlitz τ-sheaf on X over A. Its underlying module is $M = \mathbb{F}_q[\theta, t]$ and the endomorphism $\tau \colon M \to M$ is given by $(t - \theta)(\sigma \times \mathrm{id})$. For $j \colon \mathbb{A}^1 \hookrightarrow \mathbb{P}^1$ we wish to determine $j_! \underline{\mathcal{C}}^{\otimes n}$ for $n \in \mathbb{N}$.

For any $m \in \mathbb{Z}$ define $\mathcal{F}_m := \mathcal{O}_{\mathbb{P}^1}(m\infty) \otimes A$ on $X \times C$. If $\mathcal{O} := \mathcal{O}_{\mathbb{P}^1}(-1)$ denotes the ideal sheaf of ∞, then $\mathcal{F}_m = \mathcal{I}_0^{-m} \mathcal{F}_0$.

We consider $\mathcal{F}_m$ near ∞, more concretely on $\mathbb{P}^1 \smallsetminus \{0\}$. Here
$$\Gamma \left(\operatorname{Spec} \mathbb{F}_q \left[\frac{1}{\theta}, t \right], \mathcal{F}_m \right) = \theta^m \mathbb{F}_q \left[\frac{1}{\theta}, t \right].$$

(It would suffice to consider a formal neighborhood of ∞. But notationally it is actually simpler to consider $\mathbb{P}^1 \smallsetminus \{0\}$.) On $\mathbb{P}^1 \smallsetminus \{0, \infty\}$, we have

$$\theta^m f\left(\frac{1}{\theta}, t\right) \xmapsto{\tau^{\otimes n}} (t-\theta)^n \theta^{qm} f\left(\frac{1}{\theta^q}, t\right) = \theta^{(q-1)m+n} \theta^m \left(\frac{t}{\theta} - 1\right)^n f\left(\frac{1}{\theta^q}, t\right).$$

For the right-hand expression to lie in $\theta^m \mathbb{F}_q\left[\frac{1}{\theta}, t\right]$, one needs $(q-1)m + n \leq 0$. For it also to be zero at ∞ one requires $(q-1)m + n < 0$. This leads to

$$m < \frac{-n}{q-1}.$$

Lecture 4

Derived Categories and Derived Functors

In [8] we carefully develop derived categories of τ-sheaves and (quasi-)crystals and derived functors between such categories. By lack of time, this cannot be exposed in the present lecture series. So I shall confine myself to point to some issues which led in [8] to work out the derived setting in great detail.

Derived categories are the appropriate setting for derived functors. This is not particular to the case at hand. It allows one to apply standard techniques of homological algebra to deduce consequences for derived functors. For instance, spectral sequences often arise from the universal properties of the involved derived functors.

Derived functors are the main reason for introducing large ambient categories such as $\mathbf{QCrys} \supset \mathbf{Crys}$. Only in this setting functors like Rf_* can be properly defined. Therefore only in this generality theorems like proper base change and the projection formula can be established.

To be more concrete let $f \colon Y \to X$ be a morphism Then one way of defining Rf_* is via Čech resolutions. Even if the initial object resolved is coherent, the objects of the resolution are only quasi-coherent. The same holds if one works with injective resolutions. Thus a priori one only obtains a functor

$$Rf_* \colon \mathbf{D}^*\big(\mathbf{QCrys}\,(Y, A)\big) \longrightarrow \mathbf{D}^*\big(\mathbf{QCrys}\,(X, A)\big).$$

For proper f, this functor maps $\mathbf{D}^*(\mathbf{Crys}(Y, A))$ as well as $\mathbf{D}^*_{\mathrm{crys}}(\mathbf{QCrys}(Y, A))$ to $\mathbf{D}^*_{\mathrm{crys}}(\mathbf{QCrys}(X, A))$. Thus an important theorem will be the equivalence of derived categories

$$\mathbf{D}^*\big(\mathbf{Crys}\,(X, A)\big) \longrightarrow \mathbf{D}^*_{\mathrm{crys}}\big(\mathbf{QCrys}\,(X, A)\big), \tag{4.1}$$

which holds for $* \in \{b, -\}$. In the proper case it allows one to deduce, again for $* \in \{b, -\}$, a functor

$$Rf_* \colon \mathbf{D}^*\big(\mathbf{Crys}\,(Y, A)\big) \longrightarrow \mathbf{D}^*\big(\mathbf{Crys}\,(X, A)\big).$$

Note that in the present situation the equivalence (4.1) is not a simple formal result as it is in the case of quasi-coherent sheaves without an endomorphism; see [27]. The reason is that not all quasi-crystals are direct limits of crystals.

We defined crystals from τ-sheaves by a localization procedure. But one also defines derived categories from homotopy categories of complexes by a localization procedure. Thus for the theory of derived categories of quasi-crystals it is important to know that the following functor is an equivalence:

$$\mathcal{S}_{\mathrm{nilqi}}^{-1}\, \mathbf{D}^*\big(\,\mathbf{QCoh}_\tau\,\big) = \mathcal{S}_{\mathrm{nilqi}}^{-1}\, \mathbf{K}^*\big(\,\mathbf{QCoh}_\tau\,\big) \longrightarrow \mathcal{S}_{\mathrm{qi}}^{-1}\, \mathbf{K}^*\big(\,\mathcal{S}_{\mathrm{nil}}^{-1}\,\mathbf{QCoh}_\tau\,\big)$$
$$= \mathcal{S}_{\mathrm{qi}}^{-1}\, \mathbf{K}^*\big(\,\mathbf{QCrys}\,\big).$$

Another important issue concerns the computation of the derived functors Lf^* (on quasi-crystals) and Rf_* (e.g., for f proper). We define these derived functors on derived categories of an abelian category whose objects are pairs of a sheaf and an endomorphism. However we do prove that one can compute these objects by computing the derived functors on the underlying categories of sheaves and then adding the induced endomorphisms, which can also be obtained from the derived functors on sheaves without endomorphisms. For further results we refer to [8].

We conclude this section by collecting the main results on the existence of derived functors. Let $f\colon Y \to X$ be any morphism and denote by $j\colon U \hookrightarrow X$ an open immersion.

Theorem 4.1. *The exact functor f^* on* **Crys** *induces for any* $* \in \{b, +, -, \varnothing\}$ *an exact functor*

$$f^*\colon \mathbf{D}^*\big(\,\mathbf{Crys}\,(X, A)\big) \longrightarrow \mathbf{D}^*(\mathbf{Crys}(Y, A)).$$

The functor f^* on **QCrys** *induces a left derived functor*

$$Lf^*\colon \mathbf{D}^-\big(\,\mathbf{QCrys}\,(X, A)\big) \longrightarrow \mathbf{D}^-\big(\,\mathbf{QCrys}\,(Y, A)\big)$$

via resolutions inside $\mathbf{D}^-(\mathbf{QCrys}(X, A))$. *The second functor agrees with the first if restricted to* **Crys**.

In the following lecture on flatness we shall define flat crystals and very flat quasi-crystals. In Corollary 5.6 we shall see that the category $\mathbf{D}^-(\mathbf{Crys}(X, A))$ has enough flat objects. They are not flat within $\mathbf{D}^-(\mathbf{QCrys}(X, A))$. Nevertheless we have:

Theorem 4.2. *If constructed via flat resolutions within* $\mathbf{D}^-(\mathbf{Crys}(X, A))$, *the bi-functor $\otimes$ on crystals possesses a left bi-derived functor*

$$\overset{L}{\otimes}\colon \mathbf{D}^-\big(\,\mathbf{Crys}\,(X, A)\big) \times \mathbf{D}^-\big(\,\mathbf{Crys}\,(X, A)\big) \longrightarrow \mathbf{D}^-(\mathbf{Crys}(X, A)).$$

If constructed via very flat resolutions within $\mathbf{D}^-(\mathbf{QCrys}(X, A))$, *the bi-functor $\otimes$ on quasi-crystals possesses a left bi-derived functor*

$$\overset{L}{\otimes}\colon \mathbf{D}^-\big(\,\mathbf{QCrys}\,(X, A)\big) \times \mathbf{D}^-\big(\,\mathbf{QCrys}\,(X, A)\big) \longrightarrow \mathbf{D}^-\big(\,\mathbf{QCrys}\,(X, A)\big).$$

The second functor agrees with the first one if restricted to crystals.

Theorem 4.3. *Let $A \to A'$ denote a homomorphism of coefficient rings. If constructed via flat resolutions within $\mathbf{D}^-(\mathbf{Crys}(X, A))$, the functor $\_ \otimes_A A'$ on A-crystals possesses a left derived functor*

$$\_ \overset{L}{\underset{A}{\otimes}} A' : \mathbf{D}^-(\mathbf{Crys}(X, A)) \longrightarrow \mathbf{D}^-(\mathbf{Crys}(X, A')).$$

If constructed via very flat resolutions within $\mathbf{D}^-(\mathbf{QCrys}(X, A))$, the functor $\_ \otimes_A A'$ on A-quasi-crystals possesses a left derived functor

$$\_ \overset{L}{\underset{A}{\otimes}} A' : \mathbf{D}^-(\mathbf{QCrys}(X, A)) \longrightarrow \mathbf{D}^-(\mathbf{QCrys}(X, A')).$$

The second functor agrees with the first one if restricted to crystals.

Theorem 4.4. *For any $* \in \{b, +, -, \varnothing\}$ the functor f_* possesses a right derived functor*

$$Rf_* : \mathbf{D}^*(\mathbf{QCrys}(Y, A)) \longrightarrow \mathbf{D}^*(\mathbf{QCrys}(X, A)).$$

This functor can be defined via Čech resolutions. When f is proper, the subcategory $\mathbf{D}^*_{\mathrm{crys}}(\mathbf{QCrys}(\_, A))$ is preserved under f_* and thus by (4.1) it induces a functor

$$Rf_* : \mathbf{D}^*(\mathbf{Crys}(Y, A)) \longrightarrow \mathbf{D}^*(\mathbf{Crys}(X, A))$$

for $* \in \{b, -\}$.

Since $j_!$ is exact on crystals it clearly induces an exact functor

$$j_! : \mathbf{D}^*(\mathbf{Crys}(U, A)) \longrightarrow \mathbf{D}^*(\mathbf{Crys}(X, A)).$$

Combined with Rf_* and using Nagata's theorem that any morphism f of finite type has a relative compactification, i.e., that it lies in a commutative diagram

$$\begin{array}{ccc} Y & \overset{j}{\hookrightarrow} & \bar{Y} \\ {\scriptstyle f}\downarrow & \swarrow {\scriptstyle \bar{f}} & \\ X & & \end{array} \qquad (4.2)$$

where j is an open embedding and $\bar{f}$ is proper, one obtains the following.

Theorem 4.5. *For any such diagram and any $* \in \{b, -\}$ one defines an exact functor, called* direct image with proper support,

$$Rf_! := R\bar{f}_* \circ j_! : \mathbf{D}^*(\mathbf{Crys}(Y, A)) \longrightarrow \mathbf{D}^*(\mathbf{Crys}(X, A)), \qquad (4.3)$$

which, by a standard procedure, is independent of the chosen compactification. It is compatible with composition of morphisms.

We end this lecture by stating two main theorems on the above derived functors from [8, Sec. 6.7]: For the first we consider a cartesian diagram

$$
\begin{array}{ccc}
Y' & \xrightarrow{\ g'\ } & Y \\
{\scriptstyle f'}\downarrow & & \downarrow{\scriptstyle f} \\
X' & \xrightarrow{\ g\ } & X.
\end{array}
\tag{4.4}
$$

Adjunction between direct image and inverse image yields a natural transformation

$$
Lg^*Rf_* \longrightarrow Rf'_*Lg'^*
\tag{4.5}
$$

called the *base change* homomorphism.

Theorem 4.6 (Proper base change)**.** *In the cartesian diagram* (4.4), *assume that f is compactifiable. Then f' is compactifiable and there is a natural isomorphism of functors* $g^*Rf_! \cong Rf'_!g'^*$.

Theorem 4.7 (Projection formula)**.** *For compactifiable* $f\colon Y \to X$ *there is a natural isomorphism of functors* $Rf_!\_ \overset{L}{\otimes} \_ \cong Rf_!(\_ \overset{L}{\otimes} f^*\_)$.

Lecture 5

Flatness

The present lecture is mainly preparatory for the following one. There we shall define (naive and crystalline) L-functions. Their definition requires some hypotheses on the underlying crystals. This is to be expected since given an arbitrary A-module with an endomorphism there need not be a well-defined characteristic polynomial of such an endomorphism taking values in the polynomial ring over A. A sufficient condition is that the underlying module is free of finite rank over A. In the context of crystals it turns out that the proper setting to define L-functions (at least over good coefficient rings; cf. 6.7) is that of flat crystals. Flat crystals are the theme of the present lecture. In Section 5.1 we give their definition and discuss some basic results; in Section 5.2 their behavior under all (derived) functors defined so far is studied and in Section 5.3 we give a partial answer to the question to what extent a flat crystal is representable by a locally free τ-sheaf. All results stem from [8, Ch. 7].

5.1 Basics on flatness

Proposition 5.1. *The following properties for $\underline{\mathcal{F}} \in \mathbf{Crys}(X, A)$ are equivalent:*

(a) *The functor $\underline{\mathcal{F}} \otimes \_\_$: $\mathbf{Crys}(X, A) \to \mathbf{Crys}(X, A)$ is exact.*

(b) $\mathrm{Tor}_i(\underline{\mathcal{F}}, \underline{\mathcal{G}}) = 0$ *for all $i \geq 1$ and all $\underline{\mathcal{G}} \in \mathbf{Crys}(X, A)$.*

(b) $\mathrm{Tor}_1(\underline{\mathcal{F}}, \underline{\mathcal{G}}) = 0$ *for all $\underline{\mathcal{G}} \in \mathbf{Crys}(X, A)$.*

Definition 5.2. If any of the above conditions are satisfied, then $\underline{\mathcal{F}}$ is called a *flat A-crystal*.

A quasi-crystal $\underline{\mathcal{G}}$ is called *very flat* if the functor $\underline{\mathcal{G}} \otimes \_\_$: $\mathbf{QCrys}(X, A) \to \mathbf{QCrys}(X, A)$ is exact.

As remarked earlier, flat crystals need not be very flat. In the sequel we shall exclusively consider flat crystals. We introduced very flat ones only for completeness sake. Their main use is to provide $\otimes$-acyclic resolutions within $\mathbf{QCrys}$.

Flatness of crystals is a pointwise property:

Proposition 5.3. *A crystal $\underline{F}$ on X is flat if and only if $i_x^* \underline{F}$ is flat for every $x \in X$.*

Proof. For any $x \in X$ and any $\underline{G} \in \mathbf{Crys}(X, A)$, one has

$$i_x^* \operatorname{Tor}_1 \left(\underline{F}, \underline{G} \right) \cong \operatorname{Tor}_1 \left(i_x^* \underline{F}, i_x^* \underline{G} \right).$$

The assertion follows easily.   □

Definition 5.4. A τ-sheaf is called of *pullback type* if its underlying sheaf is a pullback from the first factor. A crystal is called of *pullback type* if it has a representing τ-sheaf with this property.

Example 5.5. We can now give some examples:

- Any crystal represented by a locally free τ-sheaf is flat.
- Any crystal of pullback type $\underline{F}$ is flat. This follows from Proposition 5.3 since for any $x \in X$ the τ-sheaf $i_x^* \underline{F}$ is then the pullback of a vector space on the residue field k_x of x and thus free.
- If $\underline{F}$ is a flat crystal on U, then $j_! \underline{F}$ is a flat crystal on X.

Corollary 5.6. *The category $\mathbf{Crys}(X, A)$ possesses enough flat objects.*

Proof. Let $\mathfrak{U} = \{U_i\}$ denote a finite affine cover of X and let $j_i \colon U_i \to X$ be the open embedding for i. Then for any $\underline{F} \in \mathbf{Crys}(X, A)$ the natural homomorphism $\oplus_i j_{i!} \underline{F}|_{U_i} \to \underline{F}$ is surjective. Thus it suffices to prove that each $j_{i!} \underline{F}|_{U_i}$ is the image of a flat crystal. Because $j_{i!}$ preserves flatness and is exact, in turn it suffices that each $\underline{F}|_{U_i}$ is the image of a flat crystal. But this is immediate from Lemma 3.6.   □

We state the following result without proof:

Theorem 5.7. *A crystal $\underline{F}$ is flat if and only if for all $c \in C$ and all $i \geq 1$ one has $\operatorname{Tor}_i(\underline{F}, k_c) = 0$.*

An immediate corollary is the following:

Corollary 5.8. *If A is a field, then any A-crystal is flat.*

5.2   Flatness under functors

Proposition 5.9. *Let $j \colon U \hookrightarrow X$ be an open immersion. If $\underline{F}$ and $\underline{G}$ are flat crystals, then so are*

(a) $f^* \underline{F}$,

(b) $\underline{F} \otimes \underline{G}$,

(c) $\underline{F} \otimes_A A'$,

(d) $j_! \underline{F}$.

Except for (a) all parts are rather straightforward. For (a) one may use Theorem 5.7.

Theorem 5.10. *Let $f\colon Y \to X$ be a proper morphism. Suppose $\underline{\mathcal{F}}^{\bullet} \in \mathbf{C}^b(\mathbf{Crys}(Y, A))$ is quasi-isomorphic to a bounded complex of flat crystals. Then so is $Rf_*\underline{\mathcal{F}}^{\bullet}$ within $\mathbf{C}^b(\mathbf{Crys}(X, A))$.*

The following result shows that for regular coefficient rings all objects possess flat resolutions.

Theorem 5.11. *Suppose A is regular. Then any $\underline{\mathcal{F}}^{\bullet} \in \mathbf{C}^b(\mathbf{Crys}(Y, A))$ is quasi-isomorphic to a bounded complex of flat crystals.*

Remark 5.12. Crystals of pullback type are particular examples of flat crystals. For them it is particularly easy to show that they are preserved under all of our functors, including $R^i f_*$. It is however not clear whether the complex $Rf_*\underline{\mathcal{F}}$ of a crystal $\underline{\mathcal{F}}$ of pullback type is representable by a complex in $\mathbf{D}^b(\mathbf{Crys}(X, A))$ all of whose objects are of pullback type.

5.3 Representability of flat crystals

It is a basic question to what extent flat crystals are representable by coherent τ-sheaves whose underlying sheaves are flat over $\mathcal{O}_X \otimes A$. The following results given without proof provide partial answers.

Proposition 5.13. *Suppose that $X = \operatorname{Spec} F$ for a field F and that C is regular of dimension ≤ 1. Then every flat A-crystal on X can be represented by a coherent τ-sheaf whose underlying sheaf is free.*

Theorem 5.14. *Suppose X is reduced of dimension ≤ 1 and A is artinian. Then for every flat A-crystal $\underline{\mathcal{F}}$ on X there exists an open dense embedding $j\colon U \hookrightarrow X$ such that $j^*\underline{\mathcal{F}}$ can be represented by a coherent τ-sheaf whose underlying sheaf is free.*

A particularly important result in relation to L-functions is Proposition 5.16 which provides for $x = \operatorname{Spec} k_x$ with k_x a finite field and A artinian a canonical locally free representative of a flat crystal.

Definition 5.15. A τ-sheaf $\underline{\mathcal{F}}$ is called *semisimple* if $\tau_{\mathcal{F}}\colon (\sigma \times \operatorname{id})^*\mathcal{F} \to \mathcal{F}$ is an isomorphism.

Proposition 5.16. *Let $x = \operatorname{Spec} k_x$ with k_x a finite field, let A be artinian and consider $\underline{\mathcal{F}}, \underline{\mathcal{G}} \in \mathbf{Coh}_\tau(x, A)$.*

(a) *There exists a unique direct sum decomposition $\underline{\mathcal{F}} = \underline{\mathcal{F}}_{\mathrm{ss}} \oplus \underline{\mathcal{F}}_{\mathrm{nil}}$ such that $\underline{\mathcal{F}}_{\mathrm{ss}}$ is semisimple and $\underline{\mathcal{F}}_{\mathrm{nil}}$ is nilpotent. The summands are called the* semisimple part *and the* nilpotent part *of $\underline{\mathcal{F}}$, respectively.*

(b) *The decomposition in (a) is functorial in $\underline{\mathcal{F}}$.*

(c) *Any nil-isomorphism $\underline{F} \to \underline{G}$ induces an isomorphism $\underline{F}_{ss} \to \underline{G}_{ss}$.*

(d) *The construction induces a functor $\mathbf{Crys}(x, A) \to \mathbf{Coh}_\tau(x, A)\colon \underline{F} \mapsto \underline{F}_{ss}$.*

Proof. Note first that $\tau_{\mathcal{F}}^{d_x}$ is an endomorphism $\mathcal{F} \to \mathcal{F}$, because σ^{d_x} is the identity on k_x. Moreover $\mathcal{F}$ has finite length, because $k_x \otimes A$ is artinian. Therefore, for $n \gg 0$ the subsheaves $\mathcal{F}_{ss} := \operatorname{Im} \tau_{\mathcal{F}}^{n d_x}$ and $\mathcal{F}_{\mathrm{nil}} := \operatorname{Ker} \tau_{\mathcal{F}}^{n d_x}$ are independent of n. Clearly $\tau_{\mathcal{F}}$ maps them to themselves, so they define τ-subsheaves $\underline{F}_{ss}$ and $\underline{F}_{\mathrm{nil}}$. By construction $\tau_{\mathcal{F}_{ss}}$ is surjective and $\tau_{\mathcal{F}_{\mathrm{nil}}}$ nilpotent. Since $\mathcal{F}$ and hence $\mathcal{F}_{ss}$ have finite length, $\tau_{\mathcal{F}_{ss}}$ is then also injective. Thus $\underline{F}_{ss}$ is semisimple and $\underline{F}_{\mathrm{nil}}$ nilpotent. Furthermore, the construction yields a split short exact sequence

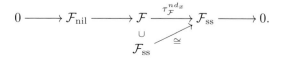

This shows that $\underline{F} = \underline{F}_{ss} \oplus \underline{F}_{\mathrm{nil}}$, proving the existence part of (a). The uniqueness follows from the fact that any semisimple τ-subsheaf of $\underline{F}$ is contained in $\underline{F}_{ss}$ and any nilpotent τ-subsheaf of $\underline{F}$ is contained in $\underline{F}_{\mathrm{nil}}$.

Part (b) follows directly from the above construction of $\mathcal{F}_{ss}$ and $\mathcal{F}_{\mathrm{nil}}$. Also (b) implies that the kernel and cokernel of any homomorphism $\underline{F}_{ss} \to \underline{G}_{ss}$ have trivial nilpotent part. This implies (c) and hence (d). $\square$

Proposition 5.17. *Let $x = \operatorname{Spec} k_x$ with k_x a finite field and let A be artinian and consider $\underline{F} \in \mathbf{Crys}(x, A)$.*

(a) *$\underline{F}_{ss}$ is the unique semisimple τ-sheaf representing $\underline{F}$.*

(b) *The functor $\mathbf{Crys}(x, A) \to \mathbf{Coh}_\tau(x, A)\colon \underline{F} \mapsto \underline{F}_{ss}$ is exact.*

(c) *$\underline{F}$ is flat if and only if the sheaf $\mathcal{F}_{ss}$ underlying $\underline{F}_{ss}$ is locally free.*

Proof. Part (a) follows from Proposition 5.16, especially from 5.16 (c). Part (b) is a consequence of 5.16 (b) and the fact that any exact sequence in $\mathbf{Crys}(x, A)$ is the image of an exact sequence in $\mathbf{Coh}_\tau(x, A)$.

For (c) note that one direction (the "if") is obvious. For the other direction, let $\underline{G}$ be a representative of the crystal $\underline{F}$ whose underlying sheaf is free; cf. Proposition 5.14. By Proposition 5.17(a) we have $\underline{G} = \underline{G}_{\mathrm{nil}} \oplus \underline{G}_{ss}$. Considering the underlying sheaves it follows that $\mathcal{G}_{ss}$ is locally free on $\mathcal{O}_{x \times C}$. By part (d) of the same proposition, we have $\underline{G}_{ss} = \underline{F}_{ss}$. $\square$

Lecture 6

The *L*-function

In this lecture we introduce two *L*-functions. A naive one for locally free τ-sheaves or τ-sheaves which are of pullback type, and a crystalline one for flat *A*-crystals, provided the ring *A* satisfies some mild hypotheses. The word naive simply refers to the fact that the naive *L*-function is not necessarily invariant under nil-isomorphism and thus it does not, in general, induce an *L*-function for crystals. In the last section we shall state a trace formula for naive *L*-functions and indicate some consequences for crystalline *L*-functions. The proof of the trace formula will be postponed to Lecture 7. For more details, see [8, Ch.'s 8 and 9].

6.1   Naive *L*-functions

As a preparation we briefly recall without proof some basic properties of the dual characteristic polynomial for endomorphisms of projective modules.

Lemma–Definition 6.1. *Let A be a commutative ring, M a finitely generated projective A-module, and $\varphi\colon M \to M$ an A-linear endomorphism.*

(a) *Let M′ be any finitely generated projective A-module such that $M \oplus M'$ is free over A. Let $\varphi'\colon M' \to M'$ be the zero endomorphism. Then the following expression is independent of the choice of M′:*

$$\det{}_A\big(\mathrm{id} - t(\varphi \oplus \varphi') \mid M \oplus M'\big) \in 1 + tA[t].$$

From now on we simply write $\det_A(\mathrm{id} - t\varphi \mid M)$ for it and call it the dual characteristic polynomial *of (M, φ).*

(b) *The assignment $(M, \varphi) \mapsto \det_A(\mathrm{id} - t\varphi \mid M)$ is multiplicative in short exact sequences.*

Lemma 6.2. *Let A be an algebra over a field k. Let k′ be a finite cyclic Galois extension of k of degree d, and σ a generator of $\mathrm{Gal}(k'/k)$. Let M be a finitely generated* *projective module over $k' \otimes_k A$ and $\varphi\colon M \to M$ an A-linear endomorphism satisfying $\varphi(xm) = {}^{\sigma}x \cdot \varphi(m)$ for all $x \in k'$ and $m \in M$. Then φ^d is $(k' \otimes A)$-linear and*

$$\det{}_A\big(\mathrm{id} - t\varphi \mid M\big) = \det{}_{k' \otimes A}\big(\mathrm{id} - t^d\varphi^d \mid M\big).$$

In particular, both sides lie in $1 + t^d A[t^d]$.

Now we return to τ-sheaves. Let X be a variety of finite type over k and let $\underline{\mathcal{F}} \in \mathbf{Coh}_{\tau}(X, A)$ be either of pullback type or locally free. For any $x \in |X|$, writing $\underline{\mathcal{F}}_x := i_x^*\underline{\mathcal{F}}$, Lemma 6.2 shows that

$$\det{}_A\big(\mathrm{id} - T\tau \mid \underline{\mathcal{F}}_x\big) \;=\; \det{}_{k_x \otimes A}\big(\mathrm{id} - T^{d_x}\tau^{d_x} \mid \underline{\mathcal{F}}_x\big) \;\in\; 1 + T^{d_x} A[T^{d_x}].$$

Definition 6.3. The *naive L-function* of $\underline{\mathcal{F}}$ over x is

$$L^{\mathrm{naive}}\big(x, \underline{\mathcal{F}}, T\big) \;:=\; \det{}_A\big(\mathrm{id} - T\tau \mid \underline{\mathcal{F}}_x\big)^{-1} \;\in\; 1 + T^{d_x} A[[T^{d_x}]].$$

As the number of points in $|X|$ of any given degree d_x is finite, we can form the product over all x within $1 + tA[[T]]$:

Definition 6.4. The *naive L-function* of $\underline{\mathcal{F}}$ over X is

$$L^{\mathrm{naive}}\big(X, \underline{\mathcal{F}}, T\big) \;:=\; \prod_{x \in |X|} L^{\mathrm{naive}}\big(x, i_x^*\underline{\mathcal{F}}, T\big) \;\in\; 1 + TA[[T]].$$

In the special case where A is reduced, L^{naive} is invariant under nil-isomorphisms. For more general A, this property may fail.

6.2   Crystalline L-functions

We first assume that A is artinian.

Definition 6.5. The *(crystalline) L-function* of a flat crystal $\underline{\mathcal{F}}$ on x is

$$L^{\mathrm{crys}}\big(x, \underline{\mathcal{F}}, T\big) \;:=\; L^{\mathrm{naive}}\big(x, \underline{\mathcal{F}}_{\mathrm{ss}}, T\big) \;\in\; 1 + T^{d_x} A[[T^{d_x}]].$$

Let now $\underline{\mathcal{F}}$ be a flat crystal on a scheme X of finite type over k. Since $L^{\mathrm{crys}}(x, \underline{\mathcal{F}}, T) \in 1 + T^{d_x} A[[T^{d_x}]]$ for any $x \in |X|$, and the number of points of any given degree d_x is finite, we can again form the product over all x within $1 + TA[[T]]$:

Definition 6.6. The *(crystalline) L-function* of a flat crystal $\underline{\mathcal{F}}$ on X is

$$L^{\mathrm{crys}}\big(X, \underline{\mathcal{F}}, T\big) \;:=\; \prod_{x \in |X|} L^{\mathrm{crys}}\big(x, i_x^*\underline{\mathcal{F}}, T\big) \;\in\; 1 + TA[[T]].$$

If $\underline{\mathcal{F}}^\bullet$ is a bounded complex of flat crystals, one defines

$$L^{\mathrm{crys}}(X,\underline{\mathcal{F}}^\bullet,T) \;:=\; \prod_{i\in\mathbb{Z}} L^{\mathrm{crys}}(X,\underline{\mathcal{F}}^i,T)^{(-1)^i}.$$

If $\underline{\mathcal{F}}^\bullet$ is quasi-isomorphic to such a complex, we use the notation $L^{\mathrm{crys}}(X,\underline{\mathcal{F}}^\bullet,T)$ as well, and mean by this the L-function of that bounded complex of flat crystals. Using the cone construction it is not difficult to show that this is well defined, that is, quasi-isomorphic bounded complexes of flat crystals have the same L-function.

We now relax our condition on A and only require that A be a *good coefficient ring*. In Remark 6.8 we give several examples of classes of rings A that are good coefficient rings. The reason for introducing this notion is that for such A one can define an L-function as follows: Changing coefficients from A to its quotient ring Q_A one obtains from a flat crystal on A a flat crystal over the artinian ring Q_A. For Q_A we know how to define a crystalline L-function. The property of being a good coefficient ring is then used to prove that the pointwise L-factors defined over Q_A have in fact coefficients in A and not just in Q_A.

We let $\mathfrak{p}_1,\dots,\mathfrak{p}_n$ denote the minimal primes of A and call

$$Q_A := A_{\mathfrak{p}_1} \oplus \dots \oplus A_{\mathfrak{p}_n}$$

the *quotient ring* of A.

Definition 6.7. We call A a *good coefficient ring* if

(a) the natural homomorphism $A \to Q_A$ is injective, and

(b) A is closed under taking finite ring extensions $A \hookrightarrow A' \hookrightarrow Q_A$ which induce bijections $\mathrm{Spec}\, A' \to \mathrm{Spec}\, A$ and isomorphisms on all residue fields.

Remark 6.8. Artin rings, regular rings and normal domains are examples of good coefficient rings. In particular, finite rings and Dedekind domains are good coefficient rings.

The following is a main result on good coefficient rings and the reason for their definition. Consider any flat $\underline{\mathcal{F}} \in \mathbf{Crys}(x,A)$ for $x = \mathrm{Spec}\, k_x$ for a finite extension k_x of k.

Lemma 6.9. *If A is a good coefficient ring, then*

$$L^{\mathrm{crys}}\big(x,\underline{\mathcal{F}}\underset{A}{\otimes}Q_A,T\big)^{-1} \in 1 + TQ_A[T]$$

has in fact coefficients in A.

The proof uses properties of flat A-crystals not developed in this lecture series. Therefore we refer to [8, §9.7].

Putting things together, we obtain:

Theorem 6.10. *Suppose that A is a good coefficient ring. Then for every complex $\underline{\mathcal{F}}^{\bullet} \in \mathbf{C}^{b}(\mathbf{Crys}^{\mathrm{flat}}(X, A))$ the L-function $L^{\mathrm{crys}}(X, \underline{\mathcal{F}}^{\bullet} \overset{L}{\otimes}_{A} Q_{A}, T)$ has coefficients in A.*

Definition 6.11. *If A is a good coefficient ring, then the* (crystalline) *L-function of $\underline{\mathcal{F}}^{\bullet} \in \mathbf{C}^{b}(\mathbf{Crys}^{\mathrm{flat}}(X, A))$ is*

$$L^{\mathrm{crys}}(X, \underline{\mathcal{F}}^{\bullet}, T) := L^{\mathrm{crys}}(X, \underline{\mathcal{F}}^{\bullet} \overset{L}{\underset{A}{\otimes}} Q_{A}, T) \in 1 + T A[[T]].$$

One has the following elementary comparison theorem between naive and crystalline L-functions. (The asserted equality is true for all pointwise L-factors and can be proved there by passing from A to Q_{A}.)

Proposition 6.12. *Suppose A is a good coefficient ring. Suppose $\underline{\mathcal{F}} \in \mathbf{Coh}_{\tau}(X, A)$ is a τ-sheaf of pullback type. Then both $L^{\mathrm{crys}}(X, \underline{\mathcal{F}}, T)$ and $L^{\mathrm{naive}}(X, \underline{\mathcal{F}}, T)$ are defined and elements of $1 + T A[[T]]$. If A is reduced then they agree.*

6.3   Trace formulas for *L*-functions

We now have the necessary definitions at our disposal to formulate the main results regarding trace formulas of τ-sheaves and crystals.

Theorem 6.13. *Let $X = \operatorname{Spec} R$ be affine and smooth of equidimension n over k and with structure morphism $s_{X} \colon X \to \operatorname{Spec} k$. Let A be artinian. Suppose the underlying sheaf of $\underline{\mathcal{F}} \in \mathbf{Coh}_{\tau}(X, A)$ is free. Let $j \colon X \hookrightarrow \overline{X}$ be a compactification of X. Then there exists $\widetilde{\underline{\mathcal{F}}} \in \mathbf{Coh}_{\tau}(\overline{X}, A)$ of pullback type such that*

(a) *$\widetilde{\underline{\mathcal{F}}} = j_{!}\underline{\mathcal{F}}$ in $\mathbf{Crys}(\overline{X}, A)$;*
(b) *$H^{i}(\overline{X}, \widetilde{\underline{\mathcal{F}}})$ is nilpotent for all $i \neq n$;*
(c) *$L^{\mathrm{naive}}(X, \underline{\mathcal{F}}, T) = L^{\mathrm{naive}}(\operatorname{Spec} k, H^{n}(\overline{X}, \widetilde{\underline{\mathcal{F}}}), T)^{(-1)^{n}}$.*

The result strongly relies on Anderson's trace formula which we state below in Theorem 7.13. It has some rather restrictive hypotheses on the base scheme and the sheaf underlying the given τ-sheaf. But formally it has the correct shape. The following result for crystals has hypotheses as general as can be expected. However the trace formula will, in general, not be an exact equality.

We write $\mathfrak{n}_{A}$ for the nilradical of A. For $f, g \in 1 + T A[[T]]$ we define $f \sim g$ to mean that there exists $h \in 1 + T\mathfrak{n}_{A}[T]$ such that $g = fh$. This defines an equivalence relation on $1 + T A[[T]]$.

Theorem 6.14. *For any morphism $f \colon Y \to X$ of schemes of finite type over k and any bounded complex $\underline{\mathcal{F}}^{\bullet}$ of flat A-crystals on Y we have*

$$L^{\mathrm{crys}}(Y, \underline{\mathcal{F}}^{\bullet}, T) \sim L^{\mathrm{crys}}(X, Rf_{!}\underline{\mathcal{F}}^{\bullet}, T).$$

An example of Deligne, see [8, Sec. 8.5], shows that for general A one cannot expect a stronger result. Note that A is reduced if and only if $\mathfrak{n}_A = 0$, and so in this case $\sim$ becomes $=$.

As we shall explain later (in Lecture 9), an example of Deligne from the mid 1970s showed that one cannot expect a stronger result.

It is worthwhile to state explicitly the following corollary:

Corollary 6.15. *Let X be a scheme of finite type over k with structure morphism $s_X \colon X \to \operatorname{Spec} k$. Then for any bounded complex $\underline{F}^\bullet$ of flat A-crystals on X we have*

$$L^{\mathrm{crys}}(X, \underline{F}^\bullet, T) \sim L^{\mathrm{crys}}(\operatorname{Spec} k, Rf_!\underline{F}^\bullet, T).$$

Since the complex $Rf_!\underline{F}^\bullet \otimes_A Q_A$ can be represented by a bounded complex of free Q_A-modules carrying some endomorphism, the right-hand side is a rational function over A, and thus so is the left-hand side.

For completeness, we also mention an important result on change of coefficients:

Theorem 6.16. *If both A and A' are good coefficient rings, then for every $\underline{F}^\bullet \in \mathbf{C}^b(\mathbf{Crys}^{\mathrm{flat}}(X, A))$ we have*

$$L^{\mathrm{crys}}\Big(X, \underline{F}^\bullet \overset{L}{\underset{A}{\otimes}} A', T\Big) \sim \lambda\big(L^{\mathrm{crys}}(X, \underline{F}^\bullet, T)\big),$$

where $\lambda \colon 1 + TA[[T]] \to 1 + TA'[[T]]$ is induced from $A \to A'$. If moreover A is artinian, then equality holds.

Lecture 7

Proof of Anderson's Trace Formula and a Cohomological Interpretation

The aim of this lecture is to give a proof of Theorem 6.13 – at least under some simplifying hypotheses. In the end we shall briefly indicate how from this result one can deduce Theorem 6.14, and that for general non-reduced A the proof of the latter result shall indeed require quite some more effort than we indicate here. The proof of the important Theorem 7.13 essentially goes back to the article [2] by G. Anderson.

Throughout this lecture, we let $X = \operatorname{Spec} R$ be an affine scheme which is smooth and of finite type over k. The ring A is an arbitrary fixed k-algebra.

7.1 The Cartier operator

Let $\Omega := \Omega_{R/k}$ be the module of Kähler differentials of R. Because R/k is smooth of equidimension n, it is a finitely generated projective R-module of rank n. Let $d\colon R \to \Omega$ denote the universal derivation as well as its extension to the de Rham complex $\bigwedge^{\bullet}\Omega$. The following result is due to Cartier; see [30, pp. 199–203].

Theorem 7.1 (Cartier)**.** *There exists an isomorphism of complexes*

$$C^{-1}\colon \left(\textstyle\bigwedge^{\bullet}\Omega, 0\right) \xrightarrow{\;\cong\;} \left(H^{\bullet}\left(\textstyle\bigwedge^{\bullet}\Omega, d\right), 0\right)$$

with zero differential, such that for all $r \in R$ and $\xi, \eta \in \bigwedge^{\bullet}\Omega$ one has

$$C^{-1}(r\xi) = r^{p}C^{-1}(\xi),$$

$$C^{-1}(dr) = r^{p-1}dr + dR,$$

$$C^{-1}(\xi \wedge \eta) = C^{-1}(\xi) \wedge C^{-1}(\eta).$$

The inverse C of C^{-1} on the highest non-vanishing exterior power $\omega := \bigwedge^{n}\Omega$ is called the Cartier operator.

For $m := \log_p(\#k) = \log_p q$ we call the m-fold iterate

$$C_q = \underbrace{C \circ C \circ \cdots \circ C}_{m}: \omega \longrightarrow \omega$$

the q-Cartier operator ($q = p^m$). It satisfies $C_q(r^q \xi) = rC(\xi)$, i.e., it is q^{-1}-linear.

7.2 Cartier sheaves

Definition 7.2. A *Cartier linear endomorphism* of a coherent sheaf $\mathcal{V}$ on $X \times C$ is an $\mathcal{O}_{X \times C}$-linear homomorphism $\kappa_\mathcal{V} \colon (\sigma \times \mathrm{id})_* \mathcal{V} \to \mathcal{V}$. The pair $\underline{\mathcal{V}} := (\mathcal{V}, \kappa_\mathcal{V})$ is then called a *Cartier sheaf on X over A*. A homomorphism of Cartier sheaves $\underline{\mathcal{V}} \to \underline{\mathcal{W}}$ is a homomorphism of the underlying sheaves $\varphi \colon \mathcal{V} \to \mathcal{W}$ compatible with the extra endomorphism κ.

We denote the category of locally free Cartier sheaves on X (over A) by $\mathbf{Cart}^{\mathrm{locfree}}(X, A)$.

As we assume that $X = \mathrm{Spec}\, R$ is affine, the above notions are expressed on modules as follows. A *Cartier linear map* on a finitely generated $(R \otimes A)$-module V is an A-linear homomorphism $\kappa_V \colon V \to V$ such that $\kappa_V((x^q \otimes a)v) = (x \otimes a) \cdot \kappa_V(v)$ for all $x \in R$, $a \in A$, and $v \in V$. For simplicity such a pair $\underline{V} := (V, \kappa_V)$ is called a *Cartier module*.

Definition 7.3. For any A define $\omega_{X,A}$ as the sheaf on $X \times C$ which is the pullback along $\mathrm{pr}_1 \colon X \times C \to X$ of the invertible sheaf on X associated to the module ω. Denote by $\kappa_{X,A} \colon (\sigma \times \mathrm{id})_* \omega_{X,A} \to \omega_{X,A}$ the endomorphism induced from C_q under this pullback.

Example 7.4. The pair $(\omega_{X,A}, \kappa_{X,A})$ is a Cartier sheaf on X over A.

If $\mathcal{F}$ is a coherent sheaf on $X \times C$, then $\mathcal{H}om_{\mathcal{O}_{X \times C}}(\mathcal{F}, \omega_{X,A})$ is again a coherent sheaf on $X \times C$. Suppose now that $\mathcal{F}$ is the underlying sheaf of a locally free τ-sheaf $\underline{\mathcal{F}}$ and let $\tilde{\tau}_{\mathcal{F}} \colon \mathcal{F} \to (\sigma \times \mathrm{id})_* \mathcal{F}$ denote the homomorphism adjoint to $\tau_{\mathcal{F}}$. The dualizing sheaf ω together with C_q allows us to assign a locally free Cartier sheaf $D(\underline{\mathcal{F}})$ on X to $\underline{\mathcal{F}}$ as follows. Its underlying sheaf is

$$D(\mathcal{F}) := \mathcal{H}om_{\mathcal{O}_{X \times C}}(\mathcal{F}, \omega_{X,A}).$$

For a section $(\sigma \times \mathrm{id})_* \varphi \in (\sigma \times \mathrm{id})_* \mathcal{H}om_{\mathcal{O}_{X \times C}}(\mathcal{F}, \omega_{X,A})$, one defines

$$\kappa_{D(\mathcal{F})}\big((\sigma \times \mathrm{id})_* \varphi\big) := \kappa_{X,A} \circ (\sigma \times \mathrm{id})_* \varphi \circ \tilde{\tau}_{\mathcal{F}}.$$

Proposition 7.5. *The functor* $\underline{\mathcal{F}} \mapsto D(\underline{\mathcal{F}})$ *induces an anti-equivalence of categories*

$$\mathbf{Coh}_\tau^{\mathrm{locfree}}(X, A) \longrightarrow \mathbf{Cart}^{\mathrm{locfree}}(X, A).$$

For arbitrary smooth schemes of finite type over k an analog of Proposition 7.5 holds. The proof is a simple patching argument by which one is reduced to the affine case.

Proof. Well-definedness and injectivity are easily verified. The main point is to prove essential surjectivity. For this one needs the adjunction between σ_* and $\sigma^!$; cf. [28, Exercise III.6.10], for the finite flat morphism $\sigma\colon X \to X$ – the ring R is finitely generated and smooth over k. It yields

$$\mathcal{H}om_{\mathcal{O}_{X\times C}}\left((\sigma \times \mathrm{id})_*\mathcal{F}, \mathcal{G}\right) \cong (\sigma \times \mathrm{id})_*\, \mathcal{H}om_{\mathcal{O}_{X\times C}}\left(\mathcal{F}, (\sigma \times \mathrm{id})^!\mathcal{G}\right).$$

Moreover the adjoint of the homomorphism $C_q\colon \sigma_*\omega \to \omega$ is an isomorphism $\omega \to \sigma^!\omega$. Details are left to the reader; cf. also [8, §7.2]. $\qquad\square$

Example 7.6. Let $X = \operatorname{Spec} k[\theta] \cong \mathbb{A}_k^1$ and $C = \operatorname{Spec} k[t] \cong \mathbb{A}_k^1$, and consider the Carlitz τ-sheaf $\mathcal{C}$ corresponding to $(k[\theta, t], (t - \theta)(\sigma \times \mathrm{id}))$.

In the case at hand, $\omega = \Omega_{k[\theta]/k} = k[\theta]\mathrm{d}\theta$ and the Cartier operator C_q, which can be remembered via $C_q(\mathrm{d}\theta/\theta) = \mathrm{d}\theta/\theta$, has the following description:

$$C_q\left(\theta^\ell\, \mathrm{d}\theta\right) = C_q\left(\theta^{\ell+1}\, \mathrm{d}\theta/\theta\right) = \begin{cases} \theta^{(\ell+1-q)/q}\, \mathrm{d}\theta & \text{if } q|(\ell + 1), \\ 0 & \text{else.} \end{cases}$$

To simplify notation, we define the expression θ^α for $\alpha \in \mathbb{Q}$ to mean 0 whenever $\alpha \in \mathbb{Q} \smallsetminus \mathbb{Z}$ and to mean the respective power for $\alpha \in \mathbb{Z}$.

We now determine $D(\mathcal{C}^{\otimes n})$. The underlying module is

$$k[\theta, t]\, \mathrm{d}\theta = \operatorname{Hom}(k[\theta, t], \omega).$$

Here an element $f(\theta, t)\, \mathrm{d}\theta$ represents the homomorphism which maps $1 \in k[\theta, t]$ to $f(\theta, t)\, \mathrm{d}\theta$ in ω. Based on this, one computes the image of $\theta^n\, \mathrm{d}\theta \in k[\theta, t]\, \mathrm{d}\theta$ as follows:

$$\kappa := D\left(\tau_{\mathcal{C}^{\otimes n}}\right)\colon \theta^\ell\, \mathrm{d}\theta \longmapsto C_q\left((t - \theta)^n\theta^\ell\, \mathrm{d}\theta/\theta\right)$$

$$= (-1)^n \sum_{\substack{i=0 \\ i \equiv \ell+n+1 \,(\mathrm{mod}\, q)}}^{n} \binom{n}{i}(-t)^i\theta^{(\ell+1+n-i-q)/q}\, \mathrm{d}\theta.$$

Let us take $\{\theta^\ell\, \mathrm{d}\theta\}_{\ell\in\mathbb{N}_0}$ as a basis of the module underlying $D(\mathcal{C}^{\otimes n})$ over $k[t]$. Then κ is $k[t]$-linear. Above we computed the image of the basis element $\theta^\ell\, \mathrm{d}\theta$. Considering the sum on the right for $\kappa(\theta^\ell\, \mathrm{d}\theta)$ we observe that the exponent ℓ essentially is divided by q, except for the added constant $\frac{1}{q}(n + 1 - q - i)$. This means that the image of $\theta^\ell\, \mathrm{d}\theta$ only involves the basis elements $\{\theta^j\, \mathrm{d}\theta\}_{j=0,1,\ldots,c+\ell/q}$ for a constant c independent of ℓ.

Thus the matrix representing κ with respect to our chosen basis has the following shape: If we draw a straight line starting at row c in the first column and with slope $-\frac{1}{q}$, then all entries below that line are zero!

7.3   Operators of trace class

Let V_0 be a k-vector space, typically of countably infinite dimension. Set $V :=$ $V_0 \otimes A$ and consider an A-linear operator $\kappa_V : V \to V$.

Definition 7.7. A k-subspace $W_0 \subset V_0$ is called a *nucleus* for κ_V if it is finite dimensional and there exists an exhaustive increasing filtration of V_0 by finite-dimensional k-vector spaces

$$W_0 \subset W_1 \subset W_2 \subset \cdots \subset V_0$$

such that $\kappa_V(W_{i+1} \otimes A) \subset W_i \otimes A$ for all $i \geq 1$. If (V, κ_V) possesses a nucleus we call it *nuclear*.

The following proposition collects some basic results which are easy consequences of the definition.

Proposition 7.8.

(a) If W_0 is a nucleus for κ_V, then for any $j \geq 0$ the exterior power $\bigwedge_k^j W_0 \subset \bigwedge_k^j V_0$ is a nucleus for $\bigwedge^j \kappa_V : \bigwedge_A^j V \to \bigwedge_A^j V$.

(b) If (V, κ_V) is nuclear, the values of the following expressions are independent of the chosen nucleus W_0:

$$\mathrm{Tr}\left(\kappa_V\right) := \mathrm{Tr}_A\left(\kappa_V|W_0 \otimes A\right) \ \in \ A,$$
$$\Delta\left(1 - T\kappa_V\right) := \det{}_A\left(1 - T\kappa_V|W_0 \otimes A\right)$$
$$= \sum_j (-1)^j \, \mathrm{Tr}\left(\textstyle\bigwedge^j \kappa_V \,\middle|\, \bigwedge_k^j W_0 \otimes A\right) T^j \ \in \ A[T].$$

They are called the trace *and the* dual characteristic polynomial *of κ_V, respectively.*

(c) Suppose $0 \to V_0' \to V_0 \to V_0'' \to 0$ is a short exact sequence of k vector spaces and that $V' := V_0' \otimes A$ is preserved by κ_V. Define $V'' := V_0'' \otimes A$ and write $\kappa_{V'}$ and $\kappa_{V''}$ for the endomorphisms induced from κ_V. If (V, κ_V) is nuclear, then so are $(V', \kappa_{V'})$ and $(V'', \kappa_{V''})$, and

$$\mathrm{Tr}\left(\kappa_V\right) = \mathrm{Tr}\left(\kappa_{V'}\right) + \mathrm{Tr}\left(\kappa_{V''}\right),$$
$$\Delta\left(1 - T\kappa_V\right) = \Delta\left(1 - T\kappa_{V'}\right) \cdot \Delta\left(1 - T\kappa_{V''}\right).$$

Note that if W_0 is a nucleus for κ_V, then $W_0 \cap V_0'$ is a nucleus for $\kappa_{V'}$.

Remark 7.9. It is unclear whether there is a reasonable theory of nuclei, trace and characteristic polynomial for pairs (V, κ_V) if the underlying module of V is not of the form $V_0 \otimes A$.

We will now see that Cartier modules provide natural examples of nuclear endomorphisms. Thus in the following we assume that V_0 is the k-vector space underlying a finitely generated R-module. (Here R is as before smooth and finitely generated over k.) Following Anderson, one introduces the following notions:

Definition 7.10. Let $r_1, \ldots, r_s$ be generators of R as a k-algebra, and let $v_1, \ldots, v_t$ be generators of V_0 as an R-module. For every integer n let $V_{0,n} \subset V_0$ denote the k-linear span of all elements admitting a representation

$$\sum_{j=1}^{t} f_j(r_1, \ldots, r_s) \, v_j$$

with polynomials $f_{ij} \in k[X_1, \ldots, X_s]$ of total degree at most n and $V_{0,-\infty} := \{0\}$. These subspaces form an exhaustive sequence of finite-dimensional k vector spaces of V_0. The function

$$\gamma \colon V \longrightarrow \mathbb{Z}^{\geq 0} \cup \{-\infty\}, \quad v \longmapsto \inf \{n \mid v \in V_{0,n} \otimes A\}$$

is called a *gauge* on V.

Definition 7.11. An A-linear operator $\kappa_V \colon V \to V$ is called *of trace class* if for every gauge γ on V there exist constants $0 \leq K_1 < 1$ and $0 \leq K_2$ such that $\gamma(\kappa_V(v)) \leq K_1 \cdot \gamma(v) + K_2$ for all $v \in V$.

Proposition 7.12. ([2, Propositions 3 and 6], [8, § 8.3]). *Let (V, κ_V) be as above.*

(a) *Any Cartier linear operator on V is of trace class with $K_1 = 1/q$.*

(b) *If κ_V is of trace class with constants K_1, K_2 for some gauge γ, then $V_{0,n}$ from 7.10 is a nucleus for κ_V for any $n \geq 1 + K_2/(1 - K_1)$.*

(c) *If κ_V is of trace class, then so is the composite of κ_V with any $(R \otimes A)$-linear endomorphism φ.*

(d) *If φ in (c) is of the form $\varphi_0 \otimes \mathrm{id}_A$ for some R-linear endomorphism φ_0 of V_0, then $\mathrm{Tr}(\kappa_V \varphi) = \mathrm{Tr}(\varphi \kappa_V)$.*

7.4 Anderson's trace formula

The following is the generalization of the main result [2, Theorem 1] of Anderson's article [2] to arbitrary artinian A. As explained in Remark 7.9, the Cartier module V requires a k-structure V_0 in order for the dual characteristic polynomial to be well defined.

Theorem 7.13 (Anderson). *Let $X = \operatorname{Spec} R$ be affine and smooth of equidimension n over k. Consider $\underline{\mathcal{F}} \in \mathbf{Coh}_\tau(X, A)$ such that $\mathcal{F} = \operatorname{pr}_1^* \mathcal{F}_0$ for a locally free coherent sheaf $\mathcal{F}_0$ on X. Set $V_0 := \operatorname{Hom}(\mathcal{F}_0, \omega_X)$ and $V = V_0 \otimes A$, so that the*

Cartier module corresponding to $D(\underline{\mathcal{F}})$ has the form (V, κ_V). Then κ_V is of trace class and

$$L^{\text{naive}}(X, \underline{\mathcal{F}}, T) = \Delta\big(1 - T\kappa_V\big)^{(-1)^{n-1}}. \tag{7.1}$$

In particular $L^{\text{naive}}(X, \underline{\mathcal{F}}, T)$ is a rational function.

Proof. We give a sketch of the proof following Anderson. Full details can be found in [2]. The proof goes by proving equation (7.1) modulo T^m for $m \in \mathbb{N}$, $m \geq 2$.

Step 1: Formula (7.1) holds modulo T^m if $m_0 := \min\{\deg(x) \mid x \in |X|\} \geq m$. In this case the left-hand side is clearly congruent to 1 modulo T^m by definition since $m_0 \geq m$. To prove the same for the right-hand side, consider first the case $m = 2$. (The case $m = 1$ is trivial.)

Let I be the ideal of R generated by the set $\{r^q - r \mid r \in R\}$. Since $m_0 \geq 2$, the ideal I must be the unit ideal: Else let $\mathfrak{m} \supset I$ be a proper maximal ideal. Then in the field $R/\mathfrak{m}$ every element would satisfy the equation $\bar{r}^q = \bar{r}$. It follows that $R/\mathfrak{m} \cong k$, contradicting the fact that $m_0 \geq m \geq 2$. Since $I = R$, we can find $r_1, \ldots, r_s, f_1, \ldots, f_s \in R$ such that

$$1 = \sum_{i=1}^{s} \big(r_i - r_i^q\big) f_i.$$

We deduce that

$$\text{Tr}\,\big(\kappa_V\big) = \text{Tr}\,\big(\kappa_V \cdot 1\big) = \text{Tr}\,\left(\sum_{i=1}^{s} \kappa_V\big(r_i - r_i^q\big) f_i\right)$$

$$= \sum_{i=1}^{s} \Big(\text{Tr}(\kappa_V r_i f_i) - \text{Tr}(r_i \kappa_V f_i)\Big) \stackrel{7.12(c)}{=} 0.$$

Hence the right-hand side of (7.1) is congruent to 1 modulo T^2.

The case $m \geq 3$ is reduced to the case $m = 2$ by regarding $\bigwedge^j V$ as a module over the ring $(R^{\otimes j})^{\Sigma_j}$ of invariants of $R^{\otimes j}$ under the natural action of the symmetric group Σ_j on j elements.

Step 2: Reduction to the case where X contains at most one point of degree less than a given m. For this one chooses an affine covering by sets U_i of the form $\text{Spec}\, R_{f_i}$ for suitable $f_i \in R$ such that each U_i contains at most one such point. Then one proves an inclusion exclusion principle for both sides. That is, in the simplest case where $X = U_1 \cup U_2$ one proves that both sides are functions f on open subsets on X such that $f(X) = f(U_1)f(U_1)/f(U_1 \cap U_2)$.

Step 3: Induction on the dimension of R. Suppose equation (7.1) is known for all smooth affine varieties of dimension less than n. Here Anderson shows that for any $f \in R$ defining a disjoint decomposition $X = \text{Spec}\, R/f \cup \text{Spec}\, R_f$ there is a corresponding decomposition of (V, κ_V) into the restriction to $\text{Spec}\, R_f$ already used in Step 2 and a suitably defined residue $\text{Res}_f(V, \kappa_V)$ of (V, κ_V) along $\text{Spec}\, R/f$. If this is established and we are in the situation of Step 2, then we choose f such that the single point of small degree lies in $\operatorname{Spec} R/f$. Then by induction hypothesis one is done.

Step 4: Initial step of the induction. One needs to prove equation (7.1) explicitly in the case where $X = \operatorname{Spec} k_x$ is the spectrum of a finite field extension k_x of k. Explicitly one has to show that

$$\Delta\big(1 - T\operatorname{Res}_{f_1} \ldots \operatorname{Res}_{f_n}\big(V, \kappa_V\big)\big)$$

where $f_1, \ldots, f_n$ is a regular sequence defining a maximal ideal $\mathfrak{m}$ of R with $x = \operatorname{Spec} R/\mathfrak{m}$ agrees with the L-factor of $\underline{\mathcal{F}}_x$. $\qquad\square$

7.5 Proof of Theorem 6.13

Here we only give a proof under the following simplifying hypotheses: The compactification $\bar{X}$ is smooth and the ideal sheaf $\mathcal{I}_0$ of a complement $i: Z \hookrightarrow \bar{X}$ to $j: X \hookrightarrow \bar{X}$ is the inverse of an ample line bundle $\mathcal{O}_{\bar{X}}(1)$. We also assume that $\mathcal{F}$ is free, say of rank r. The latter hypothesis can be easily achieved as follows. Since X is affine, the sheaf $\mathcal{F}$ corresponds to a finitely generated projective $(R \otimes A)$-module. Choose a finitely generated projective complement Q, define τ to be zero on it, and replace $\underline{\mathcal{F}}$ by its direct sum with the nilpotent τ-sheaf defined by $(Q, 0)$.

For $m \geq 0$ we define $\mathcal{F}_{0,m} := \big(\mathcal{O}_{\bar{X}}(-m)\big)^{\oplus r}$ and $\mathcal{F}_m := \mathcal{F}_{0,m} \otimes_k A$. From our construction of $j_!\underline{\mathcal{F}}$ we see that for $m \gg 0$ the endomorphism τ extends (in the present case uniquely) to $\mathcal{F}_m$ and in such a way that its inverse image along i is zero. In other words, we have

$$\tau_m: \big(\sigma \times \operatorname{id}\big)^* \mathcal{F}_m \longrightarrow \mathcal{F}_{m+1} \subset \mathcal{F}_m. \tag{7.2}$$

Let m_0 be the smallest m such that the above factorization exists. Then for any $m \geq m_0$ the resulting τ-sheaf $\underline{\mathcal{F}}_m := (\mathcal{F}_m, \tau_m)$ on $\bar{X}$ is a representative of the crystal $j_!\underline{\mathcal{F}}$.

We can now invoke Serre duality. It provides us with a canonical isomorphism

$$D\big(H^n\big(\bar{X}, \mathcal{F}_m\big)\big) \cong H^0\big(\bar{X}, D(\mathcal{F}_m)\big);$$

observe that $D(\mathcal{F}_m)$ is simply $\mathcal{F}_m^\vee \otimes \Omega_{\bar{X}/k}$. But more is true. A careful analysis yields that the above isomorphism is compatible with the endomorphism τ_m. On the left-hand side, τ_m induces a *linear* endomorphism on $H^n(\bar{X}, \mathcal{F}_m)$, and $D(\_\!\_)$ dualizes this to a linear endomorphism on $D(H^n(\bar{X}, \mathcal{F}_m))$. On the right-hand side, $D(\_\!\_)$ provides us with a Cartier linear endomorphism $D(\tau_m)$ on $D(\mathcal{F}_m)$ induced from τ_m. Taking cohomology, we obtain an induced *linear* endomorphism on $H^0(\bar{X}, D(\mathcal{F}_m))$. One can show (cf. [8, §7.4]) that Serre duality identifies the two obtained linear endomorphisms.

For $m \gg 0$ define $W_{0,m} := H^0(\bar{X}, \mathcal{H}om(\mathcal{F}_{0,m}, \Omega_{\bar{X}/k}))$, so that $W_{m,0} \otimes_k A = H^0(\bar{X}, D(\mathcal{F}_m))$. One easily verifies that the $W_{0,m}$ form an increasing exhaustive filtration of $V_0 := H^0(X, \mathcal{H}om(\mathcal{F}, \Omega_{X/k}))$. Dualizing (7.2) one deduces that $D(\tau_m) = D(\tau)$ maps $W_{0,m+1} \otimes A$ into $W_{0,m} \otimes A$. This shows that

Lemma 7.14. *For any $m \geq m_0$, the k-vector space $(W_{0,m})$ is a nucleus for $D(\underline{\mathcal{F}})$.*

Thus Anderson's trace formula tells us that

$$L^{\mathrm{naive}}(X, \underline{\mathcal{F}}, T) = \Delta\big(1 - TD(\tau)\big)^{(-1)^{n-1}}$$
$$= \det\big(1 - TD(\tau_m) \mid H^0(\bar{X}, D(\mathcal{F}_m))\big)^{(-1)^{n-1}}.$$

Serre duality, as explained above, yields

$$\det\big(1 - TD(\tau_m) \mid H^0(\bar{X}, D(\mathcal{F}_m))\big) = \det\big(1 - T\tau_{H^n(\bar{X}, \mathcal{F}_m)} \mid H^n(\bar{X}, \mathcal{F}_m)\big)$$
$$= L\big(\operatorname{Spec} k, H^n(\bar{X}, \mathcal{F}_m), T\big).$$

To complete the proof of Theorem 6.13, it simply remains to observe that the cohomology groups $H^i(\bar{X}, \mathcal{F}_m), T)$, $i \neq n$, all vanish for m sufficiently large. This follows from Serre duality and the fact that $H^i(\bar{X}, \mathcal{O}_{\bar{X}}(m) \otimes \Omega_{\bar{X}/k}^{-1}) = 0$ for $i \neq 0$ and all $m \gg 0$, since the sheaf $\mathcal{O}_{\bar{X}}(1)$ is ample on $\bar{X}$.

7.6 The crystalline trace formula for general (good) rings A

Suppose first that the ring A is a good coefficient ring and reduced and that $\underline{\mathcal{F}}$ is a flat A-crystal on some scheme X of finite type over k. Under these hypotheses the L-function $L(X, \underline{\mathcal{F}}, T)$ is well defined. Since the nil radical of A is zero, all trace formulas are exact equalities, and thus to prove them we may pass to the quotient ring Q_A of A and therefore assume that A is artinian and reduced.

Now one has several standard techniques to prove the desired trace formula in this context:

(a) If $f \colon Y \to X$ is a finite morphism, then the trace formula in Theorem 6.14 holds. Here one can in fact directly prove that for any $x \in |X|$ one has

$$\prod_{y \in f^{-1}(x)} L(y, \underline{\mathcal{F}}_y, T) = L\big(x, (f_*\underline{\mathcal{F}})_x, T\big).$$

Note that the product on the left is finite.

(b) To prove an absolute trace formula, one can decompose X into a finite disjoint union of locally closed subschemes. Suppose that X is a disjoint union

$X = U \cup Z$ with $j : U \hookrightarrow X$ and open immersion and $i : Z \hookrightarrow X$ a closed complement. Then one has the short exact sequence

$$0 \longrightarrow j_! j^* \underline{F} \longrightarrow \underline{F} \longrightarrow i_* i^* \underline{F} \longrightarrow 0.$$

Since $j_!$ and $i_* = i_!$ are exact, the spectral sequence for direct image with proper support yields, upon applying $Rf_!$, the exact triangle

$$Rs_U! j^* \underline{F} \longrightarrow Rf_! \underline{F} \longrightarrow Rs_Z! i^* \underline{F} \longrightarrow Rs_U! j^* \underline{F}[1].$$

Since for the L-functions one has

$$L(X, \underline{F}, T) = L(U, j^* \underline{F}, T) \cdot L(Z, i^* \underline{F}, T)$$

it suffices to prove an absolute trace formula for U and Z.

(c) If $f : Y = \mathbb{A}^n \to X = \mathbb{A}^{n-1}$ is the projection onto the $n-1$ first coordinates and if the trace formula is proved for $\mathbb{A}^1 \to \operatorname{Spec} k$, it follows for f. One proceeds as in (a) except that now the fibers are no longer finite. One needs to use the proper base change formula in order to deduce

$$i_x^* \, Rf_! = Rf_{|Y_x!} \, i_{Y_x!}$$

where f_{Y_x} is the restriction of f to the fiber Y_x above x and $i_{Y_x} : Y_x \hookrightarrow Y$ is the closed immersion of the fiber obtained as the pullback along $i_x : x \hookrightarrow X$.

(d) The trace formula for the structure morphism $\mathbb{A}^1 \to \operatorname{Spec} k$ can be deduced as follows. Near the generic point, i.e., on $\mathbb{A}^1$ minus finitely many closed points, one has a locally free representative due to Proposition 5.14. Using (b) it suffices to prove the formula for this open subset. But here one can simply apply Theorem 6.13 for the naive L-function.

(e) From (a)–(d) and the Noether normalization lemma (after reduction to an affine situation), one can readily deduce the absolute trace formula for any X of finite type over k.

(f) Having the absolute trace formula, one can deduce the relative trace formula for a morphism $f : Y \to X$ of finite type from proper base change as explained in a special case in (c). The key point is that one obtains the relative version as the product over the closed points $x \in |X|$ of the formula

$$\prod_{y \in |f^{-1}(y)|} L(y, \underline{F}_y, T) = L\big(x, \big(Rf_! \underline{F}\big)_x, T\big). \tag{7.3}$$

Suppose now that A is not necessarily reduced. Then (c) and (f) above will fail. The point is that in formula (7.3) we no longer have equality; it gets replaced by $\sim$ which was introduced above Theorem 6.14. Now if we take an infinite product over formulas of the type (7.3) with $\sim$ instead of $=$, we loose all control, since the infinite product over elements in $1 + T \mathfrak{n}_A[T]$ may well be a non-rational

power series. The path taken in [8, Ch. 9] to overcome this difficulty and to prove
Theorem 6.14 is rather demanding. One can stratify X so that over the finitely
many pieces of the stratification all sufficiently high twists of the initial τ-sheaf
have a locally free representative. Using this representative, one shows that on
each stratum formula (7.3) is in fact an equality for all but finitely many $x \in |X|$.

Lecture 8

Global *L*-functions for *A*-motives

So far we were mainly concerned with *L*-functions which when compared with the classical theory should be considered as local *L*-functions. In the following we present an approach, essentially due to D. Goss, to define global Carlitz type *L*-functions for arbitrary τ-sheaves over A or A-crystals with a characteristic function from their underlying scheme to the coefficient ring A (at least for certain A to be specified below). These global *L*-functions are continuous homomorphisms from $\mathbb{Z}_p$ to entire functions on $\mathbb{C}_\infty$. We state the main conjectures of Goss on these *L*-functions on meromorphy, entireness and algebraicity and indicate the proof from [3] of these conjectures which is based on the theory of [8] which was explained in the previous lectures. A different proof of Goss' conjectures was given in [45] at least for $A = \mathbb{F}_q[t]$.

Section 8.5, the last section of this lecture, consists of an extended example. We recall the explicit expression for the global Carlitz–Goss type ζ-function of the affine line. Then we derive formulas for the special values at negative integers $-n$ in terms of the cohomological formalism developed so far. These formulas can be evaluated in complexity $\mathcal{O}(\log|n|)$ by computing an explicit determinant. For $p = q$ this will also provide yet another approach on a conjecture of Goss on the distribution of zeros of the global *L*-functions evaluated at elements of $\mathbb{Z}_p$. The proof for arbitrary q was given by Sheats in [44] after previous special work for $p = q$ by Wan [51], Diaz-Vargas [12] and Poonen. Yet another approach for $p = q$ is due to Thakur [50]. We end this section with some observations and questions based on computer experiments for the ζ-functions of some other affine curves.

Throughout this lecture, we fix the following notation:

- $\bar{C}/k$ will be a smooth projective geometrically irreducible curve over the finite field k of characteristic p.

- ∞ will be a marked closed point on $\bar{C}$ and $C := \bar{C} \smallsetminus \{\infty\}$.

- $K := k(\bar{C})$ is the function field of $\bar{C}$ and $A := \Gamma(C, \mathcal{O}_{\bar{C}})$ the coordinate ring of the affine curve C.

- K_∞ is the completion of K at ∞, $\mathcal{O}_\infty \subset K_\infty$ its ring of integers, $\pi_\infty \in \mathcal{O}_\infty$ a uniformizer, k_∞ the residue field at ∞, and $d_\infty := [k_\infty : k]$ the residue degree.

Suppose X is a scheme of finite type over k with a morphism $f\colon X \to \operatorname{Spec} A$ and that $\underline{\mathcal{F}}$ is a flat A-crystal over X. (Such an f would naturally arise in the case where $\underline{\mathcal{F}}$ comes from a Drinfeld A-module or an A-motive; there one could take f as the characteristic of this object.) For every closed point $x \in |X|$, its image $\mathfrak{p}_x := f(x) \in \operatorname{Spec}(A)$ is a maximal ideal and one clearly has the divisibility $d_{\mathfrak{p}_x}|d_x$ for the residue degrees. One would like to define

$$L^{\mathrm{glob}}(X, \underline{\mathcal{F}}, s) := \prod_{x \in |X|} L\big(x, \underline{\mathcal{F}}_x, T\big)_{|T^{d_{\mathfrak{p}_x}} = \mathfrak{p}^{-s}}.$$

What is needed at this point is a good definition of $\mathfrak{p}_x^s$ and a characteristic p domain in which the exponents s will lie. Following Goss, in Section 8.1 we shall give such definitions.

An important observation toward the proof of the conjectures of Goss on L^{glob}, that goes back to [45], is that the special values of Goss L-functions agree with the L-functions of the previous lectures if one replaces $\underline{\mathcal{F}}$ by a suitable twist. The cohomological theory will prove that these special values are polynomials whose degree grows logarithmically. It was known to Goss that this would suffice to deduce the entireness (or meromorphy) of the global L-functions defined by him.

In fact, we will not completely follow Goss' approach. He defines L^{glob} as a function

$$L^{\mathrm{glob}}(X, \underline{\mathcal{F}}, \_)\colon \mathbb{Z}_p \times \mathbb{C}_\infty^* \longrightarrow \mathbb{C}_\infty.$$

Moving one copy of $\mathbb{C}_\infty$ from left to right and working with $T = z^{-1}$, we shall define a global L-function

$$L^{\mathrm{glob}}(X, \underline{\mathcal{F}}, \_)\colon \mathbb{Z}_p \longrightarrow \Big(\mathbb{C}_\infty[[T]]_{\leq c}\Big)^*,$$

where $\mathbb{C}_\infty[[T]]_{\leq c}$ denotes the ring of power Series on $\mathbb{C}_\infty$ which converge on the closed disc of radius c and the superscript $*$ denotes its units. This ring is in a natural way a Banach space, and thus in particular a topological space.

8.1   Exponentiation of ideals

The case $A = k[t]$: In this case, any maximal ideal $\mathfrak{p}$ is generated by a unique irreducible monic polynomial $f_{\mathfrak{p}} \in k[t]$, and thus a natural definition of $\mathfrak{p}^s$, at least for $n = s \in \mathbb{Z}$, is $\mathfrak{p}^n := f_{\mathfrak{p}}^n$. As can be seen in the lectures of D. Thakur, it is also natural to consider the expression $f_{\mathfrak{p}}/t^{\deg f_{\mathfrak{p}}}$, which is a 1-unit in $\mathcal{O}_\infty$. For it the expression $(f_{\mathfrak{p}}/t^{\deg f_{\mathfrak{p}}})^n$ is well defined for any $n \in \mathbb{Z}_p$. Goss defines $\mathfrak{p}^s$ for any $s = (n, z) \in \mathbb{Z}_p \times \mathbb{C}_\infty^*$, cf. [24, §8.2], by

$$\left(f_{\mathfrak{p}}/t^{\deg f_{\mathfrak{p}}}\right)^n z^{\deg \mathfrak{p}}.$$

Let us now recall the necessary definitions for arbitrary A – note that k_∞ is canonically contained in K_∞.

Definition 8.1. A continuous homomorphism $\mathrm{sign} \colon K_\infty^* \to k_\infty^*$ is a

- *sign-function* if $\mathrm{sign} = \mathrm{id}$ when restricted to k_∞^*;
- *twisted sign-function* if there exists $\sigma \in \mathrm{Gal}(k_\infty/k)$ such that $\mathrm{sign} = \sigma$ when restricted to k_∞^*.

Obviously any twisted sign function is of the form $\sigma \circ \mathrm{sign}'$ for some $\sigma \in \mathrm{Gal}(k_v/k)$ and some sign-function sign'. Any sign-function is trivial on the 1-units in $\mathcal{O}_\infty^*$. If π is a uniformizer of K_∞, then any sign-function is uniquely determined by the image of π in k_∞^*. Using that $K_\infty^* \cong \pi^{\mathbb{Z}} \times \mathcal{O}_\infty^*$ one finds that the number of sign-functions is equal to $\#k_\infty^*$, and the number of twisted sign-functions to $\#k_\infty^* \cdot d_\infty$.

We fix a sign-function sign for K_∞.

Definition 8.2. An element $x \in K_\infty^*$ is *positive* (for sign) if $\mathrm{sign}(x) = 1$. An element $x \in K^*$ is positive if its image in K_∞^* is positive. We write K_+ for the positive elements of K^* and A_+ for $A \cap K_+$.

From now on, we also fix a positive element $\pi \in K^*$ which is a uniformizer of K_∞.

The *class group* $\mathrm{Cl}(A)$ *of* A is the quotient of the group of fractional ideals $\mathcal{I}_A$ of A modulo the subgroup of principal fractional ideals $\mathcal{P}$ of A.

Definition 8.3. The *strict class group* $\mathrm{Cl}^+(A)$ of A (with respect to sign) is the quotient of the group of fractional ideals of A modulo the subgroup of principal positively generated fractional ideals $\mathcal{P}^+$ of A. (A fractional ideal is principal positively generated if it is of the form Ax for $x \in K_+$.)

In particular one has an obvious short exact sequence

$$0 \longrightarrow k_\infty^*/k^* \longrightarrow \mathrm{Cl}^+(A) \longrightarrow \mathrm{Cl}(A) \longrightarrow 0.$$

Correspondingly, one defines the *strict Hilbert class field* $H^+ \supset H$ of K as the abelian extension of K which under the Artin homomorphism has Galois group isomorphic to $\mathrm{Cl}^+(A)$. More precisely, under the reciprocity map of class field theory $H^+ \supset H \supset K$ correspond to

$$\prod_{v \text{ finite}} \mathcal{O}_v^* \times \left(\mathrm{Ker}\left(\mathrm{sign}\right)\right) \subset \prod_{v \text{ finite}} \mathcal{O}_v^* \times K_\infty^* \subset \mathbb{A}_K^*.$$

Due to our choice of π_∞, we have $\mathrm{Ker}(\mathrm{sign}) = \pi_\infty^{\mathbb{Z}} \times (1 + \mathfrak{m}_\infty)$ where $\mathfrak{m}_\infty \subset \mathcal{O}_\infty$ is the maximal ideal.

Proposition 8.4 (Goss). *Let U_1^{perf} denote the group of 1-units in the perfect closure of K_∞. There exists a unique homomorphism*

$$\langle \_ \rangle : \mathcal{I} \longrightarrow U_1^{\text{perf}}$$

such that for all $a \in K_+$ one has

$$\langle a \rangle = a \cdot \pi_\infty^{-v_\infty(a)}.$$

Proof. The idea of the proof is as follows. Let $h^+ := \#\text{Cl}^+(A)$ denote the strict class number. Then, for any fractional ideal $\mathfrak{a}$, the power $\mathfrak{a}^{h^+}$ is positively generated, and thus $\langle \mathfrak{a}^{h^+} \rangle$ is defined. To define $\langle \mathfrak{a} \rangle$, we take the unique h^+th root within U_1^{perf} – note that for $m \in \mathbb{N}$ prime to p the mth root of a 1-unit in K_∞ can simply be defined by the binomial series for $1/m$, and thus lies itself again in K_∞. However h^+ may have p as a divisor. This necessitates the use of U_1^{perf}. $\qquad\square$

The above definition and the following one of exponentiation arose in correspondence between D. Goss and D. Thakur.

Fix a d_∞th root $\pi_* \in K^{\text{alg}}$ of π_∞.

Definition 8.5. For $n \in \mathbb{Z}$ and $\mathfrak{a}$ a fractional ideal of A, one defines

$$\mathfrak{a}^n := \pi_*^{-n\deg\mathfrak{a}} \cdot \langle \mathfrak{a} \rangle^n \in K_\infty^{\text{alg}}.$$

More generally, for $s := (n, z) \in \mathbb{Z}_p \times \mathbb{C}_\infty^*$ one sets

$$\mathfrak{a}^s := z^{\deg\mathfrak{a}} \cdot \langle \mathfrak{a} \rangle^n \in K_\infty^{\text{alg}}.$$

Note that the first exponentiation is a special case of the second if one takes for s the pair (n, π_*^{-n}). Also, going through the definitions and choosing $\pi_* = \frac{1}{t}$ for $A = k[t]$, one may easily verify that in this case Definition 8.5 agrees with the exponentiation described at the beginning of this section.

Proposition 8.6 (Goss). *Let $\mathbb{V}$ be the subfield of K_∞ generated by K and all the $\mathfrak{a}^n$ for all fractional ideals $\mathfrak{a}$ of A and all $n \in \mathbb{Z}$. Then $\mathbb{V}$ is a finite extension of K. Let $\mathcal{O}_\mathbb{V}$ denote its ring of integers (over A). Then for all ideals $\mathfrak{a} \subset A$ and all $n \in \mathbb{N}$, one has $\mathfrak{a}^n \in \mathcal{O}_\mathbb{V}$.*

8.2   Definition and basic properties of the global *L*-function

Let X, f, $\underline{\mathcal{F}}$ etc. be as above. At $x \in |X|$ the local L-factor $L(x, \underline{\mathcal{F}}_x, w)$ lies in $1 + w^{d_x} A[[w^{d_x}]] \subset 1 + w^{d_{\mathfrak{p}_x}} A[[w^{d_{\mathfrak{p}_x}}]]$. This shows that the following definition makes sense:

Definition 8.7. The *global L-function* of $(X, f, \underline{\mathcal{F}})$ is defined by

$$L^{\mathrm{glob}}(X, \underline{\mathcal{F}}, \_) \colon \mathbb{Z}_p \longrightarrow 1 + T\mathbb{C}_\infty[[T]]; \quad n \longmapsto \prod_{x \in |X|} L\left(x, \underline{\mathcal{F}}_x, w\right)_{w^{d_{\mathfrak{p}x}} = T^{d_{\mathfrak{p}x}} \langle \mathfrak{p}_x \rangle^n}.$$

Remarks 8.8.

(a) It is elementary to see that there is a constant $c \in \mathbb{Q}_{>0}$, independent of $n \in \mathbb{Z}_p$, such that any of the values

$$L^{\mathrm{glob}}(X, \underline{\mathcal{F}}, n) \in 1 + T\mathbb{C}_\infty[[T]]$$

lies in the ring $\mathbb{C}_\infty[[T]]_{\leq c}$ of convergent power series around 0 on the (closed) disc $\{T \in \mathbb{C}_\infty \mid |T| \leq c\}$. Since at $T = 0$ the power series take the value 1, one can choose c such that for all $n \in \mathbb{Z}_p$ the power series $L^{\mathrm{glob}}(X, \underline{\mathcal{F}}, n)$ are bounded away from zero uniformly, i.e., for each $n \in \mathbb{Z}_p$ they are units in $\mathbb{C}_\infty[[T]]_{\leq c}$.

(b) The function $n \mapsto \langle \mathfrak{p} \rangle^n$ satisfies a very strong interpolation property. The limit $\lim_{m \to \infty} \langle \mathfrak{p} \rangle^{p^m} = 1$ is uniform in $\mathfrak{p} \in \mathbf{Max}(A)$. Moreover $\mathbb{C}_\infty[[T]]_{\leq c}$ is a Banach space under the norm given by

$$\left\| \sum_{m \geq 0} a_m T^m \right\| = \sup_{m \geq 0} |a_m| |c|^n.$$

Using the interpolation property of $n \mapsto \langle \mathfrak{p} \rangle^n$ it is not difficult to show that $L^{\mathrm{glob}}(X, \underline{\mathcal{F}}, \_)$ is a continuous function from $\mathbb{Z}_p$ into this Banach space.

(c) In Goss' formulation, the variable T is substituted by z^{-1}. Thus one has a continuous function from $\mathbb{Z}_p$ into a Banach space of convergent power series in $1/z$ around ∞ of radius $1/c$.

(d) The definition of $L^{\mathrm{glob}}(X, \underline{\mathcal{F}}, \_)$ depends on the choice of $\pi_* \in K^{\mathrm{alg}}$. This element is used to define $\langle \_ \rangle$ and its d_∞th power is a uniformizer π_∞ of K_∞ which determines our sign function.

Let us draw a first conclusion from the cohomological theory of crystals for Goss' global L-functions.

Proposition 8.9. *Suppose* $f \colon X \to \operatorname{Spec} A$ *is a morphism of finite type and* $\underline{\mathcal{F}}$ *is a flat A-crystal on* X. *Then:*

(a) *The complex* $Rf_!$ *can be represented by a bounded complex* $\underline{\mathcal{G}}^\bullet$ *of flat A-crystals on* $\operatorname{Spec} A$.

(b) *With* $\underline{\mathcal{G}}^\bullet$ *from* (a) *one has*

$$L^{\mathrm{glob}}(X, \underline{\mathcal{F}}, n) = \prod_{i \in \mathbb{Z}} L^{\mathrm{glob}}(X, \underline{\mathcal{G}}^i, n)^{(-1)^i}.$$

Part (a) follows from Theorem 5.10, and part (b) is an application of the trace formula (Theorem 6.14) to each fiber $X \times_{\operatorname{Spec} A} \mathfrak{p}$, together with the proper base change (Theorem 4.6).

Thus to study general properties of global L-functions, it suffices to consider the case $X = \operatorname{Spec} A$! We will do so from now on unless stated otherwise.

8.3   Global *L*-functions at negative integers

We now describe the link between L^{glob} and the L-function from the previous lectures. We learned this in the case $A = k[t]$ from [45].

For any ring A there is an analog of the Carlitz module, the so-called Drinfeld–Hayes module. Let $\mathcal{O}^+$ denote the ring of integers over A of $H^+ \supset K$. In [29] Hayes shows that there are $\#\mathrm{Cl}^+(A)$ sign-normalized rank 1 Drinfeld modules

$$\psi_{DH,A} \colon A \longrightarrow \mathcal{O}^+[\tau].$$

(We provide more details on these in Appendix A.3.) Denote the structure morphism $\operatorname{Spec} \mathcal{O}^+ \to \operatorname{Spec} A$ by s and the A-motive associated to $\psi_{DH,A}$ by $\mathcal{H}_A$. Thus $\mathcal{H}_A$ is a locally free τ-sheaf on $\operatorname{Spec} \mathcal{O}^+$ over A of rank 1.

It is now possible and not too hard to compute the local L-factors of the tensor powers $\mathcal{H}^{\otimes n}$, $n \in \mathbb{N}_0$; cf. [3, Lemma 3.2]. Define

$$L_X^{\mathrm{glob}}(n,T) := L^{\mathrm{glob}}\big(X, (\mathcal{O}_{X \times \operatorname{Spec} A}, \tau_{\mathrm{can}} = (\sigma \times \mathrm{id})), n\big)$$

as the global Carlitz–Goss L-function for any A-scheme X of finite type, i.e., any scheme with a structure morphism to $\operatorname{Spec} A$. We explicitly write the variable T on the left to stress that for each $n \in \mathbb{Z}_p$ the right-hand term is a function in T. Similarly, for any character χ of the abelian group $\operatorname{Gal}(H^+/K)$ define $L_X^{\mathrm{glob}}(n, \chi, T)$ as the ζ-function on $\operatorname{Spec} A$ for the character χ. Using this intuitive notation, the following formula was first observed by Goss; see also Lemma 8.22.

Theorem 8.10 (Goss). *For $n \in \mathbb{N}_0$ one has*

$$L_{\operatorname{Spec} \mathcal{O}^+}(-n, T) = L\big(\operatorname{Spec} A, s_* \mathcal{H}_A^{\otimes n}, T\big) = \prod_{\chi \in \widehat{\operatorname{Gal}(H^+/K)}} L_X^{\mathrm{glob}}\big(-n, \chi, T\big)\big|_{T = T\pi_*^n}.$$

By the symbol $(\_)|_{T = T\pi_*^n}$ we mean that one substitutes the term $T\pi_*^n$ for T.

The same computations and defining L-functions with characters for crystals, cf. [3, Corollary 3.8], yield

Theorem 8.11. *Suppose $\underline{\mathcal{F}}$ is a flat A-crystal on $\operatorname{Spec} A$. Then for $n \in \mathbb{N}_0$ one has*

$$L\big(\operatorname{Spec} A, \underline{\mathcal{F}} \otimes s_* \mathcal{H}_A^{\otimes n}, T\big) = \prod_{\chi \in \widehat{\operatorname{Gal}(H^+/K)}} L^{\mathrm{glob}}\big(\operatorname{Spec} A, \underline{\mathcal{F}}, \chi^n, -n\big)\big|_{T = T\pi_*^n}.$$

Remark 8.12. The importance of the previous theorem lies in the fact that the left-hand side is, via the trace formula (Corollary 6.15), equal to the L-function of $H^1(C_{H^+}, j_! \underline{\mathcal{F}} \otimes s_* \mathcal{H}_A^{\otimes n})$, where C_{H^+} is the smooth projective geometrically irreducible curve over k_∞ with function field H^+ and where $j \colon \operatorname{Spec} \mathcal{O}^+ \hookrightarrow C_{H^+}$ is the canonical open immersion. The cohomology can be represented by a τ-sheaf whose underlying module is free of finite rank over A. To compute this τ-sheaf one may compute the coherent cohomology of the underlying sheaf together with the endomorphism induced by τ. In doing so, one has the freedom to replace the τ-sheaf which appears as the argument of cohomology by a nil-isomorphic one.

8.4 Meromorphy and entireness

One can prove the following simple but important lemma. Its proof follows closely the method described in Section 8.5 for $\underline{\mathcal{C}}^{\otimes n}$.

Lemma 8.13. *For a locally free τ-sheaf $\underline{\mathcal{F}}$ on $\operatorname{Spec} A$ over A, the function*

$$n \longmapsto \deg_T L\big(\operatorname{Spec} A, \underline{\mathcal{F}} \otimes s_* \mathcal{H}_A^{\otimes n}, T \big)$$

is of order of growth at most $O(\log n)$.

Now for any $n \geq 0$ which is a multiple of $h^+ := \#\mathrm{Cl}^+(A)$, the characters χ^n, $\chi \in \operatorname{Gal}(\widehat{H^+}/K)$ are all trivial. Thus one deduces from the lemma and Theorem 8.11 that $L^{\mathrm{glob}}(\operatorname{Spec} A, \underline{\mathcal{F}}, -n)$ for such n is a polynomial in T whose degree grows at most like $O(\log n)$. Let $h^{(p)}$ be the prime-to-p part of h^+. By replacing $\underline{\mathcal{F}}$ by some kind of Frobenius twist on the base – this is similar to the observation made in Remark 8.20 (a) – one can show logarithmic degree growth for all negative integers which are multiples of $h^{(p)}$. These lie dense in $\mathbb{Z}_p$ and the strong uniform interpolation property of $n \mapsto \langle \mathfrak{p} \rangle^n$ yields:

Theorem 8.14. *If $\underline{\mathcal{F}}$ is a locally τ-sheaf on $\operatorname{Spec} A$ over A, then $L^{\mathrm{glob}}(\operatorname{Spec} A, \underline{\mathcal{F}}, \_)$ is a continuous function $\mathbb{Z}_p \to \mathbb{C}_\infty[[T]]^{\mathrm{ent}}$, where $\mathbb{C}_\infty[[T]]^{\mathrm{ent}}$ is the ring of entire power series with coefficients in $\mathbb{C}_\infty$ – it is in a natural way a Fréchet space, and thus in particular a metrizable topological space.*

Once the function $L^{\mathrm{glob}}(\operatorname{Spec} A, \underline{\mathcal{F}}, \_)$ is known to be entire, one deduces from Theorem 8.11 and Lemma 8.13, invoking Proposition 8.6, the following:

Theorem 8.15. *For all $n \in \mathbb{N}_0$, the values $L^{\mathrm{glob}}(\operatorname{Spec} A, \underline{\mathcal{F}}, -n)|_{T=T\pi_*^n}$ are polynomials in T with coefficients in $\mathcal{O}_V$ and their degrees in T are of growth at most $O(\log n)$.*

Definition 8.16 (Goss). Let f be a function $f \colon \mathbb{Z}_p \to 1 + T\mathbb{C}_\infty[[T]]$.

The function f is called *entire* if the image of f lies in $\mathbb{C}_\infty[[T]]^{\mathrm{ent}}$ and if f is continuous with respect to the natural topologies on $\mathbb{Z}_p$ and $\mathbb{C}_\infty[[T]]^{\mathrm{ent}}$.

The function f is called *meromorphic* if it is the quotient of two entire functions.

An entire function f is called *essentially algebraic* if for all $n \in \mathbb{N}_0$, the value $f(-n)|_{T=T\pi_*^n}$ is a polynomial whose coefficients are integral over A and lie in a fixed finite extension of K, independent of n.

A meromorphic function f is called *essentially algebraic* if it is the quotient of two entire essentially algebraic functions.

From Theorems 8.14 and 8.15 and Theorem 5.14 one deduces readily:

Theorem 8.17. *Suppose $\underline{F}$ is a flat A-crystal on $\operatorname{Spec} A$ (or a finite complex of such). Then $L^{\mathrm{glob}}(\operatorname{Spec} A, \underline{F}, \_\,)$ is meromorphic and essentially algebraic. If in addition $\underline{F}$ has a locally free representative, then $L^{\mathrm{glob}}(\operatorname{Spec} A, \underline{F}, \_\,)$ is entire and essentially algebraic.*

8.5   The global Carlitz–Goss L-function of the affine line

Throughout this section, we fix $A = k[t]$ and identify $\mathbb{A}^1 = \operatorname{Spec} A$. Recall that

$$L^{\mathrm{glob}}_{\mathbb{A}^1}(n, T) := L^{\mathrm{glob}}\big(\operatorname{Spec} A, (\mathcal{O}_{\operatorname{Spec} A \times \operatorname{Spec} A}, \sigma \times \mathrm{id}), n \big)$$
$$= \prod_{\mathfrak{p} \in \mathbf{Max}(A)} \big(1 - T^{d_{\mathfrak{p}}} \langle \mathfrak{p} \rangle^{-n} \big)^{-1}.$$

We also define for $n \in \mathbb{Z}$ the more intuitive ζ-function

$$\zeta_A(n, T) := \prod_{a \in A_+, \, \mathrm{irred.}} \big(1 - T^{\deg(a)} a^{-n} \big)^{-1}$$

$$= \prod_{\mathfrak{p} \in \mathbf{Max}(A)} \big(1 - T^{d_{\mathfrak{p}}} \mathfrak{p}^{-n} \big)^{-1} = \sum_{\mathfrak{a} \le A} T^{\deg \mathfrak{a}} \mathfrak{a}^{-n} = \sum_{d=0}^{\infty} T^d \bigg(\sum_{a \in A_{d+}} a^{-n} \bigg);$$

by A_{d+} we denote the set of monic elements of degree d in $A = k[t]$. The agreement of the second and fifth terms is immediate by expanding the Euler product. The third and fourth terms are term by term the same as the second and fifth where however we have used the exponentiation from Section 8.1.

Lemma 8.18. *For $n \in \mathbb{N}_0$ one has*

$$L^{\mathrm{glob}}_{\mathbb{A}^1}(-n, T)|_{T=T\pi_*^n} = \zeta_A(-n, T).$$

Proof.

$$L^{\mathrm{glob}}_{\mathbb{A}^1}(-n, T)|_{T=T\pi_*^n} = \prod_{\mathfrak{p} \in \mathbf{Max}(A)} \big(1 - (T\pi_*^n)^{d_{\mathfrak{p}}} \langle \mathfrak{p} \rangle^{-n} \big)^{-1}$$

$$= \prod_{\mathfrak{p} \in \mathbf{Max}(A)} \big(1 - T^{d_{\mathfrak{p}}} (1/t)^{n d_{\mathfrak{p}}} \langle \mathfrak{p} \rangle^{-n} \big)^{-1}$$

$$= \prod_{a \in A_+, \text{irred.}} (1 - T^{\deg(a)} t^{-n \deg(a)} (a/t^{\deg(a)})^{-n})^{-1}$$

$$= \prod_{a \in A_+, \text{irred.}} (1 - T^{\deg(a)} a^{-n})^{-1} \;=\; \zeta_A(-n, T). \qquad \square$$

Definition 8.19. For $n \in \mathbb{N}_0$ with q-digit expansion $n = a_0 + a_1 q + \cdots + a_r q^r$ we define $\ell(n) := a_0 + a_1 + \cdots + a_r$ to be the sum over the digits in the base q expansion of n.

Remark 8.20.

(a) As can be seen directly from the definition of $\zeta_A(n, T)$, one has $\zeta_A(n, T)^p = \zeta(pn, T^p)$.

(b) By a theorem of Lee (a student of Carlitz), one has $\sum_{a \in A_{d+}} a^n = 0$ for $d > \ell(n)/(q-1)$, so that

$$\deg_T \zeta_A(-n, T) \leq \frac{\ell(n)}{(q-1)}.$$

(c) Combining (a) and (b), for $q = p^m$ one finds

$$\deg_T \zeta_A(-n, T) \leq \frac{1}{q-1} \min \left\{ \ell(n), \ell(pn), \ell(p^2 n), \dots, \ell(p^{m-1} n) \right\}.$$

Although not stated explicitly in [44], one can easily deduce from it the following result:

Proposition 8.21. *For $n \in \mathbb{N}_0$ and q an arbitrary prime power one has*

$$\deg_T \zeta_A(-n, T) = \left\lfloor \frac{\min\{\ell(n), \ell(pn), \ell(p^2 n), \dots, \ell(p^{m-1} n)\}}{q-1} \right\rfloor.$$

Lemma 8.22. *For $n \in \mathbb{N}_0$ one has $\zeta_A(-n, T) = L(\mathbb{A}^1, \underline{C}^{\otimes n}, T)$.*

The lemma is Theorem 8.10 in its simplest case!

Proof. It suffices to show that for $f \in A_+$ irreducible and $\mathfrak{p} = (f)$ one has

$$1 - f^n T^{\deg(f)} \overset{!}{=} \det\left(1 - T\tau_{\mathcal{C}}^{\otimes n} \mid C_\mathfrak{p}^{\otimes n}\right) = \det\left(1 - T^{\deg(f)}(\tau_{\mathcal{C}}^{\otimes n})^{\deg(f)} \mid C_\mathfrak{p}^{\otimes n}\right).$$

Starting from $\tau^{\otimes n} = (t - \theta)^n (\sigma \times \text{id})$ we compute

$$\left(\tau^{\otimes n}\right)^{\deg(f)} = (t - \theta)^n (t - \theta^q)^n \cdots (t - \theta^{q^{\deg f - 1}})^n (\sigma^{\deg f} \times \text{id}).$$

On $k_\mathfrak{p} = k[\theta]/(f(\theta))$ we have $\sigma^{\deg f} = \text{id}$. Writing $\bar{\theta}$ for the image of θ in $k_\mathfrak{p}$, the roots of f are precisely the elements $\bar{\theta}^{q^i}$, $i = 0, \dots, \deg f - 1$, and so $f(t) = (t - \bar{\theta})(t - \bar{\theta}^q)^n \cdots (t - \bar{\theta}^{q^{\deg f - 1}})$. We deduce

$$\left(\tau^{\otimes n}\right)^{\deg(f)} = f(t)^n \in k_\mathfrak{p}[t],$$

which completes the proof of the lemma. $\qquad \square$

The next aim is to find for each $n \in \mathbb{N}_0$ an explicit square matrix whose dual characteristic polynomial is $\zeta_A(-n, T)$. Moreover we would like to find such a matrix of size $\deg_T(\zeta_A(-n, T))$!

The straightforward use of the representative $j_! \underline{\mathcal{C}}^{\otimes n}$ found in Example 3.12 is not suitable for this, since it yields a matrix of size $\dim H^1(\mathbb{P}^1, \mathcal{O}_{\mathbb{P}^1}(m)) = -1 - m > \frac{n}{q-1} - 1$, which is far larger than $\frac{\ell(n)}{q-1}$ or the possibly even smaller true degree given in Proposition 8.21.

Definition 8.23. For $i \in \mathbb{N}_0$ define the τ-sheaf $\underline{\mathcal{C}}^{(q^i)}$ on $\mathbb{A}^1$ over A by the pair

$$\left(k[\theta, t], \left(t^{q^i} - \theta\right)(\sigma \times \mathrm{id})\right).$$

Lemma 8.24. *The τ-sheaf $\underline{\mathcal{C}}^{(q^i)}$ is nil-isomorphic to $\underline{\mathcal{C}}^{\otimes q^i}$.*

Proof. The pullback $(\sigma^i)^* \underline{\mathcal{C}}^{(q^i)}$ is isomorphic to $\underline{\mathcal{C}}^{\otimes q^i}$ and via τ^i it is nil-isomorphic to $\underline{\mathcal{C}}^{(q^i)}$. □

Corollary 8.25. *Suppose $n = a_0 + a_1 q + \cdots + a_r q^r$ with $0 \le a_i \le q-1$ is the base q expansion of $n \in \mathbb{N}_0$. Let $g(\theta) := \prod_{i=0}^r (t^{q^i} - \theta)^{a_i}$. Then $\underline{\mathcal{C}}^{\otimes n}$ is nil-isomorphic to*

$$\underline{\mathcal{C}}^{(n)} := \bigotimes_{i=0}^r \left(\underline{\mathcal{C}}^{(q^i)}\right)^{\otimes a_i} = \left(k[\theta, t], g(\theta)(\sigma \times \mathrm{id})\right).$$

Let us now determine a representative of $j_! \underline{\mathcal{C}}^{(n)}$ for the open immersion $j : \mathbb{A}^1 \hookrightarrow \mathbb{P}^1$. Proceeding exactly as in Example 3.12, we find the following. As the underlying sheaf we take $\mathcal{O}_{\mathbb{P}^1}(-m\infty) \otimes_k A$ with $m = \lceil \frac{-1-\ell(n)}{q-1} \rceil$, and as τ_m we take $g(\theta)(\sigma \times \mathrm{id})$. Note that now

$$\mathrm{rank}_{k[t]} H^1\left(\mathbb{P}^1, \mathcal{O}_{\mathbb{P}^1}(-m\infty)\right) = \mathrm{rank}_{k[t]} H^0\left(\mathbb{P}^1, \mathcal{O}_{\mathbb{P}^1}((m-2)\infty)\right)$$

$$= m - 1 = \left\lfloor \frac{\ell(n)}{q-1} \right\rfloor =: D.$$

Suppressing $d\theta$ in the notation, we take $e_1 := 1, e_2 := \theta, \ldots, e_D := \theta^{D-1}$ as a basis over $k[t]$ of $H^0(\mathbb{P}^1, \mathcal{O}_{\mathbb{P}^1}(m - 2\infty))$. Then by Lemma 7.14 this is a nucleus in the sense of Anderson for $D(\underline{\mathcal{C}}^{(n)}) = k[\theta, t]\, d\theta$. Note that the image of e_i under $D(\tau^{(n)})$ is $\kappa(\theta^{i-1}) = C_q(g(\theta)\theta^{i-1})$. To be more explicit, we define $g_j(t) \in k[t]$ by

$$g(\theta) := \sum_{j \ge 0} g_j(t)\theta^j$$

and $g_j = 0$ for $j < 0$. Then

$$\kappa(e_i) = C_q\left(\sum_{j \ge 0} g_j(t)\theta^{j+i-1}\right) = \sum_{j \ge 0, \, j+i \equiv 0 \,(\mathrm{mod}\, q)} g_j(t)\theta^{(j+i)/q-1}$$

$$= \sum_{\ell \geq 0, j=-i+q\ell} g_{q\ell-i}(t)\theta^{\ell-1} = \sum_{\ell \geq 0, j=-i+q\ell} g_{q\ell-i}(t)e_\ell.$$

Define now M as the matrix

$$M := (m_{i,j}) := \left(g_{jq-i}\right)_{i,j=1,\ldots,D}.$$

This is the matrix representing the action of κ on the nucleus of $D(\underline{C}^{(n)})$. Hence its dual characteristic polynomial is the L-function of $\underline{C}^{(n)}$. Since A is reduced, we have shown:

Proposition 8.26.

$$\det\left(1 - TM\right) = \zeta_A\left(-n, T\right) \in 1 + TA[T].$$

Example 8.27. Let $q = 4$, $n = 181 = 2 \cdot 64 + 3 \cdot 16 + 1 \cdot 4 + 1 \cdot 1$. Thus $\ell(n) = 7$ and $D = \lfloor \frac{\ell(n)}{q-1} \rfloor = \lfloor \frac{7}{3} \rfloor = 2$. We have

$$g(\theta) = \left(t^{64} - \theta\right)^2 \left(t^{16} - \theta\right)^3 \left(t^4 - \theta\right)\left(t - \theta\right)$$
$$= \left(t^{128} - \theta^2\right)\left(t^{32} - \theta^2\right)\left(t^{16} - \theta\right)\left(t^4 - \theta\right)\left(t - \theta\right) = \sum_{j \geq 0} g_j(t)\theta^j.$$

One finds the following expressions for the coefficients g_j:

$$g_7 = 1.$$
$$g_6 = t^{16} + t^4 + t.$$
$$g_5 = t^{128} + t^{32} + t^{20} + t^{17} + t^5.$$
$$g_3 = t^{160} + t^{148} + t^{145} + t^{133} + t^{52} + t^{49} + t^{37}.$$
$$g_2 = t^{176} + t^{164} + t^{161} + t^{149} + t^{53}.$$

The matrix M is given by $\left(\begin{smallmatrix} g_3 & g_7 \\ g_2 & g_6 \end{smallmatrix}\right)$ and one finds

$$\zeta_A\left(-181, T\right) = \det\left(1 - TM\right) = 1 + T(\ldots) + T^2\left(t^{164} + t^{161} + \text{lower order terms}\right).$$

Definition 8.28. For $n \in \mathbb{N}_0$ we set $S_d(n) := \sum_{a \in A_{d+}} a^{-n}$, so that $\zeta_A(-n, T) = \sum_{d \geq 0} T^d S_d(n)$, and we define $s_d(n) := \deg_t S_d(n)$.

Using M we shall give yet another proof of the following theorem due to Wan [51], which was later reproved by Diaz-Vargas [12] and by Thakur [50].

Theorem 8.29 (Riemann hypothesis). *Let $q = p$. Then the following hold:*

(a) *For any $n \in \mathbb{N}_0$ one has*

$$\deg_T \zeta_A\left(-n, T\right) = \left\lfloor \frac{\ell(n)}{q-1} \right\rfloor \quad \text{and} \quad s_d(n) = \sum_{j=1}^{d} \deg_t \left(g_{j(q-1)}\right).$$

(b) *The sequence $(\deg_t g_{j(q-1)})_{j\geq 0}$ is strictly decreasing, and thus the Newton polygon of $\zeta_A(-n,T)$ has $\lfloor \frac{\ell(n)}{q-1} \rfloor$ distinct slopes all of width one.*

(c) *For any $m \in \mathbb{Z}_p$, the entire function $L_A(m,T)$ in T has a Newton polygon whose slopes are all of width one and thus the roots of $T \mapsto L_A(m,T)$ lie in K_∞; they are simple and of pairwise distinct valuation.*

Part (c) for $q = 4$ was proved by Poonen and for arbitrary q by Sheats in [44].

Recall that the Newton polygon of the polynomial $\deg_T \zeta_A(-n,T) \in A[T] \subset K_\infty[T]$ is the lower convex hull of the points $(d, -\deg_t(S_d(n)))_{d\geq 0}$. By (a), the slope between d and $d+1$ is $-s_{d+1}(n) - (-s_d(n)) = -\deg_t g_{d(q-1)}$. By the first assertion of (b), the sequence $\deg_t g_{j(q-1)}$ is strictly decreasing and it follows that the points $(d, -\deg_t(S_d(n)))_{d\geq 0}$ lie all on the lower convex hull and are break points, i.e., points where the slope changes. Thus the second part of (b) follows, once (a) and the first part is shown. Part (c) is a simple formal consequence of (b) as explained in [51]. To prove the theorem, our first aim will be to compute the degrees of the polynomials g_j.

Let us first recall a lemma of Lucas.

Lemma 8.30. *For integers $a_0, \ldots, a_r$ and $l_0, \ldots, l_r$ in the interval $[0, p-1]$ one has*

$$\prod_{i=0}^{r} \binom{a_i}{l_i} = \binom{\sum_{i=0}^{r} a_i p^i}{\sum_{i=0}^{r} l_i p^i}.$$

An analogous formula holds for multinomial coefficients $\binom{m}{m_1 \, m_2 \, \ldots \, m_\ell}$.

Lucas' formula follows easily from expanding both sides of the following equality by the binomial theorem:

$$(1+T)^{\sum_{i=0}^{r} a_i p^i} = \prod_{i=0}^{r} \left(1 + T^{p^i}\right)^{a_i}.$$

Lucas' lemma holds for arbitrary q; however the formulation for p implies the lemma also for all p-powers.

Note that by the lemma of Lucas the coefficient $\binom{n}{l}$ modulo p is non-zero if and only if, considering the base p expansions of n and l, each digit of the expansion of l is at most as large as the corresponding digit of n.

Lemma 8.31.

$$g_j = (-1)^j \sum_{\substack{l_0,\ldots,l_r,\, 0\leq l_i\leq a_i \\ \sum_i l_i = d-j}} \left(\prod_{i=0}^{r} \binom{a_i}{l_i}\right) t^{l_0+l_1 q+\cdots+l_r q^r} = (-1)^j \sum_{\substack{l=0 \\ \ell(l)=d-j}}^{n} \binom{n}{l} t^l.$$

For each $m \in \mathbb{N}_0$ the sum contains at most one summand of degree m in t.

Proof. The first equality in Lemma 8.31 is proved by a multiple application of the binomial theorem and collecting all the coefficients of θ^j in $\sum_{g \geq 0} g_j \theta^j = \prod_{i \geq 0} (T^{q^i} - \theta)^{a_i}$. The second equality follows from Lucas' lemma. $\qquad\square$

Set $\mu(D) = r$ and $m_{\mu(D)} = 0$ and for $j \in \{0, \ldots, D-1\}$ define $\mu(j) \in \{0, \ldots, r\}$ and $m_{\mu(j)} \in \{1, \ldots, a_{\mu(j)}\}$ by

$$(D-j)(q-1) = a_r + \cdots + a_{\mu(j)+1} + m_{\mu(j)}.$$

Lemma 8.32. *Suppose* $q = p$. *Then for each* $j \in \{0, \ldots, D\}$ *one has*

$$\deg g_j = a_r q^r + \cdots + a_{\mu(j)+1} q^{\mu(j)+1} + m_{\mu(j)} q^{\mu(j)}.$$

Proof. Note that in the case $p = q$ the coefficients $\left(\prod_{i=0}^{r} \binom{a_i}{l_i} \right)$ of g_j are all non-zero. The formulas are now immediate from the definitions of μ and m_μ. $\qquad\square$

Suppose $q = p$. Then if j increases, $D - j$ decreases and thus $\deg g_j$ is strictly decreasing in j. Thus we have proved Theorem 8.29(b) once we have proved part (a). Moreover if ℓ and ℓ' from $\{0, \ldots, D\}$ satisfy $\ell' \geq \ell + (q-1)$, then $\mu(\ell') \geq \mu(\ell) + 1$. This yields a precise result on the rate of the decrease of the $\deg_t g_j$:

Lemma 8.33. *Suppose* $p = q$. *Then for* $0 \leq \ell, \ell' \leq D-1$ *and* $\ell' \geq \ell + (q-1)$ *one has*

$$0 < \deg g_\ell - \deg g_{\ell+1} = q^{\mu(\ell)} \leq \frac{1}{q} \cdot q^{\mu(\ell')} = \frac{1}{q} \cdot \left(\deg g_{\ell'} - \deg g_{\ell'+1} \right).$$

Proof of Theorem 8.29(a). Our aim is to compute the degree in t of the coefficients of the T^i in the expression

$$\det(1 - TM) = \sum_{\pi \in \Sigma_d} \mathrm{sign}\,\pi \left(\delta_{1\pi(1)} - T m_{1\pi(1)} \right) \cdot \, \cdots \, \cdot \left(\delta_{d\pi(d)} - T m_{d\pi(d)} \right).$$

Let us expand the inner products by the distributive law. If a product contributes to T^i, then in the distributed term we need $d-j$ occurrences of terms not involving T, i.e., of 1's. The latter can only come from the diagonal. We deduce:

Up to sign, the terms contributing to T^i are those $(j \times j)$-minors of M which are symmetric, i.e., in which the same rows and columns were deleted from M.

Let J be a subset of $\{1, \ldots, d\}$ (which may be empty) and let $\pi \in \Sigma_d$ be a permutation of the set $\{1, \ldots, d\}$ which is the identity on J. Then for the pair J, π we define

$$\deg_{J,\pi} := \sum_{j \in J} \deg \left(m_{\pi(j),j} \right).$$

For fixed J, we shall show that the identity permutation is the unique one for which $\deg_{J,\pi}$ is maximal. The following lemma is the key step.

Lemma 8.34. *Suppose $q = p$ and fix $J \subset \{1, \ldots, d\}$. Then for all $\pi \in \Sigma_d \smallsetminus \{\mathrm{id}\}$ fixing J one has $\deg_{\pi,J} < \deg_{\mathrm{id},J}$. In particular $\mathrm{ord}(\det(M)) = \mathrm{ord}_{\mathrm{id}}$ and thus $\det(M)$ is non-zero.*

Proof. We only give the proof for $J = \varnothing$. The other cases are analogous. To simplify notation, we abbreviate $\deg_\pi := \deg_{\varnothing,\pi}$.

Assume, contrary to the assertion of the lemma, that for some $\pi \in \Sigma_d \smallsetminus \{\mathrm{id}\}$ we have $\deg_\pi = \max\{\deg_\tau \mid \tau \in \Sigma_d\}$. Since π is not the identity, the permutation matrix representing π has some entry above the diagonal. Let $j_0 \in \{1, \ldots, d\}$ be maximal such that $\pi(j_0) < j_0$. In row j_0 let j_1 be the column which contains the non-zero entry of the permutation matrix of π, i.e., $j_1 = \pi^{-1}(j_0)$. By the maximality of j_0 we have

$$j_1 < j_0. \tag{8.1}$$

Consider the matrix

$$\begin{pmatrix} \ddots & & & & \\ & 0 & & m_{\pi(j_0),j_0} & \\ & & \ddots & & \\ & m_{j_0,j_1} & & 0 & \\ & & & & \ddots \end{pmatrix}$$

with entries $m_{\pi(j),j}$ at $(\pi(j), j)$ for $j = 1, \ldots, d$ and zero otherwise. Define the permutation π' by

$$\pi'(j) = \pi(j) \ \text{ for } \ j \neq j_0, j_1, \quad \pi'(j_0) = j_0, \quad \pi'(j_1) = \pi(j_0).$$

Then

$$\begin{aligned}
\deg_{\pi'} - \deg_\pi &= \deg\left(m_{\pi(j_0),j_1}\right) + \deg\left(m_{j_0,j_0}\right) - \deg\left(m_{j_0,j_1}\right) - \deg\left(m_{\pi(j_0),j_0}\right) \\
&= \deg\left(g_{j_1 q - \pi(j_0)}\right) - \deg\left(g_{j_1 q - j_0}\right) + \deg\left(g_{j_0 q - j_0}\right) - \deg\left(g_{j_0 q - \pi(j_0)}\right) \\
&= \sum_{i=\pi(j_0)+1}^{j_0} \left(\deg(g_{j_1 q - i}) - \deg\left(g_{j_1 q - i+1}\right)\right) - \left(\deg(g_{j_0 q - i}) - \deg\left(g_{j_0 q - i+1}\right)\right) \\
&\geq \sum_{i=\pi(j_0)+1}^{j_0} (q-1) \geq (q-1),
\end{aligned}$$

where the last inequality follows from formula (8.1) and Lemma 8.33(a). We reach the contradiction $\deg_{\pi'} > \deg_\pi$. $\qquad\square$

For fixed J, the lemma tells us that id is the unique permutation for which $\deg_{J,\mathrm{id}}$ is maximal. Moreover from the definition of $\deg_{J,\pi}$, we see that

$$\deg_{J,\mathrm{id}} = \sum_{j \in J} \deg_t m_{jj} = \sum_{j \in J} \deg_t g_{j(q-1)}.$$

Since the degrees of the g_i are strictly decreasing, we find that among those J for which $\#J$ is fixed there is also a unique one for which $\deg_{J,\mathrm{id}}$ is maximal, namely $J = \{1, 2, \ldots, \#J\}$.

It follows that the degree of the coefficient of T^i is equal to $\deg_{\{1,2,\ldots,i\},\mathrm{id}} = \sum_{j=1,\ldots,i} \deg_t g_{j(q-1)}$. This completes the proof of Theorem 8.29(a) and thus of the theorem itself. $\qquad\square$

Let us add some further observations regarding the degrees of the coefficients $S_d(n)$. Say we write

$$\left(a_r\, a_{r-1}\, \ldots\, a_1\, a_0\right)_p$$

for the base p digit expansion of n. By the definition of $\mu(j)$ and $m_{\mu(j)}$, we have that for

$$\left(0\,0\,\ldots\,0\,\underbrace{a_{\mu(j)} - m_{\mu(j)}}_{\text{position } \mu(j)}\,a_{\mu(j)-1}\,\ldots\,a_1\,a_0\right)$$

the sum over its digits in base p is exactly j. We define

$$n_1 := \left(0\,0\,\ldots\,\underbrace{a_{\mu(p-1)} - m_{\mu(p-1)}}_{\text{pos. } \mu(p-1)}\,a_{\mu(p-1)-1}\,\ldots\,a_1\,a_0\right),$$

$$n_2 := \left(0\,0\,\ldots\,\underbrace{a_{\mu(2p-2)} - m_{\mu(2p-2)}}_{\text{pos. } \mu(2p-2)}\,a_{\mu(2p-2)-1}\,\cdots\,\underbrace{m_{\mu(p-1)}}_{\text{pos. } \mu(p-1)}\,0\,\ldots\,0\right),$$

$$n_3 := \left(0\,0\,\ldots\,\underbrace{a_{\mu(3(p-1))} - m_{\mu(3(p-1))}}_{\text{pos. } \mu(3(p-1))}\,a_{\mu(3(p-1))-1}\,\cdots\,\underbrace{m_{\mu(2(p-1))}}_{\text{pos. } \mu(2(p-1))}\,0\,\ldots\,0\right), \text{ etc.}$$

and m_d by $n = m_d + n_d + n_{d-1} + \cdots + n_1$. Thus n_1 is formed from the $p-1$ lowest digits of n; next n_2 is formed from the next $p-1$ lowest digits of n that have not been used in forming n_1, etc.

We can now rephrase Theorem 8.29(a) as follows:

$$s_1(n) = n - n_1, \quad s_2(n) = \left(n - n_1\right) + \left(n - n_1 - n_2\right), \ldots,$$
$$s_d(n) = \left(n - n_1\right) + \left(n - n_1 - n_2\right) + \cdots + \left(n - n_1 - \cdots - n_d\right), \text{ etc.}$$

and $s_\ell(n) = -\infty$ for all $\ell \geq \ell_0 + 1$ where ℓ_0 is smallest so that $n - n_1 - \cdots - n_{\ell_0}$ has digit sum less than $p - 1$ in its p-digit, or if no such $\ell_0 > 0$ exists, we set $\ell_0 = 0$.

We obtain an alternative proof of the following recursion from [50]:

Corollary 8.35 (Thakur). $s_d(n) = s_1(n) + s_{d-1}(s_1(n))$.

Proof.

$$s_d(n) = \left(n - n_1\right) + \left(n - n_1 - n_2\right) + \cdots + \left(n - n_1 - \cdots - n_d\right)$$
$$= \left(n - n_1\right) + \left(\left(n - n_1\right) - n_2\right) + \left(\left(n - n_1\right) - n_2 - n_3\right)$$
$$\quad + \cdots + \left(\left(n - n_1\right) - \cdots - n_d\right)$$

$$= s_1(n) + \big(s_1(n) - n_2\big) + \big(s_1(n) - n_2 - n_3\big) + \cdots + \big(s_1(n) - n_2 - \cdots - n_d\big)$$
$$= s_1(n) + s_{d-1}\big(s_1(n)\big).$$

Note that to define n_2, n_3, ..., n_d it is not necessary to know n. It suffices to know $n - n_1 = s_1(n)$ as defined above. $\square$

Open questions regarding the zero distributions of $\zeta_A(-n, T)$ for A different from $\mathbb{F}_q[T]$

Due to some simple but remarkable examples, Thakur [48] showed that Theorem 8.29(c) cannot hold for general A. It is known that $\zeta_A(-n, T)$ has a zero $T = 1$ if $q - 1$ divides n. Such zeros are called *trivial zeros*. For $\mathbb{F}_q[t]$ all trivial zeros are simple and if n is not divisible by $q - 1$ then $T = 1$ is not a root of $\zeta_A(-n, T)$. In [48] Thakur shows by explicit computation that for some even n, i.e., n divisible by $q - 1$, the root $T = 1$ is a double root!

Let us say that $\zeta_A(-n, T)$ has an *extra zero* at $-n$ if either $(T - 1)^2$ divides it, or if $(T - 1)$ divides it but n is not a multiple of $p - 1$.

The following patterns for negative integers $-n$ were observed when computing, using a computer algebra package, the Newton polygons for several rings A for the function $\zeta_A(-n, T) := \sum_{d \geq 0} T^d \sum_{a \in A_{d+}} a^n$ under the hypotheses $p = q$ and $d_\infty = 1$. Note that the Newton polygons all lie under or on the x-axis and start at $(0, 0)$. Moreover all slopes are less than or equal to zero.

Define $B \subset \mathbb{N}_0$ as

$$B := \{0\} \cup \{d \in \mathbb{N}_0 \mid \dim_k A_{(d+1)} > \dim A_d\}.$$

In particular, B contains every integer $d \geq 2g$, where g is the genus of A. The missing integers (in $\{0, \ldots, 2g\}$) are precisely the Weierstrass gaps for ∞.

- The x-coordinates of all the break and end points of all Newton polygons are in the set B and at every x-coordinate in B (along the Newton polygon) there is a break point.

- In particular, all slopes beyond the gth one have width 1.

- There are no extra zeros for n not divisible by $p - 1$. Thus extra zeros can only occur at horizontal slopes of width larger than one, i.e., among the first g slopes.

- Even if a slope is horizontal and of length 2, there may not be an extra zero.

- The degree in T of $\zeta_A(-n, T)$ is determined by the following rule: The number of slopes of the Newton polygon is exactly equal to $\lfloor \frac{\ell(n)}{p-1} \rfloor$.

Lecture 9

Relation to Étale Sheaves

Throughout this lecture we assume that A *is a finite k-algebra, and so in particular A is finite.* For such A, we shall set define a functor ϵ from A-crystals to constructible étale sheaves of A-modules that is an equivalence of categories. The correspondence is modeled at the Artin–Schreier sequence in étale cohomology and Deligne's [11, Fonctions L].

One reason why one is interested in finite rings A is to study geometric questions in positive characteristic via étale mod p cohomology. Another reason is the following. Suppose φ is a Drinfeld A-module and $\underline{M}(\varphi)$ its associated A-motive. Then for all finite primes $\mathfrak{p}$ of A, the $\mathfrak{p}^n$-torsion of φ provides us with a Galois representation, or a lisse étale sheaf of $(A/\mathfrak{p}^n)$-modules on the base. On the side of A-motives, $A/\mathfrak{p}^n$-torsion is described by $\underline{M}(\varphi) \otimes_A A/\mathfrak{p}^n$. Using the equivalence of categories ϵ introduced in Section 9.1 and the relation between the torsion points of φ and the above quotient of $\underline{M}(\varphi)$, the following result is straightforward.

$$\mathrm{Hom}_A\left(\varphi[\mathfrak{p}^n], A/\mathfrak{p}^n\right) \cong \epsilon\left(\underline{M}(\varphi) \otimes_A A/\mathfrak{p}^n\right).$$

That is, the dual of the module of $\mathfrak{p}^n$-torsion points is naturally associated to the motive modulo $\mathfrak{p}^n$.

In Section 9.1 we define the functor from A-crystals to constructible étale sheaves of A-modules and discuss its basic properties. Some proofs are given. In the subsequent Section 9.2 we use these results to reprove (in many but not all cases) a result of Goss and Sinnott which links properties of class groups to special values of Goss' L-functions.

9.1 An equivalence of categories

Our first aim is to define a functor

$$\epsilon\colon \mathbf{QCoh}_\tau\left(X, A\right) \longrightarrow \mathbf{\acute{E}t}\left(X, A\right).$$

For this, we consider a τ-sheaf $\underline{F}$. Using adjunction, we assume that it is given by a pair $(\mathcal{F}, \tau_{\mathcal{F}} \colon \mathcal{F} \to (\sigma \times \mathrm{id})_* \mathcal{F})$. Let $u \colon U \to X$ be any étale morphism. Pullback of τ along $u \times \mathrm{id}$ induces a homomorphism

$$(u \times \mathrm{id})^* \tau_{\mathcal{F}} \colon (u \times \mathrm{id})^* \mathcal{F} \longrightarrow (u \times \mathrm{id})^* (\sigma \times \mathrm{id})_* \mathcal{F} \cong (\sigma \times \mathrm{id})_* (u \times \mathrm{id})^* \mathcal{F}.$$

Taking global sections on $U \times C$ and observing that $\sigma \times \mathrm{id}$ is a topological isomorphism, we obtain a homomorphism of A-modules

$$(u \times \mathrm{id})^* \tau_{\mathcal{F}} \colon ((u \times \mathrm{id})^* \mathcal{F})(U \times C) \longrightarrow ((\sigma \times \mathrm{id})_* (u \times \mathrm{id})^* \mathcal{F})(U \times C)$$
$$= ((u \times \mathrm{id})^* \mathcal{F})(U \times C).$$

By slight abuse of notation, let us denote this homomorphism by τ_{et}. Then one verifies that

$$(u \colon U \to X) \longmapsto \mathrm{Ker}\left(1 - \tau_{\mathrm{et}} \colon ((u \times \mathrm{id})^* \mathcal{F})(U \times C) \longrightarrow ((u \times \mathrm{id})^* \mathcal{F})(U \times C)\right)$$

for $(u \colon U \to X)$ varying over the étale morphisms to X defines a sheaf of A-modules on the small étale site of X, denoted $\epsilon(\underline{F})$. A more concise way of defining ϵ is as follows. Let $\mathcal{F}_{\mathrm{et}}$ denote the étale sheaf associated to $\mathrm{pr}_{1*} \mathcal{F}$ by change of sites – this is what is done above, if one forgets about τ. Then $\tau_{\mathcal{F}}$ induces a homomorphism $\tau_{\mathrm{et}} \colon \mathcal{F}_{\mathrm{et}} \to \mathcal{F}_{\mathrm{et}}$ and

$$\epsilon(\underline{F}) := \mathrm{Ker}\left(\mathrm{id} - \tau_{\mathrm{et}} \colon \mathcal{F}_{\mathrm{et}} \longrightarrow \mathcal{F}_{\mathrm{et}}\right). \tag{9.1}$$

Clearly this construction is functorial in $\underline{F}$, that is, to every homomorphism $\varphi \colon \underline{F} \to \underline{G}$ it associates a homomorphism $\epsilon(\varphi) \colon \epsilon(\underline{F}) \to \epsilon(\underline{G})$. Thus it defines an A-linear functor

$$\epsilon \colon \mathbf{QCoh}_\tau (X, A) \longrightarrow \acute{\mathbf{E}}\mathbf{t}(X, A). \tag{9.2}$$

Following its construction one finds that ϵ is left exact.

Example 9.1. Let $\underline{\mathbb{1}}_{X,A}$ denote the τ-sheaf consisting of the structure sheaf $\mathcal{O}_{X \times C}$ together with its obvious τ given by

$$\tau = \sigma \times \mathrm{id} \colon \mathcal{O}_{X \times C} \longrightarrow (\sigma \times \mathrm{id})_* \mathcal{O}_{X \times C}.$$

The étale sheaf associated to $\mathcal{O}_{X \times C}$ is simply $\mathcal{O}_{X_{\mathrm{et}}} \otimes A$ with τ_{et} the morphism $(u \otimes a) \mapsto u^q \otimes a$. Therefore $\epsilon(\underline{\mathbb{1}}_{X,A}) \cong \underline{A}_X$, the constant étale sheaf on X with stalk A. In the special case $A = k$ we recover the sequence

$$0 \longrightarrow \underline{k}_X \longrightarrow \mathcal{O}_{X_{\mathrm{et}}} \overset{1-\sigma}{\longrightarrow} \mathcal{O}_{X_{\mathrm{et}}}$$

from Artin–Schreier theory.

Lemma 9.2. Let $\varphi \colon \underline{F} \to \underline{G}$ be a nil-isomorphism in $\mathbf{Coh}_\tau(X, A)$. Then the induced $\epsilon(\varphi) \colon \epsilon(\underline{F}) \to \epsilon(\underline{G})$ is an isomorphism.

Proof. Observe first that regarding τ as a homomorphism of τ-sheaves, we have $\epsilon(\tau) = \mathrm{id}$. This is so because ϵ is precisely the operation on the étale sheaf associated to $\underline{\mathcal{F}}$ of taking fixed points under τ. Clearly τ is the identity on the set of these fixed points. Having clarified this, the proof of the proposition follows immediately from applying τ to the diagram (2.2).   □

The assertion of the lemma also holds for τ-sheaves whose underlying sheaf is only quasi-coherent. The proof is however much more subtle. In [8] it is shown that ϵ factors via the category of ind-coherent τ-sheaves, i.e., τ-sheaves which can be written as inductive limits of coherent τ-sheaves. Then an argument involving direct limits reduces one to the already proved case of the lemma. In total one obtains:

Proposition 9.3. *The functor ϵ induces a unique left exact A-linear functor*

$$\bar{\epsilon} \colon \mathbf{QCrys}\,(X, A) \longrightarrow \mathbf{\acute{E}t}\,(X, A).$$

It is shown in [8] that the isomorphism ϵ defined for all pairs (X, A) (with A finite) is compatible with the formation of functors on crystals and on the étale site, respectively:

Proposition 9.4. *For $f \colon Y \to X$ a morphism, $j \colon U \hookrightarrow X$ an open immersion and $h \colon C \to C'$ a base change homomorphism, one has the following compatibilities:*

$$\epsilon \circ f^* \cong f^* \circ \epsilon, \quad \epsilon \circ \left(\_ \otimes \_\right) \cong \left(\_ \otimes \_\right) \circ \epsilon, \quad \epsilon \circ \left(\_ \otimes_A A'\right) \cong \left(\_ \otimes_A A'\right) \circ \epsilon,$$
$$\epsilon \circ f_* \cong f_* \circ \epsilon, \quad \epsilon \circ j_! \cong j_! \circ \epsilon.$$

Except for the very first compatibility, i.e., that of inverse image, the proofs are rather straightforward. Note that to the left of $\circ\,\epsilon$ the functors are functors on étale sheaves and to the right of $\epsilon\,\circ$ they are functors on A-crystals.

Let $\mathbf{\acute{E}t}_c(X, A) \subset \mathbf{\acute{E}t}(X, A)$ denote the subcategory of constructible étale sheaves. Recall that an étale sheaf of A-modules is constructible if X has a finite stratification by locally closed subsets U_i such that the restriction of the sheaf to each U_i is locally constant. This in turn means that there exists a *finite* étale morphism $V_i \to U_i$ such that the pullback to V_i is a constant sheaf on a finite A-module.

Proposition 9.5. *The image of $\mathbf{Crys}(X, A)$ under ϵ lies in $\mathbf{\acute{E}t}_c(X, A)$.*

Sketch of proof. Since being constructible is independent of the A-action, we may restrict the proof to $\mathbf{Crys}(X, k)$. We also may assume that X is reduced; cf. Theorem 3.9. Let $\underline{\mathcal{F}}$ be a coherent τ-sheaf on X over k. Since we have the functors $j_!$, f_* and f^* at our disposal we can apply noetherian induction on X in order to show that $\epsilon(\underline{\mathcal{F}})$ is constructible. Thus it suffices to fix a generic point η of X and to prove that there exists an open neighborhood U of η such that $\epsilon(\underline{\mathcal{F}}|U)$ is locally constant.

We first choose a neighborhood U of η which is regular as a scheme. By [30, Theorem 4.1.1], it suffices to show that after possibly further shrinking U one can find a τ-sheaf $\mathcal{G}$ which is nil-isomorphic to $\underline{F}$ and such that $\mathcal{G}$ is locally free and τ is an isomorphism on it. At the generic point both properties can be achieved by replacing $\underline{F}$ by $\mathrm{Im}(\tau_{\underline{F}}^m)$ for m sufficiently large. And then one shows, using $A = k$, that this extends to an open neighborhood of η. $\square$

A main theorem of [8, Ch. 10] is the following:

Theorem 9.6. *For A a finite k-algebra, the functor $\epsilon\colon \mathbf{Crys}(X, A) \to \mathbf{\acute{E}t}_c(X, A)$ is an equivalence of categories.*

Since it is compatible with all functors, the definition of flatness for both categories implies that ϵ induces an equivalence between the full subcategory of flat A-crystals and the full subcategory of flat constructible étale sheaves of A-modules:

$$\epsilon\colon \mathbf{Crys}^{\mathrm{flat}}\left(X, A\right) \xrightarrow{\cong} \mathbf{\acute{E}t}_c^{\,\mathrm{flat}}\left(X, A\right).$$

For flat A-crystals we have a definition of L-functions if X is of finite type over k. Under the same hypothesis on X, for flat constructible étale sheaves of A-modules such a definition is given in [11, Fonctions L, 2.1]. At a closed point x, it is the following:

Definition 9.7. The *L-function* of $\mathsf{F} \in \mathbf{\acute{E}t}_c^{\mathrm{flat}}(x, A)$ is

$$L\left(x, \mathsf{F}, t\right) := \det{}_A\left(\mathrm{id} - t^{d_x} \cdot \mathrm{Frob}_x^{-1} \mid \mathsf{F}_{\bar{x}}\right)^{-1} \in 1 + t^{d_x} A\left[\left[t^{d_x}\right]\right].$$

The obvious extensions to schemes X of finite type over k and to complexes representable by bounded complexes of objects in $\mathbf{\acute{E}t}_c^{\mathrm{flat}}(X, A)$ are left to the reader.

It is a basic result that ϵ is compatible with the formation of L-functions:

Proposition 9.8. *For any $\underline{F} \in \mathbf{Crys}^{\mathrm{flat}}(X, A))$ we have*

$$L\left(X, \epsilon(\underline{F}), t\right) = L^{\mathrm{crys}}\left(X, \underline{F}, t\right).$$

As a consequence of Theorem 6.14 we find:

Theorem 9.9. *Let $f\colon Y \to X$ be a morphism of schemes of finite type over k and $\mathsf{F}^\bullet \in \mathbf{D}^b(\mathbf{\acute{E}t}_c(Y, A))_{\mathrm{ftd}}$ a complex representable by a bounded complex with objects in $\mathbf{\acute{E}t}_c^{\mathrm{flat}}(X, A)$. Then one has*

$$L\left(Y, \mathsf{F}^\bullet, t\right) \sim L\left(X, Rf_!\mathsf{F}^\bullet, t\right),$$

i.e., their quotient is a unipotent polynomial.

For reduced coefficient rings A, the above result was first proved by Deligne in [11, Fonctions L, Theorem 2.2]. In [14, Theorem 1.5], Emerton and Kisin give a proof for arbitrary finite A of some characteristic p^m. By an inverse limit procedure, in [14, Corollary 1.8] they give a suitable generalization to any coefficient ring A which is a complete noetherian local $\mathbb{Z}_p$-algebra with finite residue field.

Remark 9.10. The category $\acute{\mathbf{E}}\mathbf{t}_c(X, A)$ has no duality and $f^!$, f_* and an internal Hom are either not all defined or not well behaved. Thus for the theory of A-crystals we cannot hope for this either.

9.2 A result of Goss and Sinnott

In what follows we shall use the correspondence between étale sheaves and crystals of the previous section to reprove a result of Goss and Sinnott – in many, but so far not all cases. The original proof of the result of Goss and Sinnott is based on the comparison of classical L-functions for function fields and Goss–Carlitz type L-functions for function fields. Our proof avoids all usage of classical results but uses the results from the previous section instead.

Class groups of Drinfeld–Hayes cyclotomic fields

We consider the following situation. Let K be a function field with constant field k, let ∞ be a place of K and A the ring of regular functions outside ∞. Let $H \subset H^+$ be the (strict) Hilbert class field with ring of integers $\mathcal{O} \subset \mathcal{O}^+$. By the theory of Drinfeld–Hayes modules, see [24, Ch. 7] or Appendix A.3, there exist $[H^+ : K]$ many sign-normalized rank 1 Drinfeld–Hayes modules

$$\varphi \colon A \longrightarrow \mathcal{O}^+[\tau], \quad a \longmapsto \varphi_a.$$

Let $\mathfrak{p}$ be a maximal ideal of A. Then the $\mathfrak{p}$-torsion points $\varphi[\mathfrak{p}](\bar{K})$ of φ over $\bar{K}$ form a free $(A/\mathfrak{p})$-module of rank 1 carrying an A-linear action of $\mathrm{Gal}(\bar{K}/H^+)$. If $H_{\mathfrak{p}}^+$ denotes the fixed field of the kernel of this representation, then $G := \mathrm{Gal}(H_{\mathfrak{p}}^+/H^+)$ is isomorphic to $(A/\mathfrak{p})^*$. The extension $H_{\mathfrak{p}}^+/H^+$ is totally ramified at the places of $\mathcal{O}^+$ above $\mathfrak{p}$ and unramified above all other finite places. For all places above ∞ the decomposition and inertia groups agree and are isomorphic to the subgroup $k^* \subset (A/\mathfrak{p})^*$.

$$
\begin{array}{c}
H_{\mathfrak{p}}^+ \\
\big| \, G \\
H^+ \\
\big| \\
K.
\end{array}
$$

For $K = k(t)$ and $A = k[t]$ one has $H = K$ and the Drinfeld module φ is simply the Carlitz module.

Let us denote by $\chi \colon G \to (A/\mathfrak{p})^*$ the character of G that arises from the action of $\varphi[\mathfrak{p}]$. This is the analog of the mod p cyclotomic character in classical number theory. We introduce the following notation. By $\mathrm{Jac}_{K,\mathfrak{p}}$ we denote the Jacobian variety of the smooth projective geometrically irreducible curve $C_{K,\mathfrak{p}}$ with constant field k_∞ and function field H^+, and we let $\mathrm{Cl}(H_{\mathfrak{p}}^+)$ denote the class group of the field $H_{\mathfrak{p}}^+$. Then the p-torsion subgroup of $\mathrm{Cl}(H_{\mathfrak{p}}^+)$ is isomorphic to the group of invariants under $\mathrm{Gal}(\bar{k}_\infty/k_\infty)$ of the p-torsion group $\mathrm{Jac}_{K,\mathfrak{p}}[p](\bar{k})$. For $w \in \mathbb{Z}$ (it suffices $w \in \{1, 2, \ldots, \#(A/\mathfrak{p})^*\}$) we define the χ^w components of the above groups as

$$C(w) := \big(\mathrm{Cl}(H_{\mathfrak{p}}^+) \otimes_{\mathbb{F}_p} A/\mathfrak{p}\big)_{\chi^w}, \quad \widetilde{C}(w) := \big(\mathrm{Jac}_{K,\mathfrak{p}}\,[p]\,(\bar{k}) \otimes_{\mathbb{F}_p} A/\mathfrak{p}\big)_{\chi^w}.$$

Note that $\bar{k}H^+/H^+$ is a constant field extension, while $H_{\mathfrak{p}}^+/H^+$ is purely geometric. Hence these extensions are linearly disjoint. Moreover the group G is of order prime to p and thus its action on the p-group $\mathrm{Jac}_{K,\mathfrak{p}}[p](\bar{k})$ is exact, so that

$$\widetilde{C}(w)^{\mathrm{Gal}(\bar{k}_\infty/k_\infty)} = C(w).$$

Let $\widetilde{h}_A^+$ denote the number of places of H^+ above ∞, so that $\widetilde{h}_A^+ = h_A \frac{\#k_\infty^*}{\#k^*}$. The following result (even under more general hypotheses) is due to Goss and Sinnott.

Theorem 9.11 (Goss, Sinnott). *Let $w \in \mathbb{N}$. For $a, b \in \mathbb{N}$ define $\delta_{a|b}$ to be 1 if a is a divisor of b and zero otherwise. Then the following hold:*

(a) $\dim_{A/\mathfrak{p}} \widetilde{C}(w) = \deg_T(L_{\mathrm{Spec}\,\mathcal{O}^+}(w, T) \bmod \mathfrak{p}) - \widetilde{h}_A^+ \delta_{(q-1)|w}.$

(b) $C(w) \neq 0$ *if and only if* $\mathrm{ord}_{T=1}(L_{\mathrm{Spec}\,\mathcal{O}^+}(w, T) \bmod \mathfrak{p}) > \widetilde{h}_A^+ \delta_{(q-1)|w}.$

(c) $\dim_{A/\mathfrak{p}} C(w)$ *is the multiplicity of the eigenvalue* 1 *of the action of* τ *on* $H^1(C_{K,\mathfrak{p}}, (\underline{\mathcal{M}}(\varphi)^{\otimes w})^{\mathrm{max}}) \otimes_A A/\mathfrak{p}$. *(For the superscript* max, *see Definition* 9.12.)

The proof in [25] uses congruences between L-functions of τ-sheaves and classical Hasse–Weil L-functions. This comparison is replaced by comparing the cohomology of a τ-sheaf and that of the étale sheaf associated to its mod $\mathfrak{p}$ reduction.

Before we can give the proof of the theorem, we need to introduce the concept of maximal extension of a τ-sheaf. Once this is understood, we shall present a proof of the above theorem different from that in [25].

The maximal extension of Gardeyn

The material on maximal extensions is based on work and ideas of F. Gardeyn from [16, § 2]. We follow the approach in [4]. By B we denote a k-algebra which is essentially of finite type. Typically it is equal to A or to $A/\mathfrak{n}$ for some ideal $\mathfrak{n} \subset A$. We omit almost all proofs. They can be found either in [16, § 2] or in [4, Ch. 8].

Throughout the discussion of maximal extensions, we fix an open immersion $j \colon U \hookrightarrow X$ and a complement $Z \subset X$ of U.

Definition 9.12 (Gardeyn). *Suppose* $\underline{\mathcal{F}} \in \mathbf{Coh}_\tau(U, B)$.

(a) *A coherent τ-subsheaf $\underline{\mathcal{G}}$ of $j_*\underline{\mathcal{F}}$ with $j^*\underline{\mathcal{G}} = \underline{\mathcal{F}}$ is called an* extension *of* $\underline{\mathcal{F}}$.

(b) *The union of all extensions of $\underline{\mathcal{F}}$ is denoted by $j_\#\underline{\mathcal{F}} \subset j_*\underline{\mathcal{F}}$.*

(c) *If $j_\#\underline{\mathcal{F}}$ is coherent, it is called the* maximal extension *of* $\underline{\mathcal{F}}$. *It is also denoted by* $\underline{\mathcal{F}}^{\mathrm{max}}$.

The assignment $\underline{F} \mapsto j_\# \underline{F}$ defines a functor $\mathbf{Coh}_\tau(U, B) \to \mathbf{QCoh}_\tau(X, B)$. Note that if $j_* \mathcal{F}$ is not coherent, the same holds for $j_\#(\mathcal{F}, 0) = (j_* \mathcal{F}, 0)$ – consider for instance the case $\operatorname{Spec} R \hookrightarrow \operatorname{Spec} K$ where R is a discrete valuation ring with fraction field K.

We state some basic properties:

Proposition 9.13. *Any τ-sheaf $\underline{F}$ has an extension to X which represents the crystal $j_! \underline{F}$.*

Proof. This follows from the part of the proof of Theorem 3.10 giving the existence of the crystal $j_! \underline{F}$. $\square$

The τ-sheaf $j_\# \underline{F}$ has the following intrinsic characterization modeled after the Néron mapping property:

Proposition 9.14. *Suppose $\underline{F} \in \mathbf{Coh}_\tau(U, B)$ and $\underline{G} \in \mathbf{QCoh}_\tau(X, B)$ are such that $j^* \underline{G} \cong \underline{F}$. Then $\underline{G}$ is isomorphic to $j_\# \underline{F}$ if and only if the following conditions hold:*

(a) *$\underline{G}$ is an inductive limit of coherent τ-sheaves, and*
(b) *for all $\underline{H} \in \mathbf{Coh}_\tau(X, B)$, the following canonical map is an isomorphism:*

$$\operatorname{Hom}_{\mathbf{QCoh}_\tau(X,B)} \left(\underline{H}, \underline{G} \right) \longrightarrow \operatorname{Hom}_{\mathbf{Coh}_\tau(U,B)} \left(j^* \underline{H}, \underline{F} \right).$$

Proposition 9.14 motivates the following axiomatic definition of maximal extension for crystals:

Definition 9.15. *A crystal $\underline{G} \in \mathbf{Crys}(X, B)$ is called an extension of $\underline{F}$ if $j^* \underline{G} \cong \underline{F}$. It is called a maximal extension if, in addition, the canonical map*

$$\operatorname{Hom}_{\mathbf{Crys}(X,B)} \left(\underline{H}, \underline{G} \right) \longrightarrow \operatorname{Hom}_{\mathbf{Crys}(U,B)} \left(j^* \underline{H}, \underline{F} \right)$$

is an isomorphism for all $\underline{H} \in \mathbf{Crys}(X, B)$.

Proposition 9.16. *Let $\underline{F}$ be in $\mathbf{Coh}_\tau(X, B)$. If $j_\# \underline{F}$ is coherent, then the crystal represented by $\underline{F}$ possesses a maximal extension and the latter is represented by $j_\# \underline{F}$.*

Proposition 9.17. *The functor $j_\#$ is left exact on τ-sheaves. Moreover, if one has a left exact sequence of crystals such that the outer terms have a maximal extension, then so does the central term, and the induced sequence of the maximal extensions is left exact.*

We now impose the following conditions sufficient for our intended applications. Under these, the main result on the existence of a maximal extension, Theorem 9.22, is due to Gardeyn.

(a) The ring B is finite over k or over A.
(b) X is a smooth geometrically irreducible curve over k and $U \subset X$ is dense.

Proposition 9.18. *For $\underline{F} \in \mathbf{Coh}_\tau(U, B)$ and $\underline{G}$ an extension of $\underline{F}$, the following assertions are equivalent:*

(a) *$\underline{G}$ is the maximal extension of $\underline{F}$.*

(b) *For any $x \in Z$ and $j_x \colon \operatorname{Spec} \mathcal{O}_{X,x} \hookrightarrow X$ the canonical morphism, the τ-sheaf $j_x^* \underline{G}$ is the maximal extension of $i_\eta^* \underline{F}$; here η is the generic point of X*

This proposition allows one to reduce the problem of the existence of a maximal extension to the situation where X is a discrete valuation ring. The proof is a simple patching argument.

Definition 9.19 (Gardeyn). Let $\underline{G}$ be a locally free τ-sheaf on X over B. Then $\underline{G}$ is called *good* at $x \in X$ if τ is injective on $i_x^* \underline{G}$. It is called *generically good* if it is good at the generic point of X.

Note that if $\underline{G}$ is generically good, then it is good for all x in a dense open subset.

Proposition 9.20. *If $\underline{G} \in \mathbf{Coh}_\tau(X, B)$ is an extension of $\underline{F} \in \mathbf{Coh}_\tau(U, B)$ such that $\underline{G}$ is good at all $x \in Z$, then $\underline{G} = j_\# \underline{F}$.*

The point is that after pulling back the situation to any $\operatorname{Spec} \mathcal{O}_{X,x}$ for $x \in Z$, the fact that $\underline{G}$ is good at x easily implies that it is a maximal extension. Now one can apply Proposition 9.18.

Corollary 9.21. *The unit τ-sheaf $\mathbb{1}_{X,A}$ is good at all $x \in X$. Suppose now that $\underline{G} \in \mathbf{Coh}_\tau(X, B)$ is an extension of $\underline{F} \in \mathbf{Coh}_\tau(X, B)$ such that $i_x^* \underline{G} \cong \mathbb{1}_{x,A}$ for all $x \in Z$. Then $\underline{G} = j_\# \underline{F}$ and moreover in $\mathbf{Crys}(X, B)$ the following sequence is exact:*

$$0 \longrightarrow j_! \underline{F} \longrightarrow j_\# \underline{F} \longrightarrow \oplus_{x \in Z} \mathbb{1}_{x,A} \longrightarrow 0.$$

The following are the central results on maximal extensions:

Theorem 9.22 (Gardeyn). *If $\underline{F}$ is a locally free, generically good τ-sheaf on U over B, then $j_\# \underline{F}$ is locally free.*

Theorem 9.23. *Suppose B is finite. Then $j_\# \colon \mathbf{Crys}(U, B) \to \mathbf{Crys}(X, B)$ is a well-defined functor. Moreover one has $\epsilon \circ j_\# \cong j_* \circ \epsilon$ where $\epsilon \colon \mathbf{Crys}(\ldots) \to \acute{\mathrm{Et}}_c(\ldots)$ is the functor in Theorem 9.6.*

Another simple assertion along the lines of Corollary 9.21 is the following:

Proposition 9.24. *Suppose $\underline{F} \in \operatorname{Coh}(U, A)$ has a maximal extension to X. Then the canonical homomorphism of crystals*

$$\underline{F}^{\max} \otimes_A A/\mathfrak{p} \hookrightarrow \left(\underline{F} \otimes_A A/\mathfrak{p}\right)^{\max}$$

is injective. If $\underline{F}^{\max} \otimes_A A/\mathfrak{p}$ has good reduction at all $x \in Z$, it is an isomorphism.

Proof of Theorem 9.11

Proof. It is well known that the first étale cohomology of a curve for the constant sheaf $\mathbb{F}_p$ can be expressed in terms of the p-torsion group of its Jacobian. Denoting by the superscript $^\vee$ the formation of the $\mathbb{F}_p$-dual, i.e, $\mathrm{Hom}_{\mathbb{F}_p}(\_, \mathbb{F}_p)$, one has

$$\mathrm{Jac}_{K,\mathfrak{p}}[p](\bar{k}) \cong H^1_{\mathrm{et}}(C_{K,\mathfrak{p}}/\bar{k}, \mathbb{F}_p)^\vee.$$

Both sides carry Galois actions of $\mathrm{Gal}(\bar{k}/k_\infty)$ and of G. The extension $H^+_{\mathfrak{p}}/H^+$ is totally ramified at all places above $\mathfrak{p}$. Therefore it is linearly disjoint from the constant field extension $\bar{k}H^+/H^+$, and hence the two actions commute. We tensor both sides with $A/\mathfrak{p}$ over $\mathbb{F}_p$. This allows to decompose them into isotypic components for the semisimple action of G, whenever desired. Observe that the isotypic components on the left are the groups $\tilde{C}(w)$.

To analyze the right-hand term, we introduce some notation. Let $o_\mathfrak{p}$ denote the order of $(A/\mathfrak{p})^*$. Denote by $f_\mathfrak{p}\colon C_{K,\mathfrak{p}} \to C_{H^+}$ the G-cover of smooth projective geometrically irreducible curves over k_∞ corresponding to $H^+_\mathfrak{p}/H^+$. Define $U_\mathfrak{p}$ to be $\mathrm{Spec}\,\mathcal{O}^+$ minus the finitely many places above $\mathfrak{p}$ and let $j_\mathfrak{p}$ denote the open immersion of $U_\mathfrak{p} \hookrightarrow C_{H^+}$. Over $U_\mathfrak{p}$ the representation of G on $\varphi[\mathfrak{p}]$ is unramified, and thus it defines a lisse étale sheaf of rank one over $A/\mathfrak{p}$ which we denote by $\underline{\varphi[\mathfrak{p}]}$. This sheaf and all its tensor powers become, after pullback along the finite étale cover $f_\mathfrak{p}^{-1}(U_\mathfrak{p}) \to U_\mathfrak{p}$, isomorphic to the constant sheaf $\underline{A/\mathfrak{p}}$ on $C_{K,\mathfrak{p}}$ with generic fiber $A/\mathfrak{p}$. Using simple representation theory, one deduces that

$$\left(f_{\mathfrak{p}*}\underline{A/\mathfrak{p}}\right)\big|_{U_\mathfrak{p}} \cong \bigoplus_{w\in\mathbb{Z}/o_\mathfrak{p}} \underline{\varphi[\mathfrak{p}]}^{\otimes w}.$$

From adjunction of j^* and j_* we deduce a homomorphism

$$f_{\mathfrak{p}*}\underline{A/\mathfrak{p}} \longrightarrow \bigoplus_{w\in\mathbb{Z}/o_\mathfrak{p}} j_{\mathfrak{p}*}\underline{\varphi[\mathfrak{p}]}^{\otimes w}.$$

On stalks one can verify that the map is an isomorphism: At points where the representation $\varphi[\mathfrak{p}](\bar{K})^{\otimes w}$ is ramified, the sheaf $j_{\mathfrak{p}*}\underline{\varphi[\mathfrak{p}]}^{\otimes w}$ is zero and so is the corresponding summand on the left. At the other (unramified) points above $\mathfrak{p}, \infty$, the sheaf $j_{\mathfrak{p}*}\underline{\varphi[\mathfrak{p}]}^{\otimes w}$ is lisse, as is the corresponding summand on the left. Using $H^1_{\mathrm{et}}(C_{K,\mathfrak{p}}, \_) \cong H^1_{\mathrm{et}}(C_{H^+}, f^*_\mathfrak{p}\_)$, we deduce

$$\mathrm{Jac}_{K,\mathfrak{p}}[p](\bar{k}) \otimes_{\mathbb{F}_p} A/\mathfrak{p} \cong H^1_{\mathrm{et}}\left(C_{H^+}/\bar{k}, \bigoplus_{w\in\mathbb{Z}/o_\mathfrak{p}} j_{\mathfrak{p}*}\underline{\varphi[\mathfrak{p}]}^{\otimes w}\right)^\vee.$$

Now we decompose the isomorphism into isotypic components – note that $^\vee$ changes the sign of w. This yields

$$\tilde{C}(w) \cong H^1_{\mathrm{et}}\left(C_{H^+}/\bar{k}, j_{\mathfrak{p}*}\underline{\varphi[\mathfrak{p}]}^{\otimes(-w)}\right)^\vee.$$

Our next aim is to relate the coefficient sheaf to a tensor power of the τ-sheaf $\underline{\mathcal{M}}(\varphi)$ attached to the Drinfeld module φ. We observed earlier that $\epsilon(\underline{\mathcal{M}}(\varphi) \otimes_A A/\mathfrak{p})$ on $\operatorname{Spec}\mathcal{O}^+$ is dual to $\varphi[\mathfrak{p}]$. Since $\varphi[\mathfrak{p}](\bar{K})$ is totally ramified at $\mathfrak{p}$, the same holds for tensor powers w, except if w is a multiple of $o_\mathfrak{p}$ – here $\epsilon(\underline{\mathcal{M}}(\varphi) \otimes_A A/\mathfrak{p})^{\otimes w}$ may be zero above $\mathfrak{p}$, while the representation defined by $\varphi[\mathfrak{p}](\bar{K})^{\otimes w}$ is trivial and hence lisse. Let $j\colon \operatorname{Spec}\mathcal{O}^+ \hookrightarrow C_{H+}$ denote the canonical open immersion. Using Theorem 9.23 for w not a multiple of $o_\mathfrak{p}$, we find

$$j_{\mathfrak{p}*}\underline{\varphi}[\mathfrak{p}]^{\otimes(-w)} \cong \epsilon\big(j_\#\big(\underline{\mathcal{M}}(\varphi) \otimes_A A/\mathfrak{p}\big)^{\otimes w}\big).$$

One can either use that the representation defined by $\varphi[\mathfrak{p}](\bar{K})^{\otimes w}$ is unramified at the places above ∞ if and only if $(q-1)$ divides w – the ramification group at those places is $k^* \subset A/\mathfrak{p}^* \cong G$ – or a direct computation on the side of τ-sheaves to deduce from Corollary 9.21 that

$$j_!(\underline{\mathcal{M}}(\varphi) \otimes_A A/\mathfrak{p})^{\otimes w} \hookrightarrow j_\#(\underline{\mathcal{M}}(\varphi) \otimes_A A/\mathfrak{p})^{\otimes w}$$

is an isomorphism whenever w is not a multiple of $(q-1)$ and has cokernel $\bigoplus_{\infty'|\infty} \mathbb{1}_{\infty',A/\mathfrak{p}}$ otherwise – the sum is over all places of H^+ above ∞. One can in fact also prove that $j_!\underline{\mathcal{M}}(\varphi)^{\otimes w} \hookrightarrow j_\#(\underline{\mathcal{M}}(\varphi))^{\otimes w}$ is an isomorphism for $(q-1) \nmid w$ and has cokernel $\bigoplus_{\infty'|\infty} \mathbb{1}_{\infty',A}$ otherwise. Finally we use that ϵ commutes with all functors defined for crystals, so that to compute H^1_{et} we may first compute H^1 for crystals and then apply ϵ. This yields

$$\widetilde{C}(w) \cong \epsilon\Big(H^1(C_{H+}/\bar{k}, \mathbb{1}_{C_{H+},A}) \otimes A/\mathfrak{p}\Big)\big(\operatorname{Spec}\bar{k}\big) \quad \text{for } o_\mathfrak{p}|w; \tag{9.3}$$

$$\widetilde{C}(w) \cong \epsilon\Big(\big(H^1(C_{H+}/\bar{k}, j_!\underline{\mathcal{M}}(\varphi)^{\otimes w})/\bigoplus_{\infty'|\infty}\mathbb{1}_{\operatorname{Spec}\bar{k},A}\big) \otimes A/\mathfrak{p}\Big)(\operatorname{Spec}\bar{k})$$
$$\text{for } (q-1)|w,\; o_\mathfrak{p}\nmid w \text{ or } w=0; \tag{9.4}$$

$$\widetilde{C}(w) \cong \epsilon\Big(H^1(C_{H+}/\bar{k}, j_!\underline{\mathcal{M}}(\varphi)^{\otimes w} \otimes A/\mathfrak{p}\Big)\big(\operatorname{Spec}\bar{k}\big) \quad \text{for } (q-1)\nmid w. \tag{9.5}$$

Note that without the evaluation $(\operatorname{Spec}\bar{k})$ outside ϵ we would have a sheaf on the right-hand side.

In either case, the expression inside $\epsilon(\ldots)$ is a τ-sheaf $\mathcal{G}$ on $\operatorname{Spec}\bar{k}$ over $A/\mathfrak{p}$. By Proposition 5.16 it can be written as $\mathcal{G} \cong \mathcal{G}_{\mathrm{ss}} \oplus \mathcal{G}_{\mathrm{nil}}$ where on the first summand τ is bijective and on the second nilpotent. The underlying modules in both cases are finitely generated projective over $k \otimes_k A/\mathfrak{p}$. By the theory of the Lang torsor the τ-fixed points of the first summand form free $A/\mathfrak{p}$ vector space of dimension equal to $\operatorname{rank}_{\bar{k}\otimes A/\mathfrak{p}}\mathcal{G}_{\mathrm{ss}}$; those of the second summand are clearly zero. Moreover computing ϵ in the case at hand, cf. (9.1), is precisely the operation of taking τ-fixed points – the only relevant étale morphism to $\operatorname{Spec}\bar{k}$ is the identity.

At the same time, the dual characteristic polynomial of

$$\big(H^1(C_{H+}, j_!\underline{\mathcal{M}}(\varphi)^{\otimes w} \otimes A/\mathfrak{p}\big)$$

has degree precisely equal to $\mathrm{rank}_{\bar{k}\otimes A/\mathfrak{p}}\,\mathcal{G}_{ss}$. By Theorem 8.10 and Remark 8.12 this rank is the degree of $L_{\mathcal{O}+}(w,T)$ mod $\mathfrak{p}$. Thus we have now proved Theorem 9.11 (a). One may wonder about the case $o_{\mathfrak{p}}|w$ and $w \neq 0$. There are two answers why this case is covered as well. The formal answer is that the L-functions mod $\mathfrak{p}$ for w and w' in $-\mathbb{N}_0$ are equal whenever $w \equiv w' \pmod{o_{\mathfrak{p}}}$, and so it suffices to understand the case $w = 0$. An answer obtained by looking closer at what is happening goes as follows: The places above $\mathfrak{p}$ have L-factors congruent to 1 module $\mathfrak{p}$. So it does not matter whether we leave them in or not, i.e., whether we compute via the trace formula with $H^1(C_{H^+}/\bar{k},\,j_*\underline{\mathcal{M}}(\varphi)^{\otimes w} \otimes A/\mathfrak{p})$ or with $H^1(C_{H^+}/\bar{k},\,j_{\mathfrak{p}*}\underline{\mathcal{M}}(\varphi)^{\otimes w}|_{U_{\mathfrak{p}}} \otimes A/\mathfrak{p})$.

To prove (b) and (c) observe that we obtain formulas for $C(w)$ by taking invariants under $\mathrm{Gal}(\bar{k}/k_\infty)$ in the isomorphisms (9.3) to (9.5). The effect on the right-hand sides is that we replace the curve $C_{H^+}/\bar{k}$ by $C_{H^+}/k_\infty = C_{H^+}$ and that for the resulting sheaf we compute global sections over $\mathrm{Spec}\,k_\infty$. This amounts to the same as computing the fixed points under τ of the expressions inside the brackets $\epsilon(\ldots)$. Since the τ-fixed points being non-zero is the same as the assertion that 1 is an eigenvalue of the τ-action, part (b) is now clear. Note that $\bigoplus_{\infty'|\infty}\mathbb{1}_{\mathrm{Spec}\,\bar{k},A}$ being a subcrystal of $M := H^1(C_{H^+}/\bar{k},\,j_!\underline{\mathcal{M}}(\varphi)^{\otimes w})$ in (9.4) means that $(T-1)^{\tilde{h}_A^+}$ is a factor of the L-function of M. Part (c) simply says that the dimension of the eigenspace of the eigenvalue 1 for the τ action is precisely the dimension of the space of τ-invariants, and the latter is $C(w)$. $\qquad\square$

Lecture 10

Drinfeld Modular Forms

The aim of this lecture is to give a description of Drinfeld modular forms via the cohomology of certain universal crystals on moduli spaces of rank 2 Drinfeld modules.

The basic definition of Drinfeld modular forms goes back to Goss [20, 21]. Many important contributions are due to Gekeler, e.g., [18]. Moreover, Gekeler obtains foundational results on Drinfeld modular curves in [17]. The work of Gekeler and Goss gives a satisfactory description of Drinfeld modular forms as rigid analytic functions on the Drinfeld analog of the upper half-plane. The important work [47] of Teitelbaum links this to harmonic cochains on the Bruhat–Tits tree underlying the Drinfeld symmetric space. As shown in [4], the latter provides the link to a description of modular forms via crystals.

After introducing a moduli problem for Drinfeld modules of arbitrary rank (with a full level structure) in Section 10.1, in Section 10.2 we give equations for a particular example of such a moduli space. The universal Drinfeld module on it will give rise to a crystal via Anderson's correspondence between A-modules and A-motives. Following the classical case, this crystal yields a natural candidate for a cohomological description of Drinfeld modular forms; cf. Section 10.3. The cohomological object so obtained plays the role of a motive for the space of forms of fixed weight and level. It has various realizations: Its analytic realization is directly linked to Teitelbaum's description of Drinfeld modular forms via harmonic cochains. In Section 10.4 we consider its étale realizations. They allow one to attach Galois representations to Drinfeld–Hecke eigenforms as in the classical case. Unlike in the classical case, the representations are one-dimensional! The following Section 10.5 gives some discussion of ramification properties of the Galois representations so obtained. In Section 10.6 we indicate in what sense these compatible systems of one-dimensional Galois representations associated to a cuspidal Drinfeld–Hecke eigenform arise from a (suitably defined) Hecke character. So far the nature of these characters is still rather mysterious. We conclude with Section 10.7, which contains an extended example of the computation of the crystals associated to some low-weight modular forms for $\mathbb{F}_q[t]$ and a particular level. It provides exemplary answers to many natural questions and points to open problems. Due to lack of time and space, we omit many details and refer the reader to [4].

We recall the following notation: By $C = \operatorname{Spec} A$ we denote an irreducible smooth affine curve over k whose smooth compactification is obtained by adjoining precisely one closed point ∞. We define K as the fraction field of A, K_∞ as the completion of K at ∞, and $\mathbb{C}_\infty$ as the completion of the algebraic closure of K_∞. Similarly, for any place v of K we denote by K_v the completion of K at v; by $\mathcal{O}_v$ the ring of integers of K_v, and by k_v the residue field of K_v. Often A will simply be $k[t]$. We fix a non-zero ideal $\mathfrak{n}$ of A. By $A[1/\mathfrak{n}]$ we denote the localization of A at all elements which have poles at most at $\mathfrak{n}$. The weight of a form will usually be denoted by n (or $n+2$), the letter k being taken as the name of the finite base field.

10.1   A moduli space for Drinfeld modules

Let S be a scheme over $\operatorname{Spec} A[1/\mathfrak{n}]$. Let $\underline{\varphi} := (L, \varphi)$ be a Drinfeld A-module on S of rank r, i.e., the line bundle L considered as a scheme of k-vector spaces over S is equipped with an endomorphism $\varphi \colon A \to \operatorname{End}(L), a \mapsto \varphi_a$. For any $a \in A$, the morphism $\varphi_a \colon L \to L$ is finite flat of degree $\#(A/a)^r$ and hence its kernel

$$\varphi[(a)] := \operatorname{Ker}\left(\varphi_a \colon L \to L\right)$$

is a finite flat A-module scheme over S. If furthermore all prime factors of the ideal aA are prime factors of $\mathfrak{n}$, working locally on affine charts, it follows that $d\varphi_a$ is a unit in $A[1/\mathfrak{n}]$, and thus that $\varphi_a(z)$ is a separable polynomial. This means that $\varphi[(a)]$ is étale over S. As a consequence the subscheme

$$\varphi[\mathfrak{n}] := \bigcap_{a \in \mathfrak{n} \smallsetminus \{0\}} \varphi[(a)]$$

is finite étale over S and of degree equal to $\#(A/\mathfrak{n})^r$. A level $\mathfrak{n}$-structure on $\underline{\varphi}$ is an isomorphism

$$\psi \colon \underline{(A/\mathfrak{n})}^r_S \xrightarrow{\cong} \varphi[\mathfrak{n}]$$

of finite étale group schemes over S, where $\underline{(A/\mathfrak{n})}^r_S$ denotes the constant group scheme on S with fiber $(A/\mathfrak{n})^r$.

Definition 10.1. Let $\mathcal{M}_r(\mathfrak{n})$ denote the functor on $A[1/\mathfrak{n}]$-schemes S given by

$$S \longmapsto \left\{ (\underline{\varphi}, \psi) \mid \underline{\varphi} = (L, \varphi) \text{ is a rank } r \text{ Drinfeld } A\text{-module on } S, \right.$$
$$\left. \psi \text{ is a level } \mathfrak{n}\text{-structure on } \underline{\varphi} \right\} / \cong,$$

i.e., we consider such triples up to isomorphism.

One has the following important theorem from [13]:

Theorem 10.2 (Drinfeld). *Suppose* $0 \neq \mathfrak{n} \subsetneq A$. *Then the functor* $\mathcal{M}_r(\mathfrak{n})$ *is represented by an affine scheme* $\mathfrak{M}_r(\mathfrak{n})$ *which is smooth of finite type and relative dimension* $r - 1$ *over* Spec $A[1/\mathfrak{n}]$

Remark 10.3. In [13], Drinfeld also defines level structures for levels dividing the characteristic of the Drinfeld module. Using these, he obtains a more general theorem as above: A universal Drinfeld module with level $\mathfrak{n}$-structure exists for A-schemes provided that $\mathfrak{n}$ has at least two distinct prime divisors. The universal space is regular of absolute dimension r. Its pullback to Spec $A[1/\mathfrak{n}]$ is the space $\mathfrak{M}_r(\mathfrak{n})$.

Remark 10.4. Let $(L^{\mathrm{univ}}, \varphi^{\mathrm{univ}}, \psi^{\mathrm{univ}})$ denote the universal object on $\mathfrak{M}_r(\mathfrak{n}) =$ Spec R^{univ}. Then in fact L^{univ} is the trivial bundle on Spec R^{univ}. The reason is that the image under ψ^{univ} of any non-zero element in $(A/\mathfrak{n})^r$ is a section of L which is everywhere different from the zero section, i.e., it is a nowhere vanishing global section.

For this reason, we shall in the universal situation always assume that $L^{\mathrm{univ}} = \mathcal{O}_{\mathfrak{M}_r(\mathfrak{n})}$. Moreover since $\mathfrak{M}_r(\mathfrak{n})$ is smooth and hence reduced, the universal Drinfeld module is in standard form, i.e.,

$$A \longrightarrow R^{\mathrm{univ}}[\tau], \quad a \longmapsto \varphi_a = \alpha_0(a) + \alpha_1(a)\tau + \cdots + \alpha_{r\deg(a)}\tau^{r\deg(a)}$$

with $\alpha_{r\deg a}(a) \in (R^{\mathrm{univ}})^*$.

Exercise 10.5. Let φ be a Drinfeld A-module of rank r in standard form with $L = \mathbb{G}_a$, i.e., a ring homomorphism

$$A \longrightarrow R[\tau], \quad a \longmapsto \varphi_a = \alpha_0(a) + \alpha_1(a)\tau + \cdots + \alpha_{r\deg(a)}\tau^{r\deg(a)}$$

for some A-algebra R. Suppose $s \in R$ is an a-torsion point which is non-zero on any component of R, i.e., $\alpha_0(a)s + \alpha_1(a)s^q + \cdots + \alpha_{r\deg(a)}s^{q^{r\deg(a)}} = 0$. Show that $a \mapsto s\varphi_a s^{-1}$ defines an isomorphic Drinfeld A-module such that 1 is an a-torsion point.

10.2 An explicit example

To make the above example more explicit, we consider the following special case: Let $A = k[t]$ and $\mathfrak{n} = (t)$. Then any Drinfeld A-module of rank r over an affine basis Spec R in standard form is described by the image of $t \in A$ in $R[\tau]$. This image is a polynomial of degree r which we denote by

$$\alpha_0 + \alpha_a \tau + \cdots + \alpha_r \tau^r$$

with $\alpha_r \in R^*$. We assume that R is an $A[1/t]$-algebra, so that also $\alpha_0 \in R^*$, and hence φ_t is a separable polynomial. Thus over a finite étale extension of R the solution set of $\varphi_t = 0$ is an $\mathbb{F}_q$-vector space of dimension r. Suppose we have a basis of t-torsion points $s_1, \ldots, s_r$ already defined over $\operatorname{Spec} R$. We trivialize the bundle L via the section s_1. This means that we have $s_1 = 1$ on $L(\operatorname{Spec} R)$ which is isomorphic via s_1 to $\mathbb{G}_a(\operatorname{Spec} R) = (R, +)$.

The set of all t-torsion points is thus the set $\sum_{i=1}^r s_i \alpha_i$, where the α_i range over all elements of k. Since these points are precisely the roots of φ_t, we find that

$$\varphi_t(z) = c \cdot \prod_{\underline{\alpha} \in k^r} \left(z - \sum_{i=1}^r s_i \alpha_i \right). \tag{10.1}$$

Recall the following result from [24, 1.3.7]:

Proposition 10.6 (Moore determinant). *Suppose $w_1, \ldots, w_r$ lie in an $\mathbb{F}_q$-algebra. Then*

$$\begin{vmatrix} w_1 & w_1^q & w_1^{q^2} & \cdots & w_1^{q^{r-1}} \\ w_2 & w_2^q & w_2^{q^2} & \cdots & w_2^{q^{r-1}} \\ \vdots & \vdots & \vdots & \ddots & \vdots \\ w_r & w_r^q & w_r^{q^2} & \cdots & w_r^{q^{r-1}} \end{vmatrix} = \prod_{i=1}^r \prod_{(\ell_{i-1}, \ldots, \ell_1) \in k^{i-1}} \left(w_i + \ell_{i-1} w_{i-1} + \cdots + \ell_1 w_1 \right).$$

By the theory of the Moore determinant, we obtain

$$\varphi_t(z) = c \cdot \begin{vmatrix} z & z^q & z^{q^2} & \cdots & z^{q^r} \\ s_r & s_r^q & s_r^{q^2} & \cdots & s_r^{q^r} \\ \vdots & \vdots & \vdots & \ddots & \vdots \\ s_2 & s_2^q & s_2^{q^2} & \cdots & s_2^{q^r} \\ 1 & 1 & 1 & \cdots & 1 \end{vmatrix} \Bigg/ \begin{vmatrix} s_r & s_r^q & \cdots & s_r^{q^{r-1}} \\ \vdots & \vdots & \ddots & \vdots \\ s_2 & s_2^q & \cdots & s_2^{q^{r-1}} \\ 1 & 1 & \cdots & 1 \end{vmatrix}. \tag{10.2}$$

Since the constant term of φ_t, i.e., the coefficient of z, is θ, the image of t under $A[1/t] \to R$, we can solve for c by computing the coefficient of z on the right-hand side. It yields

$$\theta = c \cdot \begin{vmatrix} s_r^q & s_r^{q^2} & \cdots & s_r^{q^r} \\ \vdots & \vdots & \ddots & \vdots \\ s_2^q & s_2^{q^2} & \cdots & s_2^{q^r} \\ 1 & 1 & \cdots & 1 \end{vmatrix} \Bigg/ \begin{vmatrix} s_r & s_r^q & \cdots & s_r^{q^{r-1}} \\ \vdots & \vdots & \ddots & \vdots \\ s_2 & s_2^q & \cdots & s_2^{q^{r-1}} \\ 1 & 1 & \cdots & 1 \end{vmatrix}$$

$$= c \cdot \begin{vmatrix} s_r & s_r^q & \cdots & s_r^{q^{r-1}} \\ \vdots & \vdots & \ddots & \vdots \\ s_2 & s_2^q & \cdots & s_2^{q^{r-1}} \\ 1 & 1 & \cdots & 1 \end{vmatrix}^{q-1}.$$

Proposition 10.7. *Let*

$$
R = k \left[\theta^{\pm 1}, s_2, \ldots, s_r,
\begin{vmatrix}
s_r & s_r{}^q & \cdots & s_r{}^{q^{r-1}} \\
\vdots & \vdots & \ddots & \vdots \\
s_2 & s_2{}^q & \cdots & s_2{}^{q^{r-1}} \\
1 & 1 & \cdots & 1
\end{vmatrix}^{-1}
\right]
$$

for indeterminates $\theta, s_2, \ldots, s_r$, and let $\varphi\colon k[t] \to R$ be the rank r Drinfeld module where φ_t is defined by (10.2). Then $\mathfrak{M}_r(t) \cong \operatorname{Spec} R$ and the universal triple is

$$
(\varphi, \mathbb{G}_{a,R}, \psi)
$$

where $\psi\colon (k[t]/(t))^r \to \varphi[t]$ is defined by mapping the ith basis vector on the left to s_i with the convention that $s_1 = 1$.

Proof. Let (L', φ', ψ') be a Drinfeld module with a full level t-structure on a scheme $S = \operatorname{Spec} R'$. Assume first that S is affine. As in the above case we may take the section $\psi'(1, 0, \ldots, 0)$ of L to trivialize it. By the construction of R, there is a homomorphism from $R \to R'$ over $\mathbb{F}_q[\theta^{\pm 1}]$-algebras sending s_i to the torsion point $s_i' := \psi'(0, \ldots, 0, 1, 0, \ldots, 0)$, where 1 occurs in the ith place. The s_i' determine in the same way as the s_i the function φ_t'. Hence (L', φ', ψ') is the pullback of (L, φ, ψ) under the morphism $\operatorname{Spec} R' \to \operatorname{Spec} R$. Moreover the morphism $R \to R'$ with this property is unique: The element s_1' determines a unique isomorphism $L \to \mathbb{G}_a$. With respect to the coordinates of $\mathbb{G}_a$ given by $s_1' = 1$, the sections $s_2, \ldots, s_r$ are uniquely determined from (L', φ', ψ') and hence $R \to R'$ is unique.

Now, let S be arbitrary. Fix an affine cover $\{\operatorname{Spec} R_i\}_i$. By the preceding paragraph we have unique morphisms $\operatorname{Spec} R_i \to \operatorname{Spec} R$. However by the uniqueness it also follows that on any affine subscheme of $\operatorname{Spec} R_i \cap \operatorname{Spec} R_j$, the two restriction to this subscheme agree. This in turn means that the local morphisms patch to a morphism $S \to \operatorname{Spec} R$ under which (L', φ', ψ') is the pullback of (L, φ, ψ). The uniqueness is true locally, hence also globally. This completes the proof of the representability of the functor $\mathcal{M}_r(\mathfrak{n})$. $\qquad\square$

Remark 10.8. Geometrically $\operatorname{Spec} R$ can be described as follows: It is the affine space $\mathbb{A}^{r-1}_{k[\theta^{\pm 1}]}$ with all the $\#k^{r-1}$ hyperplanes with coordinates in k removed. To see this, observe that $s_1 = 1$ and, by applying the Moore determinant, one has

$$
\begin{vmatrix}
s_r & s_r{}^q & \cdots & s_r{}^{q^{r-1}} \\
\vdots & \vdots & \ddots & \vdots \\
s_2 & s_2{}^q & \cdots & s_2{}^{q^{r-1}} \\
1 & 1 & \cdots & 1
\end{vmatrix}
= \prod_{i=1}^{r} \prod_{(\ell_{i-1}, \ldots, \ell_1) \in k^{i-1}} (s_i + \ell_{i-1} s_{i-1} + \cdots + \ell_1 s_1).
$$

Proposition 10.9. *We keep the notation of Proposition* 10.7. *The t-motive on* Spec R *corresponding to the universal Drinfeld module is isomorphic to the pair*

$$
\underline{F} := \left(R[t]^r, \tau = \begin{pmatrix} 0 & 0 & \cdots & 0 & \frac{t-\theta}{\alpha_r} \\ 1 & 0 & \cdots & 0 & \frac{-\alpha_1}{\alpha_r} \\ 0 & 1 & \cdots & 0 & \frac{-\alpha_2}{\alpha_r} \\ \vdots & \vdots & \ddots & \vdots & \vdots \\ 0 & 0 & \cdots & 1 & \frac{-\alpha_{r-1}}{\alpha_r} \end{pmatrix} (\sigma_R \times \mathrm{id}_t) \right)
$$

where

$$
\alpha_i = \theta \cdot \left. \begin{vmatrix} s_r & \cdots & s_r^{q^{i-1}} & s_r^{q^{i+1}} & \cdots & s_r^{q^r} \\ \vdots & \ddots & \vdots & \vdots & \ddots & \vdots \\ s_2 & \cdots & s_2^{q^{i-1}} & s_2^{q^{i+1}} & \cdots & s_2^{q^r} \\ 1 & \cdots & 1 & 1 & \cdots & 1 \end{vmatrix} \middle/ \begin{vmatrix} s_r & s_r^q & \cdots & s_r^{q^{r-1}} \\ \vdots & \vdots & \ddots & \vdots \\ s_2 & s_2^q & \cdots & s_2^{q^{r-1}} \\ 1 & 1 & \cdots & 1 \end{vmatrix}^q \right.
$$

for $i = 0, \dots, r$.

Proof. The shape of τ is determined as in Example 1.10. The formulas for the coefficients result easily from (10.2) by first eliminating c and then expanding the determinant in the numerator of (10.2) according to the first row. □

Let us, for some computations below, describe the case $n = 2$ in greater detail. For simplicity, we write $s := s_1$. In this case,

$$
R = k[\theta^{\pm 1}, s, (s^q - s)^{-1}],
$$

$$
\underline{F} = \left(R[t]^2, \ \tau = \begin{pmatrix} 0 & (t/\theta - 1)(s - s^q)^{q-1} \\ 1 & (s - s^{q^2})(s - s^q)^{-1} \end{pmatrix} (\sigma_R \times \mathrm{id}_t) \right).
$$

Substituting $u := s^q - s$ and observing that $s - s^{q^2} = u + u^q$, we obtain

$$
R_u := k[\theta^{\pm 1}, u^{\pm 1}], \quad \underline{F} = \left(R_u[t]^2, \ \tau = \begin{pmatrix} 0 & (t/\theta - 1)u^{q-1} \\ 1 & 1 + u^{q-1} \end{pmatrix} (\sigma_{R_u} \times \mathrm{id}_t) \right).
$$

$$
\tag{10.3}
$$

The introduction of u corresponds to a cover Spec $R \to$ Spec R_u of degree q. The space Spec R_u is a moduli space for Drinfeld modules with a level $\Gamma_1(t)$-structure, where $\Gamma_1(t)$ is the set of matrices $\begin{pmatrix} a & b \\ c & d \end{pmatrix} \in \mathrm{GL}_2(k[t])$ such that $a, d \equiv 1 \pmod{t}$ and $c \equiv 0 \pmod{t}$. Note that the moduli correspond to a choice of two t-torsion points $1, u$ where u is only determined up to adding a multiple of 1. (Due to our choice of coordinates for the line bundle underlying the rank 2 Drinfeld module, the first torsion point is 1.)

In the sequel, the symmetric powers $\mathrm{Sym}^n \underline{F}$ and their extension by zero to a compactification of $\mathfrak{M}_r(\mathfrak{n})$ will play an important role. We make this explicit in the setting of (10.3). Here a smooth compactification of $\operatorname{Spec} R_u = \mathbb{G}_{m/k[\theta^{\pm 1}]}$ is $\mathbb{P}^1_{k[\theta^{\pm 1}]}$. One simply has to extend $\underline{\mathcal{F}}$ to 0 and ∞. (Taking symmetric powers is compatible with this extension process.)

At $u = 0$, the matrix $\begin{pmatrix} 0 & (t/\theta-1)u^{q-1} \\ 1 & 1+u^{q-1} \end{pmatrix}$ specializes to $\begin{pmatrix} 0 & 0 \\ 1 & 1 \end{pmatrix}$, i.e., the extension is defined but not zero. At $u = \infty$ specializing the matrix leads to poles. To analyze the situation, we introduce $v = 1/u$, so that the matrix describing τ becomes $\begin{pmatrix} 0 & (t/\theta-1)v^{1-q} \\ 1 & 1+v^{1-q} \end{pmatrix}$. Next we multiply the standard basis e_1, e_2 of $R_u[t]^2$ by v. Then the action of τ for this new basis is given by

$$v^{-1}\begin{pmatrix} 0 & (t/\theta-1)v^{1-q} \\ 1 & 1+v^{1-q} \end{pmatrix}(\sigma_{R_u} \times \mathrm{id}_t)v = v^{-1}\begin{pmatrix} 0 & (t/\theta-1)v^{q-1} \\ 1 & 1+v^{q-1} \end{pmatrix}v^q$$

$$= \begin{pmatrix} 0 & (t/\theta-1) \\ v^{1-q} & v^{1-q}+1 \end{pmatrix}.$$

The following result summarizes the above discussion.

Proposition 10.10. *Consider $R_u = k[\theta^{\pm 1}, u^{\pm 1}]$ as an algebra over $A[1/\theta] = k[\theta^{\pm 1}]$.*

(a) *The moduli space of rank 2 Drinfeld modules with level $\Gamma_1(t)$-structure is isomorphic to $\operatorname{Spec} R_u$ as a scheme over $\operatorname{Spec} A[1/\theta]$.*

(b) *A relative smooth compactification of $\operatorname{Spec} R_u$ is the projective line $\mathbb{P}^1_{A[1/\theta]}$.*

(c) *The A-motive attached to the universal Drinfeld A-module over $\operatorname{Spec} R_u$ is given by*

$$\underline{\mathcal{F}} = \left(R_u[t]^2, \ \tau = \begin{pmatrix} 0 & (t/\theta-1)u^{q-1} \\ 1 & 1+u^{q-1} \end{pmatrix}(\sigma_{R_u} \times \mathrm{id}_t) \right).$$

(d) *$j_{\#}\underline{\mathcal{F}} := \left(\mathcal{O}^{\oplus 2}_{\mathbb{P}^1_{k[\theta^{\pm 1}]}}(-1\cdot[\infty]), \tau \right)$ and $j_!\underline{\mathcal{F}} := \left(\mathcal{O}^{\oplus 2}_{\mathbb{P}^1_{k[\theta^{\pm 1}]}}(-2\cdot[\infty]-1\cdot[0]), \tau \right)$ are a coherent extension of $\underline{\mathcal{F}}$ to $\mathbb{P}^1_{A[1/\theta]}$ and an extension by zero, respectively. Moreover one has a canonical monomorphism $j_!\underline{\mathcal{F}} \hookrightarrow j_{\#}\underline{\mathcal{F}}$ whose cokernel is a skyscraper sheaf supported on $\{0,\infty\}$.*

To compute the cohomology of $j_!\underline{\mathcal{F}}$ it suffices to compute that of $j_{\#}\underline{\mathcal{F}}$, since the discrepancy is easy to describe by the cokernel of $j_!\underline{\mathcal{F}} \hookrightarrow j_{\#}\underline{\mathcal{F}}$. This simplifies the computation of the cohomology of the crystal $j_! \operatorname{Sym}^n \underline{\mathcal{F}}$ significantly, since the coherent cohomology under $\mathbb{P}^1_{k[\theta^{\pm 1}]} \to \operatorname{Spec} k[\theta^{\pm 1}]$ (with the induced τ) is much easier to carry out for $\operatorname{Sym}^n j_{\#}\underline{\mathcal{F}}$ than for $\operatorname{Sym}^n j_!\underline{\mathcal{F}} = j_! \operatorname{Sym}^n \underline{\mathcal{F}}$.

10.3 Drinfeld modular forms via cohomology

Let us return to the general situation over $\mathfrak{M}_r(\mathfrak{n})$. *Assume that $r = 2$. For $r > 2$, the material below has not been carried out.* Only recently Pink has constructed compactifications of the moduli spaces $\mathfrak{M}_r(\mathfrak{n})$ with good properties; cf. [42] (and also [43]).

Let

$$f \colon \mathfrak{M}_r(\mathfrak{n}) \longrightarrow \operatorname{Spec} A[1/n] \tag{10.4}$$

denote the structure morphism and define

$$\underline{\mathcal{S}}_n(\mathfrak{n}) := R^1 f_! \operatorname{Sym}^n (j_! \underline{\mathcal{F}}).$$

(For $i \neq 1$ one has $R^i f_! \operatorname{Sym}^n(j_! \underline{\mathcal{F}}) = 0$.) Here are some basic facts on $\underline{\mathcal{S}}_n(\mathfrak{n})$:

(a) By general theory, since $\underline{\mathcal{F}}$ is flat as a crystal, we deduce that $\underline{\mathcal{S}}_n(\mathfrak{n})$ is flat as an A-crystal. This implies that on an open subscheme of $A[1/n]$ it has a free representative. However one has better representability results:

(b) Since $\underline{\mathcal{F}}$ is of pullback type (it is a τ-sheaf on an affine scheme),

$$R^i f_! \operatorname{Sym}^n (j_! \underline{\mathcal{F}}) = 0$$

is of pullback type. From this one can deduce that it has a free representative as a τ-sheaf. This representative does have, however, the disadvantage that the action of τ may be highly nilpotent.

(c) Using Gardeyn's theory of maximal extensions, one can also construct a representing τ-sheaf whose underlying sheaf is locally free *and* on which τ^{lin} is injective. We denote it by $\underline{\mathcal{S}}_n(\mathfrak{n})^{\mathrm{max}}$.

Our first aim is to give an interpretation for the *analytic realization* of $\underline{\mathcal{S}}_n(\mathfrak{n})$. For this, fix a homomorphism $k[t] \to A$ such that $\operatorname{Im}(t) \in A \smallsetminus k$. Denote by $K_\infty\{t\}$ the entire power series over K_∞. Define

$$\underline{\mathcal{S}}_n^{\mathrm{an}}(\mathfrak{n}) := \left(\underline{\mathcal{S}}_n(\mathfrak{n}) / K_\infty \right) \otimes_{K_\infty[t]} K_\infty\{t\} \quad \text{and} \quad M_n^B(\mathfrak{n}) := \left(\underline{\mathcal{S}}_n^{\mathrm{an}}(\mathfrak{n}) \right)^\tau$$

where $(\underline{\mathcal{S}}_n(\mathfrak{n})/K_\infty)$ is the base change of $\underline{\mathcal{S}}_n(\mathfrak{n})$ under $\operatorname{Spec} K_\infty \to \operatorname{Spec} A(1/n)$, and the superscript τ indicates that one takes τ-fixed points. In the simplest case $A = k[t]$, the pair defining $\underline{\mathcal{S}}_n^{\mathrm{an}}(\mathfrak{n})$ is a free sheaf on the rigid analytic $\mathbb{A}^1$ and τ defines a semilinear endomorphism on it of which one could think of as a system of differential equations. Then $M_n^B(\mathfrak{n})$ is the solution set of this system. It is not hard to see that $M_n^B(\mathfrak{n})$ is independent of the chosen representative of the crystal $\underline{\mathcal{S}}_n(\mathfrak{n})$. From this it is not hard to see that it is free over A of rank at most the rank of $\underline{\mathcal{S}}_n(\mathfrak{n})^{\mathrm{max}}$.

Denote by $S_n(\Gamma(\mathfrak{n}))$ the space of Drinfeld modular forms of full level $\mathfrak{n}$ and by $F_\mathfrak{n}$ the ray class group of F with conductor $\mathfrak{n}$. One of the central results of [4] is the following:

Theorem 10.11. *There is a Hecke-equivariant isomorphism*

$$\left(M_n^B(\mathfrak{n}) \right)^\vee \otimes_A \mathbb{C}_\infty \xrightarrow{\cong} S_n\left(\Gamma(\mathfrak{n}) \right)^{\mathrm{Gal}(F_\mathfrak{n}/F)}.$$

In the above theorem the cardinality of $\mathrm{Gal}(F_{\mathfrak{n}}/F)$ describes the number of connected components of $\mathcal{M}_2(\mathfrak{n})$ over the algebraic closure of K. In an adelic description of Drinfeld modular forms no such exponent is necessary.

We recall for the convenience of the reader the definition of the Hecke action on the crystal $\underline{S}_n(\mathfrak{n})$. (This induces the action on $M_n^B(\mathfrak{n})$.) The action on $S_n(\Gamma(\mathfrak{n}))^{\mathrm{Gal}(F_{\mathfrak{n}}/F)}$ can be similarly defined. In the case where A does not have class number one, this definition should only be given adelically. Since we have not developed the corresponding language, we will not give the definition and simply refer to [4]. To define the Hecke action, let us for any prime $\mathfrak{p}$ not dividing $\mathfrak{n}$ denote by $\mathfrak{M}_2(\mathfrak{n}, \mathfrak{p})$ the moduli scheme for quadruples (L, φ, ψ, C) where (L, φ, ψ) is a Drinfeld A-module (L, φ) of rank r with full level $\mathfrak{n}$-structure ψ and C is a cyclic $\mathfrak{p}$-torsion subscheme of L (in the sense of Drinfeld if $\mathfrak{p}$ is the characteristic of the base scheme). Consider

$$\mathfrak{M}_2(\mathfrak{n}) \xleftarrow{\pi_1} \mathfrak{M}_2(\mathfrak{n}, \mathfrak{p}) \xrightarrow{\pi_2} \mathfrak{M}_2(\mathfrak{n})$$

with $\pi_1((L, \varphi, \psi, C)) = (L, \varphi, \psi)$ and $\pi_2((L, \varphi, \psi, C)) = (L/C, \varphi/C, \psi/C)$. Denote by $\mathcal{G}$ the τ-sheaf $\mathrm{Sym}^n \underline{\mathcal{F}}$ and by $\mathcal{G}_{\mathfrak{p}}$ the nth symmetric power of the A-motive on $\mathfrak{M}_2(\mathfrak{n}, \mathfrak{p})$ associated to its tautological Drinfeld module. By the universal property of $\mathfrak{M}_2(\mathfrak{n})$ it follows that there are canonical isomorphisms

$$\pi_2^* \mathcal{G} \cong \mathcal{G}_{\mathfrak{p}} \cong \pi_1^* \mathcal{G}. \tag{10.5}$$

Adjunction yields a natural homomorphism

$$\mathcal{G} \longrightarrow \pi_{1*} \pi_1^* \mathcal{G}. \tag{10.6}$$

Since π_2 is finite flat of degree $\deg \mathfrak{p} + 1$ there also is a trace homomorphism

$$\mathrm{Tr}\colon \pi_{2*} \pi_2^* \mathcal{G} \longrightarrow \mathcal{G}. \tag{10.7}$$

The above isomorphisms and homomorphisms extend to $\overline{\mathfrak{M}_2(\mathfrak{n})}$ for $j_! \mathcal{G}$ and

$$j\colon \mathfrak{M}_2(\mathfrak{n}) \hookrightarrow \overline{\mathfrak{M}_2(\mathfrak{n})}$$

a compactification over $\mathrm{Spec}\, A(1/n)$. In analogy to (10.4), we denote the structure homomorphism $\mathfrak{M}_2(\mathfrak{n}, \mathfrak{p})$ by $f_{\mathfrak{p}}$. Then the following chain of homomorphisms defines the Hecke operator $T_{\mathfrak{p}}$:

$$R^1 f_! \mathcal{G} \xrightarrow{(10.6)} R^1 f_! \pi_{1*} \pi_1^* \mathcal{G} \xrightarrow{\text{can. isom.}} R^1 f_{\mathfrak{p}!} \pi_1^* \mathcal{G} \xrightarrow{(10.5)} R^1 f_{\mathfrak{p}!} \pi_2^* \mathcal{G}$$

$$\xrightarrow{\text{can. isom.}} R^1 f_! \pi_{2*} \pi_2^* \mathcal{G} \xrightarrow{(10.7)} R^1 f_! \mathcal{G}.$$

This is the standard way to make a correspondence act on cohomology. It also applies to the definition of Drinfeld modular forms as global sections of a suitable line bundle (which depends on the weight) and agrees there with other common definitions of the Hecke operator $T_{\mathfrak{p}}$.

Sketch of proof of Theorem 10.11. The proof given in [4] has its basic structure modeled at the classical proof by Shimura. Some details seem to be quite different however. Here we shall only give a rough sketch of the individual steps of the proof.

(a) By a theorem of Teitelbaum, the right hand side of the isomorphism in Theorem 10.11 is isomorphic to the space of harmonic cochains $C^{\mathrm{har}}(\Gamma(\mathfrak{n}), V_n(\mathbb{C}_\infty))$ on the Bruhat-Tits tree for PGL_2. These are $\Gamma(\mathfrak{n})$-equivariant function on the edges of this tree into a certain $\mathbb{C}_\infty[\Gamma(\mathfrak{n})]$-module $V_n(F_\infty)$. In fact, the $\Gamma(\mathfrak{n})$-module can be naturally obtained by coefficient change from a $F(\Gamma(\mathfrak{n}))$-module $V_n(F)$ defined already over F. Bet even more is true: One can define a local system $V_n(A)$ of free A-modules on the edges of the tree that naturally carries a $\Gamma(\mathfrak{n})$-action, such that this local system is an A-structure from the local system given by $V_n(F)$. Concretely, for every edge of the tree, the local system $V_n(A)$ is given by a projective A-submodule of $V_n(F)$ of rank equal to $\dim_F V_n(F)$ and such that under the action of $\Gamma(\mathfrak{n})$ on the tree there is a corresponding compatible action on these A-submodules of $V_n(F)$. An $V_n(A)$-valued $\Gamma(\mathfrak{n})$-invariant harmonic cochain is now a map which to any edge of the tree assigns a value in the A-module defined for this edge and such that the values are $\Gamma(\mathfrak{n})$-equivariant. Thus we have $S_n(\Gamma(\mathfrak{n})) \cong C_{\mathrm{har}}(\Gamma(\mathfrak{n}), V_n(A)) \otimes_A \mathbb{C}_\infty$. It this suffices to prove that there is a natural Hecke-equivariant isomorphism

$$\left(M_n^B(\mathfrak{n})\right)^\vee \xrightarrow{\cong} C_{\mathrm{har}}(\Gamma(\mathfrak{n}), V_n(A))^{\mathrm{Gal}(F_\mathfrak{n}/F)}.$$

(b) Let us now consider the right-hand side $M_n^B(\mathfrak{n})$. It was obtained by base change of the crystal $\underline{S}_n(\mathfrak{n})$ to K_∞, passing to analytic coefficients and then taking τ-invariants. Parallel to the algebraic theory of A-crystals over an algebraic base including the functors defined there, one can develop a theory of crystals over rigid analytic spaces and with analytic coefficients. Moreover one can define a natural rigidification functor from the algebraic to the rigid analytic setting which is compatible with all functors. This allows one to recover $\underline{S}_n(\mathfrak{n})/F_\infty \otimes_{F_\infty[t]} F_\infty\{t\}$ as follows: Denote by $\mathfrak{M}_2(\mathfrak{n})^{\mathrm{an}}_{/F_\infty}$ the rigidification of the scheme $\mathfrak{M}_2(\mathfrak{n})$ after base change from $A[1/n]$ to F_∞. This rigid analytic space is (after finite extensions of the base, e.g., from F_∞ to $F_{\mathfrak{n},\infty}$) isomorphic to $\Gamma(\mathfrak{n})\backslash\Omega^{\mathrm{Gal}(F_\mathfrak{n}/F)}$ where Ω is the Drinfeld symmetric space of dimension one over F_∞. In particular this rigid analytic curve has a module interpretation. Let $\mathrm{Sym}^n \mathcal{F}^{\mathrm{an}}$ denote the rigid τ-sheaf on $\Gamma(\mathfrak{n})\backslash\Omega$ with $F_\infty\{t\}$-coefficients associated to $\mathrm{Sym}^n \mathcal{F}$. The sheaf $\mathcal{F}^{\mathrm{an}}$ can also be obtained purely from the universal analytic Drinfeld module over $\Gamma(\mathfrak{n})\backslash\Omega$. Extension by zero leads to an extension by zero in the rigid setting where it is however important that one rigidifies an algebraic compactification. Let us denote by $H^1_{\mathrm{an},c}$ the cohomology with compact support on this rigid analytic site of τ-sheaves (or crystals). Then there is a natural isomorphism

$$M^B(\mathfrak{n}) \cong H^1_{\mathrm{an},c}\left(\overline{\mathfrak{M}_2(\mathfrak{n})}^{\mathrm{an}}_{/F_\infty}, j_! \operatorname{Sym}^n \underline{\mathcal{F}}^{\mathrm{an}}\right)^\tau.$$

Thus it now suffices to construct a natural isomorphism

$$\left(H^1_{\mathrm{an},c}\left(\Gamma(\mathfrak{n})\backslash \Omega^*, j_! \operatorname{Sym}^n \underline{\mathcal{F}}^{\mathrm{an}}\right)^\tau\right)^\vee \xrightarrow{\cong} C_{\mathrm{har}}\left(\Gamma(\mathfrak{n}), V_n(A)\right) \qquad (10.8)$$

where $\Omega^* = \Omega \cup \mathbb{P}^1(F)$ ("suitably topologized") and where the base of Ω on the left is sufficiently large and lies between F and $\mathbb{C}_\infty$. Moreover one needs to prove the Hecke compatibility of this isomorphism if extended in a natural way to its $\# \operatorname{Gal}(F_\mathfrak{n}/F)$-fold sum.

(c) To prove the isomorphism (10.8), one computes the left-hand side using an explicit Čech cover of $\Gamma(\mathfrak{n})\backslash \Omega^*$. The cover is obtained as follows: There is a well-known reduction map $\Omega \to \mathcal{T}$ where $\mathcal{T}$ is the Bruhat–Tits tree for $\mathrm{PGL}_2(F_\infty)$. The cover is equivariant with respect to an action of $\Gamma(\mathfrak{n})$, and there is an induced reduction map

$$\Gamma(\mathfrak{n})\backslash \Omega \longrightarrow \Gamma(\mathfrak{n})\backslash \mathcal{T}.$$

There is a finite number of orbits of vertices $\Gamma(\mathfrak{n})v_i$, $i = 1, \ldots, n$, as well as of edges $\Gamma(\mathfrak{n})e_j$, $j \in \{1, \ldots, m\}$, in $\Gamma\backslash\mathcal{T}$ on which the action of $\Gamma(\mathfrak{n})$ is free. If these orbits are removed the remaining graph becomes a disjoint union of subgraphs c_ℓ which contain no loops and are in a 1-1 correspondence with the cusps of $\Gamma(\mathfrak{n})\backslash\Omega$. The preimage of a closed ε-neighborhood of any $\Gamma(\mathfrak{n})v_i$ is an affinoid subset $\mathfrak{U}_i \subset \Gamma(\mathfrak{n})\backslash\Omega$, a disc minus q open subdiscs. The preimage of a closed ε-neighborhood of any c_ℓ after adding one puncture is an affinoid subset $\mathfrak{W}_\ell \subset \Gamma(\mathfrak{n})\backslash\Omega^*$ of the cusp. The preimage of any edge orbit $\Gamma(\mathfrak{n})e_j$ minus closed $\varepsilon/2$-neighborhoods at each end is an annulus $\mathfrak{V}_j \subset \Gamma(\mathfrak{n})\backslash\Omega$. Using the *uniformizability* of the universal Drinfeld module over the $\mathfrak{U}_i$ and the $\mathfrak{V}_j$ and the fact that $\operatorname{Sym}^n \underline{\mathcal{F}}^{\mathrm{an}}$ is extended by zero to the cusps, one finds

$$H^0_{\mathrm{an},c}\left(\mathfrak{V}_j, j_! \operatorname{Sym}^n \underline{\mathcal{F}}^{\mathrm{an}}\right)^\tau \cong V_n(A)^\vee|_{e_j}, \quad H^0_{\mathrm{an},c}\left(\mathfrak{W}_\ell, j_! \operatorname{Sym}^n \underline{\mathcal{F}}^{\mathrm{an}}\right)^\tau = 0,$$

and if $e_{j'}$ is any edge neighboring v_i then

$$H^0_{\mathrm{an},c}(\mathfrak{U}_j, j_! \operatorname{Sym}^n \underline{\mathcal{F}}^{\mathrm{an}})^\tau \cong V_n(A)^\vee|_{e_{j'}}.$$

The Čech complex is particularly simple, since any triple intersections of distinct sets of the covering are empty and any non-empty double intersections are given by a small annulus on some $\mathfrak{V}_j$ where the double intersection is by intersecting $\mathfrak{V}_j$ and some adjacent $\mathfrak{U}_i$. By explicit inspection, one can show that the Čech complex is dual to the stable complex given by Teitelbaum to compute harmonic cochains. This yields the asserted isomorphism (10.8).

(d) Finally, one verifies Hecke equivariance by comparing explicit formulas for Hecke operators on the Čech cover and on harmonic cochains. $\qquad \square$

10.4   Galois representations associated to Drinfeld modular forms

One can also study the étale realizations of the crystal $\underline{S}_n(\mathfrak{n})$. For this we fix a maximal ideal $\mathfrak{p}$ of the coefficient ring A. The functor ϵ yields the inverse system of étale sheaves

$$\left\{\epsilon\big(\underline{S}_n(\mathfrak{n}) \otimes_A A/\mathfrak{p}^m\big)\right\}_{m\in\mathbb{N}}$$

on $\operatorname{Spec} A(\mathfrak{n})$. Its inverse limit is an étale $A_{\mathfrak{p}}$-sheaf $\underline{S}_n(\mathfrak{n})_{\text{et},\mathfrak{p}}$ over $\operatorname{Spec} A[1/\mathfrak{n}]$. We shall discuss later that it can be ramified at a finite number of places of $\operatorname{Spec} A[1/\mathfrak{n}]$. However we have the following result:

Theorem 10.12. *The étale sheaf* $\epsilon\big(\underline{S}_n(\mathfrak{n}) \otimes_{A[1/\mathfrak{n}]\otimes A}(F \otimes A/\mathfrak{p}^m)\big)$ *has rank* $s_n(\mathfrak{n}) :=$ $\dim S_n(\mathfrak{n}) \cdot \#\operatorname{Gal}(F_{\mathfrak{n}}/F)$.

Sketch of proof. Recall that the functor from crystals to étale sheaves is compatible with all functors. Thus the sheaf given in the theorem is isomorphic to

$$H^1_{\text{et},c}\big(\mathfrak{M}_2(\mathfrak{n})/F, \operatorname{Sym}^n \underline{\mathcal{F}} \otimes_A A/\mathfrak{p}^m\big) \cong H^1_{\text{et},c}\big(\mathfrak{M}_2(\mathfrak{n})/F, \operatorname{Sym}^n \varphi\big[\mathfrak{p}^m\big]^{\vee}\big),$$

where φ is the universal Drinfeld module on $\mathfrak{M}_2(\mathfrak{n})/F$. By a result of Gekeler [17], the modular curve $\overline{\mathfrak{M}_2(\mathfrak{n})}/F$ is ordinary for any $\mathfrak{n}$. By a result of Pink [41], there is a Grothendieck–Ogg–Shafarevich type formula for the $\mathbb{F}_p$-dimension of the cohomology of étale $\mathbb{F}_p$-sheaves on curves, provided the curve has an ordinary cover over which the monodromy of the étale sheaf is unipotent. In the case at hand, by Gekeler's result we can take the cover $\overline{\mathfrak{M}_2(\mathfrak{n}\mathfrak{p}^m)}$. The formula of Pink shows that

$$\dim_{\mathbb{F}_p} \epsilon\big(\underline{S}_n(\mathfrak{n}) \otimes_{A[1/\mathfrak{n}]\otimes A}(F \otimes A/\mathfrak{p}^m)\big) = \dim_{\mathbb{F}_p} A/\mathfrak{p}^m s_n(\mathfrak{n})$$

for any m. From this the theorem follows easily – the proof is essentially the same as the proof that the $\mathfrak{n}$-torsion of a Drinfeld module away from the characteristic is equal to $A/\mathfrak{n}^r$. $\square$

It follows that $\underline{S}_n(\mathfrak{n})_{\text{et},\mathfrak{p}}/F$ defines a continuous homomorphism

$$\rho_{A,\mathfrak{n}} \colon \operatorname{Gal}\big(F^{\text{sep}}/F\big) \longrightarrow \operatorname{GL}_{s_n(\mathfrak{n})}\big(A_{\mathfrak{p}}\big).$$

Moreover $\underline{S}_n(\mathfrak{n})_{\text{et},\mathfrak{p}}$ carries the Hecke action induced from $\underline{S}_n(\mathfrak{n})$.

In [4] the following result is proved:

Theorem 10.13. *For any prime* $\mathfrak{q}$ *different from* $\mathfrak{n}$ *and* $\mathfrak{p}$, *the actions of* $T_{\mathfrak{q}}$ *and of* $\operatorname{Frob}_{\mathfrak{q}}$ *are the same on the reduction of* $\underline{S}_n(\mathfrak{n})_{\text{et},\mathfrak{p}}$ *from* $\operatorname{Spec} A[1/\mathfrak{p}]$ *to* $\operatorname{Spec} A/\mathfrak{q}$.

Since the Hecke operators commute among each other and since the $\operatorname{Frob}_{\mathfrak{q}}$ where $\mathfrak{q}$ runs through all maximal ideal prime ideals of $\operatorname{Spec} A[1/\mathfrak{n}\mathfrak{p}]$ are dense in $\operatorname{Gal}(F^{\text{sep}}/F)$, we deduce:

Corollary 10.14. *The image of $\rho_{A,\mathfrak{n}}$ is abelian.*

The crystal $\underline{\mathcal{S}}_n(\mathfrak{n})$ has given rise to two realizations, namely an analytic one and for each maximal ideal of A an étale one:

$$\left(\underline{\mathcal{S}}_n(\mathfrak{n})_{/K_\infty}^{\mathrm{an}}\right)^\tau \leftarrow - - \underline{\mathcal{S}}_n(\mathfrak{n}) - - \to \underline{\mathcal{S}}_n(\mathfrak{n})_{\mathrm{et},\mathfrak{p}}.$$

In each case, the realizations inherits a Hecke action. As examples show (see [35]), the Hecke action may not be semisimple. So we pass in both cases to the semisimplification and decompose $\underline{\mathcal{S}}_n(\mathfrak{n})$ into Hecke eigenspaces (if necessary after inverting some elements in the coefficient ring A). This yields a correspondence between Hecke eigensystems of Drinfeld modular forms and simple abelian Galois representations. Before we give the precise statement from [4], we recall the following classical theorem due to Goss:

Theorem 10.15 (Goss). *Let f be a Hecke eigenform of weight n and level $\mathfrak{n}$ with Hecke eigenvalues $a_\mathfrak{p}(f)$ for all $\mathfrak{p}$ not dividing $\mathfrak{n}$. Then all $a_\mathfrak{p}(f)$ are integral over A and the field $F_f := F(a_\mathfrak{p}(f) \mid \mathfrak{p} \in \mathrm{Spec}\, A[1/\mathfrak{n}])$ is a finite extension of F.*

Denote by $\mathcal{O}_f$ the ring of integers of F_f. For any prime $\mathfrak{P}$ of $\mathcal{O}_f$, the completion at $\mathfrak{P}$ will be $\widehat{\mathcal{O}_f}^{\mathfrak{P}}$.

Theorem 10.16 (B.). *Let f be as above and suppose that f is cuspidal. Then there exists a system of Galois representations*

$$\rho_{f,\mathfrak{P}} \colon \mathrm{Gal}(F^{\mathrm{sep}}/F) \to \mathrm{GL}_1(\widehat{\mathcal{O}_f}^{\mathfrak{P}})_{\mathfrak{P} \in \mathbf{Max}(\mathcal{O}_f)}$$

uniquely characterized by the condition that, for each fixed $\mathfrak{P}$, one has for almost all $\mathfrak{q}$ prime to $\mathfrak{P}\mathfrak{n}$ the equation

$$\rho_{f,\mathfrak{P}}\left(\mathrm{Frob}_\mathfrak{q}\right) = a_\mathfrak{q}(f),$$

so that the right-hand side is independent of the prime $\mathfrak{P}$.

Remark 10.17.

(a) It is not clear whether there is a theory of old and new forms for Drinfeld modular forms. So one cannot proceed as in the classical case.

(b) There are various counterexamples to a strong multiplicity one theorem, by Gekeler and Gekeler–Reversat; e.g., [19, Example 9.7.4] for an example in weight 2.

(c) As far as we know there are no counterexamples to multiplicity one for $S_n(\Gamma_0(\mathfrak{p}))$ for fixed n and a prime $\mathfrak{p}$ of $\mathrm{Spec}\, A$.

(d) Despite the results of the following section, the ramification locus of the system $(\rho_{f,\mathfrak{P}})_{\mathfrak{P} \in \mathbf{Max}(\mathcal{O}_f)}$ is rather mysterious.

10.5   Ramification of Galois representations associated to Drinfeld modular forms

From a result of Katz [31] one easily deduces the following.

Theorem 10.18. *Let $\underline{S}_n(\mathfrak{n})^{\max}$ be the maximal extension in the sense of Gardeyn representing the same-named crystal. Define D as the support of the cokernel of the injective homomorphism $\tau^{\mathrm{lin}} \colon (\sigma \times \mathrm{id})^* \underline{S}_n(\mathfrak{n})^{\max} \to \underline{S}_n(\mathfrak{n})^{\max}$. Let f be a Drinfeld modular form of level $\mathfrak{n}$ and weight n, and let $\mathfrak{P}$ be a maximal ideal of $\mathcal{O}_f$ and $\mathfrak{p}$ its contraction to A. Then $\rho_{f,\mathfrak{P}}$ is unramified at all primes $\mathfrak{q}$ of A such that $(\mathfrak{q}, \mathfrak{p})$ is not in D. Moreover for such $\mathfrak{q}$ one has*

$$\det\left(1 - z\rho_{f,\mathfrak{P}}\left(\mathrm{Frob}_{\mathfrak{q}}\right)\right) = 1 - za_{\mathfrak{q}}(f),$$
$$\det\left(1 - z\rho_{A,\mathfrak{n}}\left(\mathrm{Frob}_{\mathfrak{q}}\right)\right) = \det\left(1 - zT_{\mathfrak{q}}^{\mathrm{ss}}|S_n(\mathfrak{n})\right),$$

where $T_{\mathfrak{q}}^{\mathrm{ss}}$ is the semisimplification of $T_{\mathfrak{q}}$ acting on the analytic space of modular forms $S_n(\mathfrak{n})$ of weight n and level $\mathfrak{n}$.

Remark 10.19. After replacing the coefficient ring A by a larger ring A' which is a localization of A at a suitable element, one can in fact decompose $\underline{S}_n(\mathfrak{n})$ into components corresponding to generalized eigenforms under the Hecke action. By enlarging A' to a ring A'', one may furthermore assume that A'' contains all the Hecke eigenvalues of all eigenforms. Then, over A'', to any eigenform f one has a corresponding subcrystal of $\underline{S}_f$ of $\underline{S}_n(\mathfrak{n}) \otimes_A A''$. Katz' criterion then applies to the Gardeyn maximal model of $\underline{S}_f$. This gives, in theory, a precise description of the ramification locus of $\rho_{f,\mathfrak{P}}$ – provided that $\mathfrak{P}$ is in $\mathbf{Max}(A'')$ – given by a divisor D_f on $\mathrm{Spec}\, A \times \mathrm{Spec}\, A''$. The definition of D in the previous theorem is coarser. It gives a bound on ramification for all eigenforms f simultaneously.

While Galois representations $\rho_{f,\mathfrak{P}}$ for eigenforms f of level $\mathfrak{n}$ tend to be ramified at the places above $\mathfrak{n}$, it is not clear how the additional places of ramification are linked to $\mathfrak{P}$. In an abstract sense, the answer is that this link is given by D, or more precisely by D_f. Concretely, we do not know how to determine D_f from f or D from n and $\mathfrak{n}$. Some explicit examples are given below. One clue to the ramification locus is given by Theorems 10.21 and 10.22. They describe a link between places of bad reduction of modular curves and the ramification of modular forms. The following result might serve as a motivation.

Let K be a local field of characteristic p with ring of integers $\mathcal{O}$ and residue field k. Let $\mathcal{A}/\mathcal{O}$ be an abelian scheme with generic fiber A/K of dimension g and special fiber A/k. The p^n-torsion subscheme of $\mathcal{A}$ (or A) is denoted by $\mathcal{A}[p^n]$ (or $A[p^n]$, respectively) and for any field $L \supset K$ we write $A[p^n](L)$ for the group of L-valued points of $A[p^n]$. Consider the p-adic Tate module

$$\mathrm{Tate}_p\, A := \varprojlim_n A\left[p^n\right]\left(K^{\mathrm{sep}}\right)$$

of A. The module underlying $\mathrm{Tate}_p\, A$ is free over $\mathbb{Z}_p$. One defines the *p-rank of A* as $\mathrm{rank}_p A := \mathrm{rank}_{\mathbb{Z}_p} \mathrm{Tate}_p\, A$. It satisfies $0 \leq \mathrm{rank}_p A \leq g$. The action of $G_K := \mathrm{Gal}(K^{\mathrm{sep}}/K)$ on $\mathrm{Tate}_p\, A$ is $\mathbb{Z}_p$-linear and thus with respect to some $\mathbb{Z}_p$-basis it yields a Galois representation

$$\rho_{A,p}\colon G_K \longrightarrow \mathrm{Aut}_{\mathbb{Z}_p}\left(\mathrm{Tate}_p\, A\right) \cong \mathrm{GL}_{\mathrm{rank}_p A}(\mathbb{Z}_p).$$

By Hensel's Lemma any p^n-torsion point of $\mathcal{A}/k$ will lift to a unique p^n-torsion point of A/K. Thus

$$\mathrm{rank}_p\, \mathcal{A}/k \leq \mathrm{rank}_p A,$$

i.e., the *p*-rank can only decrease under reduction. The following result from [6] links the ramification of $\rho_{A,p}$ to the *p*-rank:

Theorem 10.20. *The p-rank is invariant under reduction if and only if the action of G_K on $\mathrm{Tate}_p\, A$ is unramified.*

Let us, after this short interlude come back to the ramification of Drinfeld modular forms: Recall that for a Hecke eigenform f, we denote by $a_{\mathfrak{q}}(f)$ the Hecke eigenvalue of f at $\mathfrak{q}$. Let $J_{\mathfrak{n}}$ denote the Jacobian of the Drinfeld modular curve for level $\mathfrak{n}$. Then the following two results are shown in [6] (in a more precise form):

Theorem 10.21. *Suppose that f is a doubly cuspidal Drinfeld–Hecke eigenform of weight 2 and level $\mathfrak{n}$. Then for a prime $\mathfrak{q}$ not dividing $\mathfrak{n}$ the following results hold:*

(a) *If $a_{\mathfrak{q}}(f) \neq 0$, then for any (or for all) $\mathfrak{P} \in \mathbf{Max}(\mathcal{O}_f)$ the representation $\rho_{f,\mathfrak{P}}$ is unramified at $\mathfrak{q}$.*

(b) *If $a_{\mathfrak{q}}(f) = 0$, then the Jacobian $J_{\mathfrak{n}}$ has non-ordinary reduction modulo $\mathfrak{q}$.*

Note that in the case of weight 2 there is a representation ρ_f from G_K into GL_1 over a finite field $\mathbb{F}$ contained in $\mathcal{O}$ such that $\rho_{f,\mathfrak{P}} = \rho_f \otimes_{\mathbb{F}} \mathcal{O}_{\mathfrak{P}}$ for all $\mathfrak{P} \in \mathbf{Max}(\mathcal{O}_f)$. This is similar to the case of classical weight 1 modular forms, and it explains why the condition in (a) is independent of $\mathfrak{P}$.

Suppose now that f has weight larger than two, and consider a representation $\rho_{f,\mathfrak{P}}$ for $\mathfrak{P} \in \mathbf{Max}(\mathcal{O}_f)$ over $\mathfrak{p} \in \mathbf{Max}(A)$. As $\rho_{f,\mathfrak{P}}$ is associated to a Hecke character (see 10.25) and because it is known that such have square free levels, it follows that $\rho_{f,\mathfrak{P}}$ is ramified at $\mathfrak{q}$ if and only if its reduction mod $\mathfrak{P}$ is so. As in the case of classical modular forms, the reduction mod $\mathfrak{P}$ is congruent to the representation of a form of weight 2 and level $\mathfrak{np}$. To the latter one can apply the previous result. This yields:

Theorem 10.22. *Suppose f is a doubly cuspidal Drinfeld–Hecke eigenform of weight $n \geq 3$ and level $\mathfrak{n}$. Let $\mathfrak{P}$ be in $\mathbf{Max}(\mathcal{O}_f)$ with contraction $\mathfrak{p} \in \mathbf{Max}(A)$. Then for a prime $\mathfrak{q}$ not dividing $\mathfrak{n}\mathfrak{q}$ the following hold:*

(a) *If $a_{\mathfrak{q}}(f) \neq 0 \pmod{\mathfrak{P}}$, then the representation $\rho_{f,\mathfrak{P}}$ is ramified at $\mathfrak{q}$.*

(b) *If $a_{\mathfrak{q}}(f) = 0 \pmod{\mathfrak{P}}$, then the Jacobian $J_{\mathfrak{np}}$ has non-ordinary reduction modulo $\mathfrak{q}$.*

Note that in known examples, e.g., [6], the places of ramification of $\rho_{f,\mathfrak{P}}$ which are prime to the level $\mathfrak{n}$ do typically depend on $\mathfrak{P}$ unlike in the case of weight 2.

Question 10.23. For classical modular forms it is simple to list all the primes which are ramified for the associated Galois representations. By the theory of new forms these primes are those dividing the minimal level associated to the modular form together with the place p (or the places above p) if one considers p-adic Galois representations.

Because of this simplicity one wonders if there is also a simple recipe in the case of Drinfeld modular forms. Numerical data seem insufficient to make any predictions. This deserves to be studied more systematically. Because of Theorem 10.21 this question is directly linked to the reduction behavior of Drinfeld modular curves (at primes of good reduction!) and their associated Jacobians.

10.6   Drinfeld modular forms and Hecke characters

In [7] we introduce a notion of Hecke character that was more general than previous definitions due to Gross [26] and others. Our motivation was a question of Serre and independently Goss which asked whether Drinfeld modular forms are linked to Hecke characters. In [6] this question was answered in the affirmative. We will briefly indicate this result.

Definition 10.24. Let F be a global function field over $\mathbb{F}_p$. A homomorphism

$$\chi \colon \mathbb{A}_F^* \longrightarrow \overline{\mathbb{F}_p(t)}^*,$$

where $\mathbb{A}_F^*$ denotes the ideles of F and $\overline{\mathbb{F}_p(t)}$ is discretely topologized, is a *Hecke character (of type Σ)* if

(a) χ is continuous (i.e., trivial on a compact open subgroup of $\mathbb{A}_F^*$) and

(b) there exists a finite subset $\Sigma = \{\sigma_1, \ldots, \sigma_r\}$ of field homomorphisms $\sigma_i \colon F \hookrightarrow \overline{\mathbb{F}_p(t)}$ and $n_i \in \mathbb{Z}$ for $i = 1, \ldots, r$, such that

$$\chi(\alpha) = \sigma_1(\alpha)^{n_1} \cdot \ldots \cdot \sigma_r(\alpha)^{n_r},$$

where on the left we regard F^* as being diagonally embedded in $\mathbb{A}_F^*$.

Note that for any compact open subgroup $U \subset \mathbb{A}_F^*$ the coset space $F^* \backslash \mathbb{A}_F^* / U$ admits a surjective degree map to $\mathbb{Z}$ whose kernel is finite and may be interpreted as a class group. By an observation of Goss [22], the Hecke characters as defined above have square free conductors.

The main result on Hecke characters and modular forms is the following.

Theorem 10.25. *For any cuspidal Drinfeld–Hecke eigenform f with eigenvalues $(a_\mathfrak{p}(f))_{\mathfrak{p}\in\mathbf{Max}(A)}$ there exists a unique Hecke character*

$$\chi_f : \mathbb{A}_F^* \longrightarrow K_f^*$$

such that

$$a_\mathfrak{p}(f) = \chi_f\big(1,\ldots,1,\underbrace{\varpi_\mathfrak{p}}_{\text{at }\mathfrak{p}},1,\ldots,1\big)$$

for almost all $\mathfrak{p}\in\mathbf{Max}(A)$.

Unfortunately, the set Σ_f for the Hecke character χ_f, as required by Definition 10.24, is rather mysterious. The proof of the theorem sheds no light on it. What is however not so hard to see is that the ramification divisor D_f introduced in 10.19 is equal to $\bigcup_{\sigma\in\Sigma_f}\mathrm{Graph}(\sigma)$ where the $\sigma\in\Sigma_f$ are viewed as morphisms of algebraic curves.

Example 10.26. The following Hecke characters are taken from [6]. The are associated to Drinfeld modular forms.

Let $F = \mathbb{F}_q(\theta)$, let n be in $\{2,\ldots,p\}$ and consider $\sigma_n : \mathbb{F}_q(\theta) \longrightarrow \mathbb{F}_q(t) : \theta \mapsto (1-n)t$. . Define

$$U := \big(1 + \theta\mathbb{F}_q[[\theta]]\big) \times \prod_{\nu\nmid 0,\infty} \mathcal{O}_v^* \times \mathbb{F}_q\Big(\big(\tfrac{1}{\theta}\big)\Big)^*.$$

Then the natural homomorphism

$$\mathbb{F}_q(\theta)^* \xrightarrow{\simeq} \mathbb{A}_F^*/U$$

is an isomorphism. Hence there exists a unique Hecke character

$$\chi_n : \mathbb{A}_F^* \to \overline{\mathbb{F}_p(t)}^*$$

such that χ_n is trivial on U and such that it agrees with σ_n on $\mathbb{F}_q(\theta)$.

Remark 10.27. The Hecke character χ_f provides a compact way of storing essential information about the cuspidal Drinfeld–Hecke eigenform f. To explain this, suppose that $A = \mathbb{F}_q[t]$, that the weight of f is n and that we have computed its Hecke eigenvalues $a_\mathfrak{p}(t)$ for many primes $\mathfrak{p}$ not dividing $\mathfrak{n}$. The conductor of χ_f consists of those primes $\mathfrak{p}\nmid\mathfrak{n}$ for which $T_\mathfrak{p}$ acts as zero and some primes dividing $\mathfrak{n}$. Having computed many eigenvalues, we may thus hope to know the prime-to-$\mathfrak{n}$ part of the conductor of χ_f and thus a lower bound for the conductor $\mathfrak{m}_f \subset A$ of χ_f, by which we mean the largest square-free ideal such that χ_f is trivial on the group U_f of all ideles congruent to 1 modulo $\mathfrak{m}_f$. Suppose furthermore that we know the coefficient field K_f of f. For theoretical reasons the character χ_f is trivial on F_∞^* (the image of the decomposition group at ∞ of any $\rho_{f,\mathfrak{P}}$ is trivial).

If the weight n is 2, then χ_f is of finite order and in particular Σ is empty. Knowing a bound on $\mathfrak{m}_f$ and that χ_f is trivial on F_∞^*, by computing sufficiently many Hecke eigenvalues, we can completely determine χ_f as a function on $F^* \backslash \mathbb{A}_F^* / U_f F_\infty^*$.

If the weight n is larger than 2, it is necessary to find the embeddings $\sigma_i \colon \mathbb{F}_q(\theta) \to K_f$, $i = 1, \ldots, r$ and their exponents n_i. Denote by $b_i \in K_f$ the image of θ under σ_i. If g is any element of $A = \mathbb{F}_q[\theta]$ which is congruent to 1 modulo $\mathfrak{m}_f$, then T_g acts on f as $\prod_i g(b_i)^{n_i}$. At the same time, if $(f) = \prod_{\mathfrak{p}} \mathfrak{p}^{m_{\mathfrak{p}}}$ for exponents $m_{\mathfrak{p}} = \operatorname{ord}_{\mathfrak{p}}(g) \in \mathbb{N}_0$, then we have the equation

$$\prod_{\mathfrak{p}} \left(a_{\mathfrak{p}}(f) \right)^{m_{\mathfrak{p}}} = \prod_i g(b_i)^{n_i}.$$

The number r, the exponents n_i and the b_i can be determined by the following algorithm. Let n run through the positive integers and let (n_i) run through all (unordered) partitions of n. For each partition, determine the solution set of

$$\prod_{\mathfrak{p}} \left(a_{\mathfrak{p}}(f) \right)^{m_{\mathfrak{p}}} = \prod_i g(x_i)^{n_i},$$

while g runs through many polynomials congruent to 1 modulo $\mathfrak{m}_f$. The algorithm terminates if a solution is found. The algorithm will terminate because of the above theorem. If it terminates and if sufficiently many g have been tested, the solution can assumed to be correct. In all explicitly known cases one has $r = 1$ and $n_1 = 1$; but this may be due to the fact that not so many examples are known.

10.7 An extended example

In this section we will carry out the explicit computation of the cohomology of certain crystals associated to Drinfeld cusp forms of low weight and level $\Gamma_1(t)$. We consider $R_u = k[\theta^{\pm 1}, u^{\pm 1}]$ as an algebra over $A[1/\theta] = k[\theta^{\pm 1}]$ as in Proposition 10.10 and let f be the morphism of the corresponding schemes. Let furthermore $\bar{f} \colon \mathbb{P}^1_{A[1/\theta]} \to \operatorname{Spec} A[1/\theta]$ be the relative compactification of $\operatorname{Spec} R_u$. We consider the τ-sheaf

$$\underline{\mathcal{F}} = \left(R_u[t]^2, \ \tau = \begin{pmatrix} 0 & (t/\theta - 1) u^{q-1} \\ 1 & 1 + u^{q-1} \end{pmatrix} (\sigma_{R_u} \times \operatorname{id}_t) \right)$$

and wish to compute $R^1 \bar{f}_*$ of

$$\operatorname{Sym}^n j_\# \underline{\mathcal{F}} := \operatorname{Sym}^n \left(\mathcal{O}^{\oplus 2}_{\mathbb{P}^1_{A[1/\theta]}} (-1 \cdot [\infty]), \tau \right)$$

as a crystal. Abbreviating $b := (t/\theta - 1)u^{q-1}$ and $c := 1 + u^{q-1}$, the endomorphism $\mathrm{Sym}^n \tau$ on $\mathrm{Sym}^n R_u[t]^2 = R_u[t]^{n+1}$ is given by

$$\alpha_n(\sigma \times \mathrm{id}), \quad \text{where} \quad \alpha_n := \begin{pmatrix} & & & & & & b^n \\ & & & & \ddots & & \vdots \\ & & & b^3 & \cdots & b^3 c^{n-3}\binom{n}{3} \\ & & b^2 & b^2 c\binom{3}{2} & \cdots & b^2 c^{n-2}\binom{n}{2} \\ & b & bc\binom{2}{1} & bc^2\binom{3}{1} & \cdots & bc^{n-1}\binom{n}{1} \\ 1 & c & c^2 & c^3 & \cdots & c^n \end{pmatrix}.$$

We denote the corresponding basis of $R[u]^{n+1}$ by $e_j, j = 0, \dots, n$.

Next recall that we can compute the cohomology of a coherent sheaf $\mathcal{G}$ on $\mathbb{P}^1_S$ (over any affine base S) as follows: Let $\mathbb{A}^1_S \subset \mathbb{P}^1_S$ be the standard affine line contained in $\mathbb{P}^1_S$. Let $\mathcal{O}_{\infty,S}$ be the affine coordinate ring of the completion of $\mathbb{P}^1_S$ along the section $\infty \times S$ at ∞ and let $K_{\infty,S}$ be the ring obtained from $\mathcal{O}_{\infty,S}$ by inverting the section at ∞. Then one has the short exact sequence

$$0 \longrightarrow H^0(\mathbb{P}^1_S, \mathcal{G}) \longrightarrow H^0(\mathbb{A}^1_S, \mathcal{G}) \oplus \mathcal{G}_{|\mathcal{O}_{\infty,S}} \longrightarrow \mathcal{G}_{|K_{\infty,S}} \longrightarrow H^1(\mathbb{P}^1_S, \mathcal{G}) \longrightarrow 0.$$

The sequence is obtained as the direct limit over U over the sequences for the computation of Čech cohomology where $\mathbb{P}^1$ is covered $\mathbb{A}^1$ and a second affine set U containing $\infty \times S$. We apply this to the base $\mathrm{Spec}\, \mathcal{A}$ with $\mathcal{A} = k[\theta^{\pm 1}, t]$ and the sheaf $\mathrm{Sym}^n \mathcal{F}$. Disregarding τ we obtain the short exact sequence

$$0 \longrightarrow \mathcal{A}[u]^{n+1} \oplus \left(\frac{1}{u}\right)^n k\left[\left[\frac{1}{u}\right]\right] \otimes_k \mathcal{A}^{n+1} \longrightarrow k\left(\left(\frac{1}{u}\right)\right) \otimes_k \mathcal{A}^{n+1} \longrightarrow \mathrm{Coker}_n \longrightarrow 0.$$

This shows that $\mathrm{Coker}_n = H^1(\mathbb{P}^1, \mathrm{Sym}^n j_\# \mathcal{F})$ is a free $\mathcal{A}$-module with basis $\{u^{-i} e_j \mid i = 1, \dots, n-1; j = 0, \dots, n\}$. Let us write the elements of the cokernel as

$$u^{-1} v_1 + \cdots + u^{1-n} v_{n-1},$$

where the v_j are column vectors over $\mathcal{A}$ of length $n + 1$. (We can write them in the basis $e_0, \dots, e_n$.) Applying τ to the summands $u^{-i} v_i$ yields

$$\tau(u^{-i} v_i) = u^{-iq} \alpha_n v_i = \left(u^{-i(q-1)} \alpha_n\right) u^{-i} v_i.$$

Now observe that α_n lies in $\mathcal{A}[u^{q-1}]$. Thus $u^{-i(q-1)}$ shifts the pole order at $u = 0$ (and $u = \infty$) by multiples of $(q-1)$. We define matrices $\alpha_{n,i} \in M_{(n+1)\times(n+1)}(\mathcal{A})$ so that

$$\alpha_n = \sum_{i \geq 0} \alpha_{n,i} u^{i(q-1)}.$$

Assumption 10.28. We now assume that the weight n lies in the interval $\{0, \ldots, q\}$.

Because the exponent of u^{-i} lies in $\{-1, \ldots, 1-n\}$, the absolute value of the difference of two such i is at most $q-2$. Therefore at most one summand in

$$\tau(u^{-i} v_i) = \left(\sum_{i' \geq 0} u^{(i'-i)(q-1)} \alpha_{n,i'} \right) u^{-i} v_i$$

is non-zero in Coker_n, namely that for $i' = i$. Let $\beta = \left(\frac{t}{\theta} - 1 \right)$ and abbreviate $x = u^{q-1}$. We enumerate the rows and columns by $r, s \in \{0, \ldots, n\}$. We let $\widetilde{r} := n - r$, so that this variable counts rows from the bottom starting at zero. Then the (r, s)-coefficient of α_n is

$$c^{r+s-n} b^{n-r} \binom{s}{n-r} = c^{s-\widetilde{r}} b^{\widetilde{r}} \binom{s}{\widetilde{r}} = (1+x)^{s-\widetilde{r}} x^{\widetilde{r}} \beta^{\widetilde{r}} \binom{s}{\widetilde{r}}$$

$$= \sum_{\ell=0}^{s-\widetilde{r}} x^{\ell+\widetilde{r}} \beta^{\widetilde{r}} \binom{s-\widetilde{r}}{\ell} \binom{s}{\widetilde{r}}.$$

The (r, s)-coefficient of $\alpha_{n,i}$ is the coefficient of x^i in the previous line. Thus it is the summand for $\ell = i - \widetilde{r}$, i.e.,

$$\beta^{\widetilde{r}} \binom{s-\widetilde{r}}{i-\widetilde{r}} \binom{s}{\widetilde{r}} = \beta^{\widetilde{r}} \frac{(s-\widetilde{r})!}{(i-\widetilde{r})!(s-i)!} \frac{s!}{(s-\widetilde{r})!\widetilde{r}!}$$

$$= \beta^{\widetilde{r}} \frac{i!}{(i-\widetilde{r})!\widetilde{r}!} \frac{s!}{(s-i)!i!} = \beta^{\widetilde{r}} \binom{i}{i-\widetilde{r}} \binom{s}{i}.$$

Let w_i be the transpose of the row vector $\left(0, \ldots, 0, \binom{i}{0} \beta^i, \binom{i}{1} \beta^{i-1}, \ldots, \binom{i}{i} \beta^0 \right)$ and let x_i be the row vector $\left(0, \ldots, 0, \binom{i}{i}, \binom{i+1}{i}, \ldots, \binom{n}{i} \right)$. Then $\alpha_{n,i} = w_i \otimes x_i$ and so

$$\tau(u^{-i} v_i) = u^{-i} w_i \cdot (x_i v_i).$$

We deduce the following. As a τ-sheaf, Coker_n is the direct sum of the sub-$\mathcal{A}$-modules W_i spanned by $u^{-i} e_j$, $j = 0, \ldots, n$. The τ-submodule W_i contains itself the τ submodule $\mathcal{A} u^{-i} w_i$, and because the image of W_i under τ is contained in $\mathcal{A} u^{-i} w_i$, it is nil-isomorphic to W_i. Thus we find that

$$\oplus_{j=1}^{n-1} \left(\mathcal{A} u^{-i} w_i, \tau_{|\mathcal{A} u^{-i} w_i} \right) \longrightarrow \left(\mathrm{Coker}_n, \tau \right)$$

is a nil-isomorphism. We compute the τ-action on $u^{-i} w_i$:

$$\tau(u^{-i} w_i) = u^{-i} w_i (x_i \cdot (\sigma \times \mathrm{id}) w_i) = (u^{-i} w_i) \sum_{\ell=0}^{\min\{i, n-i\}} \binom{i}{\ell} \binom{n-\ell}{i} \widetilde{\beta}^\ell$$

with $\widetilde{\beta} = \left(\frac{t}{\theta^q} - 1\right)$. Set

$$\gamma_{n,i} := \sum_{\ell=0}^{\min\{i,n-i\}} \binom{i}{\ell}\binom{n-\ell}{i}\beta^\ell \quad \text{and} \quad \mathcal{L}_{n,i} := \left(k[\theta^{\pm 1}, t], \gamma_{n,i}(\sigma \times \mathrm{id})\right).$$

Then $(\mathcal{A}u^{-i}w_i, \tau_{|\mathcal{A}w_i}) \cong \sigma^* \mathcal{L}_{n,i}$ and we have thus shown that as A-crystals we have:

Proposition 10.29. *Suppose* $0 \leq n \leq q$. *Then*

$$\underline{S}_{n+2}\big(\Gamma_1(t)\big) = R^1 \bar{f}_* \, \mathrm{Sym}^n j_\# \mathcal{F} \cong \bigoplus_{i=1}^{n-1} \mathcal{L}_{n,i}.$$

Moreover $\mathcal{L}_{n,i} = \mathcal{L}_{n,n-i}$ *and* $\underline{S}_{n+2}(\Gamma_1(t)) = 0$ *for* $n = 0, 1$.

In Remark 10.33 we shall compare the above formula for $\gamma_{n,i}$ to a similar expression in [35, Formula (7.3)].

Example 10.30. For $i = 1$ and $2 \leq n \leq q$ one has $\gamma_{n,1} = \binom{1}{0}\binom{n}{1}\beta^0 + \binom{1}{1}\binom{n-1}{1}\beta = 1 + (n-1)\frac{t}{\theta}$. The corresponding ramification divisor in the sense of Remark 10.19 is defined by $\theta = (1-n)t$ on $\mathrm{Spec}\, k[\theta^{\pm 1}, t]$. This leads to the Hecke character described in Example 10.26.

Example 10.31. Next we compute the local L-factors of $\mathcal{L}_{n,1}$. The base scheme is $X := \mathbb{A}_k \setminus \{0\}$ in the coordinate θ. Let $\mathfrak{p}$ be a place of X defined by the irreducible polynomial $h(\theta) \in k[\theta]$. We normalize it so that $h(0) = 1$. This is possible because $\mathfrak{p} \neq 0$. The residue field at $\mathfrak{p}$ is $k_{\mathfrak{p}} = k[\theta]/(h(\theta)) = k[\bar{\theta}]$ with $\bar{\theta}$ a root of h over $\bar{k}$. In $k_{\mathfrak{p}}[t]$ we have

$$h(t) = \left(1 - \frac{t}{\bar{\theta}}\right) \cdot \left(1 - \frac{t}{\bar{\theta}^q}\right) \cdot \ \cdots \ \cdot \left(1 - \frac{t}{\bar{\theta}^{q^{\deg h - 1}}}\right).$$

Thus

$$\det(1 - \tau_{\mathcal{L}_{n,1}} T^{\deg \mathfrak{p}}) = \det\left(1 - \left(1 + (n-1)\frac{t}{\bar{\theta}}\right) \cdot \left(1 + (n-1)\frac{t}{\bar{\theta}^q}\right)\right.$$
$$\left. \cdot \ \cdots \ \cdot \left(1 + (n-1)\frac{t}{\bar{\theta}^{q^{\deg h - 1}}}\right) T^{\deg h}\right)$$

$$= 1 - h\big((1-n)t\big)T^{\deg h}.$$

Thus, if we denote by $g_{n,1}$ the cuspidal Drinfeld–Hecke eigenform corresponding to $\mathcal{L}_{n,1}$, then its eigenvalue at $\mathfrak{p} = (h)$ is $h\big((1-n)t\big)$. A similar but more involved computation yields the eigenvalue system for the form $g_{n,i}$ corresponding to $\mathcal{L}_{n,i}$ and any $1 \leq i \leq n - 1$.

Example 10.32. For classical as well as Drinfeld modular forms, their automorphic weight, say n, is the exponent of the automorphy factor in the transformation formulas for the action of the congruence subgroup defined by the level of the form. In the classical case there is naturally a weight attached to the motive associated with a cuspidal Hecke eigenform. This *motivic weight* is the exponent of the complex absolute values of the roots of the characteristic polynomials defined by the Hecke action. By the proof of the Ramanujan–Petersson conjecture due to Deligne, this weight is $(n-1)/2$. This weight occurs in the formula for the absolute values of the pth Hecke eigenvalue of a classical modular form f:

$$\left| a_p(f) \right|_{\mathbb{C}} \le 2p^{(n-1)/2}.$$

It is therefore natural to also ask for a motivic weight of a cuspidal Drinfeld–Hecke eigenform. Does it exist and how is it related to the weight that occurs in the exponent of the automorphy factor in the transformation law of the form? In the Drinfeld modular case, the characteristic polynomial arising from Hecke operators at $\mathfrak{p}$ is one-dimensional. Therefore the motivic weight of a Drinfeld–Hecke eigenform f is (if it exists) the exponent $q \in \mathbb{Q}$ such that

$$v_\infty\left(\left| a_{\mathfrak{p}}(f) \right| \right) = -q \deg \mathfrak{p} \quad \text{for almost all } \mathfrak{p} \in \mathbf{Max}\,(A).$$

This weight is modeled after Anderson's definitions of purity and weights for t-motives [1, 1.9 and 1.10].

The τ-sheaves $\underline{\mathcal{L}}_{n,i}$ defined in Proposition 10.29 possess a motivic weight. It is equal to $\deg_t \gamma_{n,i}$, since by computations as in the previous example one shows that $v_\infty(|a_{\mathfrak{p}}(g_{n,i})|) = -\deg \gamma_{n,i} \cdot \deg \mathfrak{p}$. For $q = p$ the formulas in Remark 10.33 yield $\deg_{n,i} = \min\{n - i, i\}$. Thus, for a given n, any weight in $\{1, 2, \ldots, \lfloor \frac{n}{2} \rfloor\}$ occurs. For $q \ne p$, the possible weights are more difficult to analyze because the expression $\binom{i}{m}\binom{n-i}{m}$ can vanish for $m = \min\{i, n - i\}$ (and also for many values less than this minimum – see Lemma 8.30.

We expect but have no proof that the motives for all cuspidal Drinfeld–Hecke eigenforms are pure, i.e., that they have a motivic weight. If this is true, it can be shown from the cohomological formalism in [8] that this weight lies, for a given n, in the interval $\{0, \ldots, \lfloor \frac{n}{2} \rfloor\}$. That the range is optimal is shown by the above examples. Moreover the example shows that it is not possible to compute the motivic weight from the automorphic weight.

Remark 10.33. In [35, Formula (7.3)] a differently looking formula is given from which the Hecke eigenvalue systems for the forms $g_{n,i}$ are computed (for primes of degree one). From the following claim it follows that the formulas given there agree with those here (and thus with those given in [4]).

Claim:

$$\gamma_{n,i} = \sum_{m \ge 0} \binom{i}{m}\binom{n-i}{m}\left(\frac{t}{\theta} \right)^m,$$

where we recall that $\gamma_{n,i} := \sum_{\ell=0}^{\min\{i,n-i\}} \binom{i}{\ell}\binom{n-\ell}{i}\beta^\ell$ with $\beta = \frac{t}{\theta} - 1$. The proof follows by the use of generating series in a standard fashion. The key steps are

$$\sum_{n \ge 0} \gamma_{n,i} X^n = \sum_{\ell \ge 0} \binom{i}{\ell}(\beta X)^\ell \sum_{n \ge i+\ell} \binom{n-\ell}{i} X^{n-\ell}$$

$$= \sum_{\ell \geq 0} \binom{i}{\ell} (\beta X)^{\ell} (1 - X)^{-i} X^i = \left(\frac{1 + \beta X}{1 - X} \right)^i X^i$$

$$= \left(1 + \frac{\frac{t}{\theta} X}{1 - X} \right)^i X^i = \sum_{m \geq 0} \binom{i}{m} \left(\frac{t}{\theta} X \right)^m (1 - X)^{-m} X^i$$

$$= \sum_{m \geq 0} \binom{i}{m} \left(\frac{t}{\theta} \right)^m \sum_{n \geq i+m} \binom{n - i}{m} X^n$$

$$= \sum_{n \geq 0} X^n \sum_{m \geq 0} \binom{i}{m} \binom{n - i}{m} \left(\frac{t}{\theta} \right)^m.$$

Appendix
Further Results on Drinfeld Modules

In this appendix we collect as a reference some further results on Drinfeld modules which were used in parts of the lecture notes. Throughout the appendix we denote by ι the canonical embedding $A \hookrightarrow \mathbb{C}_\infty$.

A.1 Drinfeld A-modules over $\mathbb{C}_\infty$

Important examples of Drinfeld A-modules are obtainable over $\mathbb{C}_\infty$ via a uniformization theory modeled after that of elliptic curves.

For Drinfeld modules over $X = \operatorname{Spec} \mathbb{C}_\infty$ one has the following result of Drinfeld [13]:

Theorem A.1 (Drinfeld). *Let φ be a Drinfeld module over $\mathbb{C}_\infty$ with $d\varphi \colon A \hookrightarrow \mathbb{C}_\infty$ equal to ι. Then there exists a unique entire function*

$$
e_\varphi \colon \mathbb{C}_\infty \longrightarrow \mathbb{C}_\infty, \quad x \longmapsto x + \sum_{i \geq 1} a_i x^{q^i} \quad \left(a_i \in \mathbb{C}_\infty \right)
$$

such that for all $a \in A$ the following diagram is commutative:

$$
\begin{array}{ccc}
\mathbb{C}_\infty & \xrightarrow{\ e_\varphi\ } & \mathbb{C}_\infty \\
{\scriptstyle x \mapsto ax}\big\downarrow & & \big\downarrow{\scriptstyle x \mapsto \varphi_a(x)} \\
\mathbb{C}_\infty & \xrightarrow{\ e_\varphi\ } & \mathbb{C}_\infty
\end{array}
\tag{A.9}
$$

Moreover e_φ is a k-linear epimorphism and its kernel is a projective A-module of rank equal to the rank of the Drinfeld A-module which is discrete in $\mathbb{C}_\infty$.

Conversely, to any discrete projective A-submodule $\Lambda \subset \mathbb{C}_\infty$ of rank r one can associate a unique exponential function

$$
e_\Lambda(x) = x + \sum_{i \geq 1} a_i x^{q^i} \left(= x \prod_{\lambda \in \Lambda \setminus \{0\}} \left(1 - \frac{x}{\lambda} \right) \right),
$$

whose set of roots is the divisor Λ, *and a unique Drinfeld A-module* φ_Λ *of rank* r *over* $\mathbb{C}_\infty$ *such that* (A.9) *commutes for* $\varphi = \varphi_\Lambda$ *and* $e_\varphi = e_\Lambda$. *The characteristic of* φ_Λ *is the canonical inclusion* $A \hookrightarrow \mathbb{C}_\infty$.

Moreover, two Drinfeld A-modules φ_Λ *and* $\varphi_{\Lambda'}$ *are isomorphic (over* $\mathbb{C}_\infty$) *if and only if there is a scalar* $\lambda \in \mathbb{C}_\infty^*$ *such that* $\lambda\Lambda = \Lambda'$. *They are isogenous if there exists* $\lambda \in \mathbb{C}_\infty^*$ *and* $a \in A \setminus \{0\}$ *such that* $a\Lambda' \subset \lambda\Lambda \subset \Lambda'$.

In particular there exist Drinfeld A-modules of all ranks $r \in \mathbb{N} = \{1, 2, 3, \ldots\}$. For any rank $r \geq 2$ there exist infinitely many non-isomorphic Drinfeld A-modules of rank r over $\mathbb{C}_\infty$. In the rank 1 case, the number of isomorphism classes of Drinfeld A-modules (over $\mathbb{C}_\infty$) is equal to the number of projective A-modules of rank 1, i.e., to the cardinality of the class group $\mathrm{Cl}(A) = \mathrm{Pic}(A)$ of A. In fact within each isomorphism class there is one representative which is defined over the class field H of K with respect to ∞. There will be more on this in Appendix A.3.

Indication of proof of Theorem A.1. Let φ be given and fix $a \in A \setminus k$ and write $\varphi_a = a + \sum_{j=1}^{t} u_j \tau^j$ ($t = r \deg a$). Then (A.9) yields the recursion

$$a_i\left(a^{q^i} - a\right) = \sum_{j=1}^{t} u_j a_{i-j}^{q^j}$$

for the coefficients of e_φ, where we set $a_i = 0$ for $i < 0$. Let v denote the valuation on $\mathbb{C}_\infty$ such that $v(\pi) = 1$ for π a uniformizer π of F at ∞. Let $C := \min_{j=1,\ldots,t} v(u_j)$. Then from $v(a) < 0$ one deduces that

$$\frac{v(a_i)}{q^i} \geq \left(\frac{C}{q^i} - v(a)\right) + \min_{j=1,\ldots,t} \frac{v(a_{i-j})}{q^{i-j}}.$$

Choose $0 < \theta < -v(a)$ so that for $i \gg 0$ one has $\frac{C}{q^i} - v(a) \geq \theta$. Setting $B_i := \min_{j=1,\ldots,t} \frac{v(a_{i-j})}{q^{i-j}}$ it follows that there is some $i_0 > 0$ such that for all $i \geq i_0$ one has

$$B_{i+1} \geq B_i \quad \text{and} \quad B_{i+t} \geq B_i + \theta.$$

Thus (B_i) converges to ∞ and hence $\lim_{i\to\infty} \frac{v(a_i)}{q^i} = \infty$. This shows that e_φ has infinite radius of convergence, i.e., that it is entire. (The uniqueness of e_φ for a follows from the initial condition $de_\varphi = 1$.)

To show that e_φ is independent of a, write $e_\varphi = \sum_{i\geq 0} a_i \tau^i$ as a formal power series in τ over $\mathbb{C}_\infty$. Then for any $b \in A$ the expression $e_\varphi^{-1} \varphi_b e_\varphi$ is again such a power series. Since the image of A under φ is commutative, it follows that $e_\varphi^{-1} \varphi_b e_\varphi$ commutes with $a = e_\varphi^{-1} \varphi_a e_\varphi$ as an element in the formal non-commutative power series ring $\mathbb{C}_\infty[[\tau]]$ over τ. One deduces that $e_\varphi^{-1} \varphi_b e_\varphi$ is a constant and by taking derivatives that $b = e_\varphi^{-1} \varphi_b e_\varphi$. Hence (A.9) commutes for any $b \in A$.

That $\mathrm{Ker}(e_\varphi)$ is an A-module is immediate from the commutativity of (A.9). The discreteness follows as in complex analysis from the entireness and non-constancy. That the roots have multiplicity one is deduced from the root at 0 having multiplicity one. The rank of $\mathrm{Ker}(e_\varphi)$ as an A-module is obtained by considering torsion points (introduced in the following section) and exploiting their relation to the rank. All further assertions are rather straightforward. $\qquad\qquad\Box$

A.2   Torsion points and isogenies of Drinfeld modules

Theorem A.1 indicates that Drinfeld modules are characteristic p analogs of elliptic curves. This suggests that torsion points of Drinfeld modules carry an interesting Galois action. Formally one defines modules of torsion points (or torsion schemes) as follows.

Fix a non-zero ideal $\mathfrak{a} \subset A$ and a Drinfeld A-module φ on X. For any $a \in A \smallsetminus \{0\}$ the kernel of $\varphi_a \colon L \to L$ is a finite flat group scheme over X of rank equal to $q^{r \deg a}$. (Passing to local coordinates, it suffices to verify this for Drinfeld modules of standard type – where it is rather easy.) From this one deduces that the $\mathfrak{a}$-torsion scheme of φ, defined as

$$\varphi[\mathfrak{a}] := \bigcap_{a \in \mathfrak{a}} \mathrm{Ker}\left(\varphi_a\right),$$

is a finite flat A-module scheme over X of rank $q^{r \deg(\mathfrak{a})}$. If $\mathfrak{a}$ is prime to the characteristic of φ, then $\varphi[\mathfrak{a}]$ is finite étale over X. In the case $X = \mathrm{Spec}\, F$ for a field F, the finite group scheme $\varphi[\mathfrak{a}]$ becomes trivial over a finite Galois extension L of F. As an A-module, the group $\varphi[\mathfrak{a}](L)$ is isomorphic to $(A/\mathfrak{a})^r$ provided that $\mathfrak{a}$ is prime to the characteristic of φ. The Galois action and the A-action commute on $\varphi[\mathfrak{a}]$. This yields the following first result:

Theorem A.2. *Let φ be a Drinfeld A-module of rank r on $\mathrm{Spec}\, F$ for a field F with algebraic closure $\overline{F}$. Suppose $\mathfrak{a}$ is prime to the characteristic of φ. Then the action of $\mathrm{Gal}(\overline{F}/F)$ on $\varphi[\mathfrak{a}](\overline{F})$ induces a representation*

$$\mathrm{Gal}\left(\overline{F}/F\right) \longrightarrow \mathrm{GL}_r\left(A/\mathfrak{a}\right).$$

If $\mathfrak{a}$ is a non-zero principal ideal generated by $a \in A$, then $\varphi[\mathfrak{a}] = \mathrm{Ker}(\varphi_a)$. If one regards φ_a as an isogeny $\varphi \to \varphi$, then $\varphi[(a)]$ is the kernel of this isogeny. More generally, for any non-zero ideal $\mathfrak{a}$ of A, principal or not, there exists an isogeny $\psi \colon \varphi \to \varphi'$ for a suitable Drinfeld module φ' such that $\varphi[\mathfrak{a}] = \mathrm{Ker}(\psi)$, as subgroup schemes of φ:

Suppose first that $X = \mathrm{Spec}\, F$ for a field F containing the finite field k. We follow [24, Sec. 4.7]. Given $\mathfrak{a}$ one considers the left ideal $\mathfrak{a}_\varphi := \sum_{a \in \mathfrak{a}} F\{\tau\}\varphi_a$ of $F\{\tau\}$. Since all left ideals are principal, the ideal has a monic generator $\varphi^{\mathfrak{a}}$ that satisfies $F\{\tau\} \cdot \varphi^{\mathfrak{a}} = \mathfrak{a}_\varphi$. Since $\mathfrak{a}_\varphi \cdot \varphi_b \subset \mathfrak{a}_\varphi$ for all $b \in A$, for any $b \in A$ there is a unique $\varphi'_b \in F\{\tau\}$ such that $\varphi'_b \varphi_{\mathfrak{a}} = \varphi_{\mathfrak{a}} \varphi_b$.

Exercise A.3.

(a) $b \mapsto \varphi'_b$ defines a Drinfeld A-module $\varphi' : A \to F\{\tau\}$ such that $\varphi_\mathfrak{a}$ is an isogeny $\varphi \to \varphi'$. Moreover $\mathrm{Ker}(\varphi_\mathfrak{a}) = \varphi[\mathfrak{a}]$ as subgroup schemes of $\mathbb{G}_{a/F}$.

(b) Denote φ' from (a) by $\mathfrak{a} * \varphi$. Then $(\mathfrak{a}, \varphi) \mapsto \mathfrak{a} * \varphi$ defines an action of the monoid of non-zero ideals of A under multiplication on the set of Drinfeld A-modules, i.e. $(\mathfrak{a}\mathfrak{b}) * \varphi = \mathfrak{a} * (\mathfrak{b} * \varphi)$. The action preserves the rank and the height of φ. It is trivial on principal ideals, and thus it induces an action of the class group $\mathrm{Cl}(A)$ of A on the set of Drinfeld A-modules.

Suppose next that R is an integrally closed domain with fraction field F and φ is a Drinfeld A-module over $\mathrm{Spec}\, R$ in standard form. Using the normality of R, one can verify that $\varphi_\mathfrak{a}$ and $\mathfrak{a} * \varphi$ constructed above have coefficients in R. From there it is easy to verify all assertions of the exercise over R as well. The case that R is a normal domain applies in particular to the case of Drinfeld moduli schemes with a level structure. For the general we refer to [34, Sec. 3.5].

In [13, Prop. 2.3], Drinfeld gives a more general criterion for the existence of an isogeny whose kernel is a finite A-subgroup scheme of $\mathbb{G}_{a/F}$. The criterion can easily be extended to the case where $X = \mathrm{Spec}\, R$ with R a normal domain.

For later reference, we provide some explicit formulas: Suppose $\mathfrak{a} = (a)$ is a principal non-zero ideal of A. Then

$$\varphi_{(a)} = \mu_\varphi(a)^{-1} \varphi_a \quad \text{for } \mu_\varphi(a) \text{ the leading coefficient of } \varphi_a. \tag{A.10}$$

From this, a short computation yields the following formula for the leading term of $(a) * \varphi$:

$$\mu_{(a)*\varphi}(b) = \mu_\varphi(b) \cdot \mu_\varphi(a)^{1-q^{\deg b}} = \mu_\varphi(b)^{q^{\deg a}} \quad \forall b \in A. \tag{A.11}$$

In the particular case where the base is $\mathrm{Spec}\,\mathbb{C}_\infty$, the characteristic is the canonical embedding $\iota : A \hookrightarrow \mathbb{C}_\infty$, and $r = 1$, one can say more. By Theorem A.1 any such module is given by a rank 1 A-lattice in $\mathbb{C}_\infty$. Homotheties induce isomorphisms of Drinfeld modules and any two lattices are isogenous up to rescaling. Making all identifications explicit, one obtains the following result:

Proposition A.4 (Drinfeld, Hayes). *The set of isomorphism classes of Drinfeld A-modules of rank 1 over $\mathbb{C}_\infty$ with $d\varphi = \iota$ is a principal homogeneous space under the $*$-operation of $\mathrm{Cl}(A)$.*

A.3 Drinfeld–Hayes modules

In this section we collect some results on Drinfeld–Hayes modules. The reader is advised to recall the definitions of a sign-function and the corresponding strict class group from Definitions 8.1 and 8.3. As in (A.10), the leading term of a Drinfeld module φ is denoted μ_φ. We fix a sign-function sign throughout this section.

Definition A.5. A rank 1 sign-normalized Drinfeld module or simply a *Drinfeld–Hayes module (for* sign) is a rank 1 Drinfeld module φ over $\mathbb{C}_\infty$ with $d\varphi = \iota$ whose leading term μ_φ agrees with the restriction of a twisted sign function (sign) to $A \subset K_\infty$.

A good reference for the following result is [24, Theorem 7.2.15].

Theorem A.6 (Hayes). *Every rank 1 Drinfeld module over $\mathbb{C}_\infty$ with characteristic ι is isomorphic to a Drinfeld–Hayes module.*

Indication of proof. Define the $\mathbb{Z}$-graded ring $\mathrm{gr}(K_\infty) = \oplus_{i \in \mathbb{Z}} M^i/M^{i-1}$ using the filtration $M^i := \{x \in K_\infty : v_\infty(x) \geq -i\}$. Let L be a subfield of $\mathbb{C}_\infty$ and let $\varphi \colon A \to L\{\tau\}$ be a rank 1 Drinfeld module with $d\varphi = \iota$.

Sublemma A.7. *There exists a unique map $\lambda_\varphi \colon \mathrm{gr}(K_\infty) \to L^*$ with the following properties:*

(a) *For all $a \in A$ with image $\bar{a}$ in $M^{v_\infty(a)}/M^{v_\infty(a)-1} \subset \mathrm{gr}(K_\infty)$ one has $\lambda_\varphi(\bar{a}) = \mu_\varphi(a)$.*

(b) *For all $\alpha, \beta \in \mathrm{gr}(K_\infty)$ one has $\lambda_\varphi(\alpha\beta) = \lambda_\varphi(\alpha)^{q^{\deg \beta}} \lambda_\varphi(\beta)$.*

One first observes that $\mu_\varphi(ab) = \mu_\varphi(a)^{q^{\deg b}} \mu_\varphi(b)$ for any $a, b \in A$. Then one uses property (b) to extend the definition of λ_φ on the images $\bar{a}$ for $a \in A$ to all of $\mathrm{gr}(K_\infty)$. One also observes that λ_φ is the identity on k and a Galois automorphism when restricted to k_∞. The details are left to the reader.

The next result, whose proof we leave again to the reader, describes the change of λ_φ under isomorphism.

Sublemma A.8. *Suppose $\varphi' = \alpha\varphi\alpha^{-1}$ for some $\alpha \in L^*$. Then*

$$\lambda_{\varphi'}(x) = \lambda_\varphi(x)\alpha^{(1-q^{-\deg(x)})}.$$

Now, given φ, choose $\alpha \in \mathbb{C}^*$ such that $\alpha^{q-1} = \lambda_\varphi(\pi)$, so that $\varphi' = \alpha\varphi\alpha^{-1}$ satisfies $\lambda_{\varphi'}(\pi) = 1$. Because $\lambda_{\varphi'}$ is given by some $\sigma \in \mathrm{Gal}(k_\infty/k)$ when restricted to k_∞, one deduces that φ' is sign-normalized. □

Denote by $\mathcal{M}^1_{\mathrm{sign}}(\mathbb{C}_\infty)$ the set of sign-normalized rank 1 modules over $\mathbb{C}_\infty$. Since the number of isomorphism classes of Drinfeld A-modules of rank 1 over $\mathbb{C}_\infty$ with characteristic ι is finite and equal to the class number of A, and since the number of choices for α in the previous proof is finite, the set $\mathcal{M}^1_{\mathrm{sign}}(\mathbb{C}_\infty)$ is finite. Recall the action of fractional ideals $\mathfrak{a}$ on Drinfeld modules from the paragraphs above Proposition A.4. The following is from [24, §7.2]:

Theorem A.9. *The action of $\mathcal{I}_A$ on rank 1 Drinfeld A-modules preserves the sign-normalization and thus defines a well-defined action*

$$\mathcal{I}_A \times \mathcal{M}^1_{\mathrm{sign}}(\mathbb{C}_\infty) \longrightarrow \mathcal{M}^1_{\mathrm{sign}}(\mathbb{C}_\infty), \quad (\mathfrak{a}, \varphi) \longmapsto \mathfrak{a} * \varphi.$$

The action of $\mathcal{P}^+$ is trivial. Under the induced action

$$\mathrm{Cl}^+(A) \times \mathcal{M}^1_{\mathrm{sign}}(\mathbb{C}_\infty) \longrightarrow \mathcal{M}^1_{\mathrm{sign}}(\mathbb{C}_\infty), \quad ([\mathfrak{a}], \varphi) \longmapsto \mathfrak{a} * \varphi,$$

the set $\mathcal{M}^1_{\mathrm{sign}}(\mathbb{C}_\infty)$ becomes a principal homogeneous space under $\mathrm{Cl}^+(A)$, i.e., the action is simply transitive and the stabilizer of any element is trivial.

Indication of proof. It is easy to see that the $*$-action preserves $\mathcal{M}^1_{\mathrm{sign}}(\mathbb{C}_\infty)$: For an ideal I, the $*$-action on the leading coefficient is of the form $y \mapsto y^{q^d}$ for q^d the rank of the set $\varphi[I](\mathbb{C}_\infty)$.

For a principal ideal (a) the $*$-action on the leading term was determined in (A.11). Thus if φ is sign-normalized and if a is positive the effect on leading terms is trivial because of $\mu_\varphi(a) = \mathrm{sign}(a) = 1$. In particular $\mathcal{P}^+$ acts trivially.

Next we show that $\mathrm{Cl}(A)^+$ acts faithfully. Suppose $\mathfrak{a} * \varphi = \varphi$. Since this implies in particular that $\mathfrak{a} * \varphi$ is isomorphic to φ, we deduce that $\mathfrak{a}$ is principal, say equal to (a). In this case we can again use formula (A.11). It implies that for all $b \in A$ we must have

$$\mu_\varphi(a)^{1-q^{\deg b}} = 1.$$

Since the gcd of the $\deg b$ is 1, it follows that $\mu_\varphi(a)^{q-1} = 1$, i.e., $\alpha := \mu_\varphi(a) \in k^*$. But then $a\alpha^{-1}$ is a positive generator of $\mathfrak{a}$ and the faithfulness of the action is shown.

Finally by determining the cardinalities of $\mathrm{Cl}(A)^+$ and of $\mathcal{M}^1_{\mathrm{sign}}(\mathbb{C}_\infty)$ the proof will be complete: By definition $\#\mathrm{Cl}(A)^+ = \#\mathrm{Cl}(A) \cdot \#(\mathcal{P}/\mathcal{P}^+)$. But all elements of k^*_∞ occur as signs of some $\alpha \in K$ and principal ideals αA are positively generated if and only if $\alpha \in k^*$. Hence $\#\mathrm{Cl}(A)^+ = \#\mathrm{Cl}(A) \cdot \#k^*_\infty/\#k^*$. Next we observe that all isomorphism classes of rank 1 Drinfeld A-modules over $\mathbb{C}_\infty$ of characteristic ι are represented in $\mathcal{M}^1_{\mathrm{sign}}(\mathbb{C}_\infty)$. We count the number of times a class occurs in $\mathcal{M}^1_{\mathrm{sign}}(\mathbb{C}_\infty)$. If φ is a Drinfeld–Hayes module for sign, and if the same holds for $\alpha\varphi\alpha^{-1}$, then one shows that $\alpha \in k^*_\infty$. Moreover the two are equal if and only if $\alpha \in k^*$. Hence the cardinality of $\mathcal{M}^1_{\mathrm{sign}}(\mathbb{C}_\infty)$ is equal to the number of isomorphism classes of rank 1 Drinfeld A-modules over $\mathbb{C}_\infty$ of characteristic ι times $\#k^*_\infty/\#k^*$, and thus equal to $\#\mathrm{Cl}(A)^+ = \#\mathrm{Cl}(A)\cdot\#k^*_\infty/\#k^*$ by Theorem A.4. $\qquad\square$

One now argues as in the case of CM elliptic curves to deduce that every $\varphi \in \mathcal{M}^1_{\mathrm{sign}}(\mathbb{C}_\infty)$ is defined over H^+. Let $\widetilde{H} \subset \mathbb{C}_\infty$ be the field of definition of ψ. Since $\mathrm{Aut}(\mathbb{C}_\infty/K)$ preserves $\mathcal{M}^1_{\mathrm{sign}}(\mathbb{C}_\infty)$, the extension $\widetilde{H}/K$ is finite. Considering the infinite place it follows that $\widetilde{H}/K$ is separable. Using that the automorphisms commute with the $*$-operation, one shows that the extension is abelian. The $*$-action also shows that $\widetilde{H}$ is independent of φ.

Next one shows that φ has its coefficients in $\widetilde{\mathcal{O}}$, the normal closure of A in $\widetilde{H}$. The Drinfeld module has potentially good reduction everywhere. But the leading coefficient is a unit, and thus the Drinfeld module can be reduced without twist.

This allows one to use reduction modulo any prime of $\widetilde{\mathcal{O}}$ as a tool. It is not hard to see that to test equality of sign-normalized rank 1 Drinfeld A-modules it suffices to test it modulo any prime of $\widetilde{\mathcal{O}}$. This implies that the inertia group at any finite place is trivial. In particular the Artin symbol $\sigma_{\mathfrak{p}}$ is defined at any prime of $\widetilde{\mathcal{O}}$. One deduces the following Shimura type reciprocity law:

Theorem A.10. *If σ_I denotes the Artin symbol of a fractional ideal I of A, then $\sigma_I \varphi = I * \varphi$. Thus $\widetilde{H} = H^+$. Moreover every Drinfeld–Hayes module is defined over the ring of integers $\mathcal{O}^+$ of H relative to $A \subset K$.*

The reciprocity identifies the Galois action with the $*$-action on $\mathcal{M}_{\mathrm{sign}}^1(\mathbb{C}_\infty)$ (it is rather trivial to see that any type Galois action preserves sign normalization). One easily deduces:

Corollary A.11. $\mathrm{Gal}(H^+/H) \cong \mathbb{F}_\infty^*/\mathbb{F}^*$ *is totally and tamely ramified at ∞. It is unramified outside ∞.*

Remark A.12. One can show that any rank 1 Drinfeld module with characteristic ι can be defined over the ring of integers of the Hilbert class field H. However this representative within the isomorphism class is less canonical and its leading coefficient has no simple description.

A.4   Torsion points of Drinfeld–Hayes modules

Fix a sign-normalized rank 1 Drinfeld module φ over $\mathbb{C}_\infty$. For $I \in \mathcal{I}_A$, let φ_I denote the isogeny $\varphi \to I * \varphi$. Recall that $\varphi[I]$ denotes the A-module of I-torsion points of φ. Denote by $\mathcal{M}_{I,\mathrm{sign}}^1(\mathbb{C}_\infty)$ the set of isomorphism classes of pairs (φ, λ) where $\varphi \in \mathcal{M}_{\mathrm{sign}}^1(\mathbb{C}_\infty)$ and λ is a primitive I-torsion point of φ. Let $\mathcal{I}_A(I)$ denote the set of fractional ideals prime to I.

Theorem A.13. *Let $I \subset A$ be an ideal. The action of $\mathcal{I}_A$ on $\mathcal{M}_{I,\mathrm{sign}}^1(\mathbb{C}_\infty)$ given by*

$$J * (\varphi, \lambda) = (J * \varphi, \varphi_J(\lambda))$$

is well defined and transitive. The stabilizer of any pair (φ, λ) is the subgroup $\mathcal{P}_I^+ \subset \mathcal{P}^+$ of positively generated fractional ideals prime to I. The set $\mathcal{M}_{I,\mathrm{sign}}^1(\mathbb{C}_\infty)$ is a principal homogeneous space under the induced action of $\mathrm{Cl}(A, I) := \mathcal{I}_A(I)/\mathcal{P}_I^+$. One has an exact sequence

$$0 \longrightarrow (A/I)^* \longrightarrow \mathrm{Cl}(A, I) \longrightarrow \mathrm{Cl}^+(A) \longrightarrow 0.$$

The field $H^+(\varphi[I])$ is the ray class field of K of conductor I at the finite places and which is totally split at ∞ above H^+. One has $\mathrm{Gal}(H^+(\varphi[I])/K) \cong \mathrm{Cl}(A, I)$. Let σ_J denote the Galois automorphism which under the Artin reciprocity map corresponds to $J \in \mathcal{I}_A(I)$. The Shimura type reciprocity law reads: For every $J \in \mathcal{I}_A(I)$,

$$\sigma_J(\varphi, \lambda) = J * (\varphi, \lambda).$$

One has the following ramification properties:

Theorem A.14. *Let $P \subset A$ be a prime ideal and $\lambda \in \varphi[P]$ be a primitive element. The extension $H^+(\varphi[P^i])/H^+$ is totally ramified at P and unramified at all other finite places. It is Galois with Galois group $\mathrm{Gal}(H^+(\varphi[P^i])/H^+) \cong (A/P^i)^*$. The action of this group on $\varphi[P]$ is given by the character*

$$\sigma \longmapsto \left(\frac{\sigma(\lambda)}{\lambda} \right).$$

The extension H^+/H is totally ramified at ∞ and unramified at all other places. It is Galois with $\mathrm{Gal}(H^+/H) \cong \mathbb{F}_\infty^*/\mathbb{F}_q^*$. The decomposition D_∞ group at ∞ in $\mathrm{Gal}(H^+(\varphi[P^i])/H)$ is isomorphic to $\mathbb{F}_\infty^*$. The action of α in $\mathbb{F}_q^* \cong D_\infty \cap \mathrm{Gal}(H^+(\varphi[P])/H^+)$ on $\varphi[P]$ is given by

$$\sigma_\alpha(\lambda) = \mu_\psi(\alpha)^{-1}(\lambda) = \alpha^{-1}(\lambda).$$

Bibliography

[1] Anderson, G., *t-motives*, Duke Math. J. **53** (1986), 457–502.

[2] Anderson, G., *An elementary approach to L-functions mod p*, J. Number Theory **80(2)** (2000), 291–303.

[3] Böckle, G., *Global L-functions over function fields*, Math. Ann. **323(4)** (2002), 737–795.

[4] Böckle, G., *An Eichler–Shimura isomorphism over function fields between Drinfeld modular forms and cohomology classes of crystals*, preprint 2004, http://www.uni-due.de/arith-geom/boeckle/preprints.html

[5] Böckle, G., *Arithmetic over function fields: a cohomological approach*, in: Number Fields and Function Fields – Two Parallel Worlds, 1–38, Progr. Math., 239, Birkhäuser, Boston, MA, 2005.

[6] Böckle, G., *Hecke characters associated to Drinfeld modular forms*, preprint.

[7] Böckle, G., *Algebraic Hecke characters and compatible systems of abelian mod p Galois representations over global fields*, Manuscripta Math. **140** (2013), no. 3-4, 303–331.

[8] Böckle, G., Pink, R., *Cohomological theory of crystals over function fields*, Tracts in Math. **5**, European Mathematical Society, 2009.

[9] Deligne, P., *Cohomologie à support propre et construction du foncteur $f^!$*, Appendix of: Hartshorne, R., Residues and Duality, Lecture Notes in Math. **20**, Springer, Berlin, 1966, 404–421.

[10] Deligne, P., *La conjecture de Weil, I*, Inst. Hautes Études Sci. Publ. Math. **43** (1974), 273–307.

[11] Deligne, P. et al., *Cohomologie Étale*, Séminaire de Géométrie Algébrique du Bois Marie, SGA4$\frac{1}{2}$, Lecture Notes in Math. **569**, Springer, Berlin, 1977.

[12] Diaz-Vargas, J., *Riemann hypothesis for $\mathbb{F}_q[t]$*, J. Number Theory **59 (2)** (1996), 313–318.

[13] Drinfeld, V.G., *Elliptic modules*, Math. USSR-Sb. **23(4)** (1974), 561–592.

[14] Emerton, M., Kisin, M., *A Riemann–Hilbert correspondence for unit F-crystals*, Astérisque **293** (2004).

[15] Freitag, E., Kiehl, R., *Étale cohomology and the Weil conjecture*, Ergeb. Math. Grenzgeb. **13**, Springer, Berlin, 1988.

[16] Gardeyn, F., *The structure of analytic τ-sheaves*, J. Number Theory **100** (2003), 332–362.

[17] Gekeler, E.-U., *Drinfeld modular curves*, Lecture Notes in Math. **1231**, Springer, Berlin, 1986.

[18] Gekeler, E.-U., *On the coefficients of Drinfeld modular formes*, Invent. Math. **93** (1988), 667–700.

[19] Gekeler, E.-U., Reversat, M., *Jacobians of Drinfeld modular forms*, J. reine angew. Math. **476** (1996), 27–93.

[20] Goss, D., *π̄-adic Eisenstein series for function fields*, Compos. Math. **41** (1980), 3–38.

[21] Goss, D., *Modular forms for $\mathbb{F}_r$*, J. reine angew. Math. **317** (1980), 16–39.

[22] Goss, D., *L-series of Grössencharakters of type A_0 for function fields*, in: *p*-adic Methods in Number Theory and Algebraic Geometry, Contemp. Math. **133**, Amer. Math. Soc., 1992, 119–139.

[23] Goss, D., *L-series of t-motives and Drinfeld Modules*, in: The Arithmetic of Function Fields, Goss, D., et al. (eds.), Proceedings of the Workshop at Ohio State University, 1991, Walter de Gruyter, Berlin, 1992, 313–402.

[24] Goss, D., *Basic Structures of Function Field Arithmetic*, Ergeb. Math. Grenzgeb. **35**, Springer, Berlin, 1996.

[25] Goss, D., Sinnott, W., *Class-groups of function fields*, Duke Math. J. **528** (1985), 507–516.

[26] Gross, B., *Algebraic Hecke characters for function fields*, Seminar on Number Theory, Paris 1980–1981, Progr. Math. **22**, Birkhäuser, Basel, 1982, 87–90.

[27] Hartshorne, R., *Residues and Duality*, Lecture Notes in Math. **20**, Springer, Berlin, 1966.

[28] Hartshorne, R., *Algebraic Geometry*, Graduate Texts in Math. **52**, Springer, New York, 1977.

[29] Hayes, D., *A brief introduction to Drinfel'd modules*, in: The Arithmetic of Function Fields, Goss, D. et al. (eds.), Ohio State University Math. Res. Inst. Publications, de Gruyter, Berlin, 1992.

[30] Katz, N., *Nilpotent connections and the monodromy theorem: applications of a theorem of Turittin*, Publ. Math. IHÉS **39** (1970), 175–232.

[31] Katz, N., *p-adic properties of modular schemes and modular varieties*, in: Modular Functions of One Variable III, Lecture Notes in Math. **350**, Springer, Berlin, 1973, 69–191.

[32] Kiehl, R., Weissauer, R., *Weil conjectures, perverse sheaves and ℓ-adic Fourier transform*, Ergeb. Math. Grenzgeb. **42**, Springer, Berlin, 2001.

[33] Lafforgue, V., *Valeurs spéciales des fonctions L en caractéristique p*, J. Number Theory **129(10)** (2009), 2600–2634.

[34] T. Lehmkuhl, Compactification of Drinfeld Modular Surfaces, Memoirs AMS **197** (2009), no. 921.

[35] Li, W., Meemark, Y., Hecke operators on Drinfeld cusp forms, J. Number Theory **128** (2008), 1941–1965.

[36] Lütkebohmert, W., On compactification of schemes, *Manuscripta Math.* **80(1)** (1993), 95–111.

[37] Manin, Y.I., Panchishkin A.A., *Introduction to Modern Number Theory*, Springer, Berlin, 2005.

[38] Milne, J., *Étale Cohomology*, Princeton Math. Series **33**, Princeton University Press, Princeton, 1980.

[39] Nagata, M., *Imbedding of an abstract variety in a complete variety*, J. Math. Kyoto Univ. **2** (1962), 1–10.

[40] Nagata, M., *A generalization of the imbedding problem of an abstract variety in a complete variety*, J. Math. Kyoto Univ. **3** (1963), 89–102.

[41] Pink, R., *Euler–Poincaré Formula in equal characteristic under ordinariness assumptions*, Manuscripta Math. **102(1)** (2000), 1–24.

[42] Pink, R., *Compactification of Drinfeld modular varieties and Drinfeld modular forms of arbitrary rank*, Manuscripta Math. **140** (2013), no. 3-4, 333–361.

[43] Pink, R., Schieder S., *Compactification of a Drinfeld period domain over a finite field*, J. Algebraic Geom. **23** (2014), no. 2, 201–243.

[44] Sheats, J., *The Riemann hypothesis for the Goss zeta function for $\mathbb{F}_q[T]$*, J. Number Theory **71(1)** (1998), 121–157.

[45] Taguchi, Y., Wan, D., *L-functions of φ-sheaves and Drinfeld modules*, J. Amer. Math. Soc. **9(3)** (1996), 755–781.

[46] Taguchi, Y., Wan, D., *Entireness of L-functions of φ-sheaves on affine complete intersections*, J. Number Theory **63(1)** (1997), 170–179.

[47] Teitelbaum, J., *The Poisson kernel for Drinfeld modular curves*, J. Amer. Math. Soc. **4(3)** (1991), 491–511.

[48] Thakur, D., *On characteristic p zeta functions*, Compos. Math. **99(3)** (1995), 231–247.

[49] Thakur, D., *Function Field Arithmetic*, World Scientific, 2004.

[50] Thakur, D., *Power sums with applications to multizeta and zeta zero distribution for $\mathbb{F}_q[t]$*, Finite Fields Appl. **15** (2009), 534–552.

[51] Wan, D., *On the Riemann hypothesis for the characteristic p zeta function*, J. Number Theory **58 (1)** (1996), 196–212.

[52] Weibel, C., *An Introduction to Homological Algebra*, Cambridge University Press, Cambridge, 1994.

On Geometric Iwasawa Theory and Special Values of Zeta Functions

David Burns and Fabien Trihan

On Geometric Iwasawa Theory and Special Values of Zeta Functions

David Burns and Fabien Trihan

with an appendix by Francesc Bars

Introduction

Having succumbed to the requests of the organisers of the Research Programme on Function Field Arithmetic that was held in 2010 at the CRM in Barcelona, we present here a survey of some recent results concerning certain aspects of the Iwasawa theory of varieties over finite fields.

Such a survey is almost inevitably incomplete and in addition, due to a lack of appropriate expertise, we have felt it necessary to omit any discussion of results related to characteristic-p-valued L-functions. The very significant recent work of Taelman in, for example, [72] and [73] is therefore a particularly important omission. But, more generally, we would like to apologize to all authors of the many recent advances in this area that we do not consider in any detail here.

We shall focus on discussing results related to the Iwasawa theory of higher direct images of flat, smooth $\mathbb{Z}_p$-sheaves on separated schemes and to the Selmer modules of (certain classes of) abelian varieties in each case over possibly non-commutative, compact p-adic Lie extensions of global function fields. Indeed, our primary focus is on discussing the results of the first-named author in [20], of Vauclair and the second-named author in [80] and of Lai, Longhi, Tan and the second-named author in [55], but in each case we have also tried to describe (to the best of our knowledge) connections to the work of others.

In particular, we shall make much use of the approach to non-commutative Iwasawa theory that was introduced by Coates, Fukaya, Kato, Sujatha and Venjakob in [29], and also of several important aspects of the extensive theory that has recently been developed (by a variety of authors) in that area. We shall also try to explain the interest of studying main conjectures in this non-commutative setting by discussing a selection of rather concrete consequences of the main result of [20] that concern the leading terms and values of the associated (equivariant) Zeta functions.

In a little more detail, the main contents of these notes are as follows. In § 1 we quickly review some relevant background material concerning algebraic K-theory, the Iwasawa algebras of (possibly non-abelian) compact p-adic Lie groups and the theory of pro-coverings and perfect complexes of pro-sheaves.

In § 2 we discuss results concerning the Iwasawa theory of higher direct images of flat, smooth sheaves, focusing on the natural main conjecture of non-commutative Iwasawa theory in this context that was proved by the first author in [20]. The latter result in fact constitutes a natural non-commutative generalization of the main conjecture that was proved by Emerton and Kisin in [35], and hence also of the main conjecture proved by Crew in [30], and we shall discuss the key ideas that are involved in its proof in § 3.

In § 4 and § 5 we discuss the (precise statement and) proof of the main conjecture of non-commutative Iwasawa theory for semistable abelian varieties over ·certain unramified p-adic Lie extensions that is given in [80].

In § 6 we discuss the main results of [55] concerning the Iwasawa theory of Selmer modules of constant ordinary abelian varieties over certain $\mathbb{Z}_p$-power extensions of global function fields, and in § 7 we discuss the key ideas underlying the proof of these results.

In § 8, § 9 and § 10 we then try to give some indication of the interest of the above results by discussing a selection of the consequences of the main result stated in § 2 that are proved in [20] and are motivated, in large part, by the general philosophy of 'equivariant Tamagawa number conjectures' that was introduced by Flach and the first-named author in [21].

Finally, in an appendix to these notes, Francesc Bars discusses certain recent results of himself, Bandini and Longhi which show how several of the key results that are discussed above can, or could possibly, be extended to the setting of Galois extensions of global function fields for which the associated Iwasawa algebras are not noetherian.

Many of the results that are stated in § 2 and § 8, respectively in § 4 and § 6, were discussed in lecture courses that were given by the first-, respectively second-named author at the CRM in Barcelona in early 2010, and both authors are very grateful to the institute for providing this opportunity. They are also very grateful indeed to Francesc Bars for the wonderful hospitality shown to them throughout their visits to the CRM, for his generous encouragement and, especially, for providing us with the appendix to these notes. The authors are also grateful to Ignazio Longhi for useful discussions and, in addition, the second-named author wishes to acknowledge financial support received from EPSRC.

1 Preliminaries

In this section we shall, for the reader's convenience, recall some basic background material relating to relative algebraic K-theory, refined Euler characteristics, the Iwasawa algebras of certain kinds of non-commutative p-adic Lie groups and relevant aspects of the theory of pro-coverings and perfect complexes of pro-sheaves.

1.1 Relative algebraic K-theory and Iwasawa algebras

Throughout these notes, modules are always to be understood, unless explicitly stated otherwise, as left modules.

For any associative, unital, left noetherian ring R, we write $D(R)$ for the derived category of R-modules. We also write $D^-(R)$, $D^+(R)$ and $D^{\mathrm{p}}(R)$ for the full triangulated subcategories of $D(R)$ comprising complexes that are isomorphic to an object of the categories $C^-(R)$, $C^+(R)$ and $C^{\mathrm{p}}(R)$ of bounded above complexes of projective R-modules, bounded below complexes of injective R-modules and bounded complexes of finitely generated projective R-modules.

For any homomorphism $R \to R'$ of rings as above, we write $K_0(R, R')$ for the relative algebraic K_0-group that is defined in terms of explicit generators and relations by Swan in [71, p. 215]. We recall in particular that this group fits into a canonical exact sequence of abelian groups of the form

$$K_1(R) \longrightarrow K_1(R') \xrightarrow{\partial_{R,R'}} K_0(R, R') \longrightarrow K_0(R) \longrightarrow K_0(R'). \qquad (1.1)$$

Here, for any ring A, we write $K_1(A)$ for its Whitehead group and $K_0(A)$ for the Grothendieck group of the category of finitely generated projective A-modules, and the first and last arrows in (1.1) denote the homomorphisms that are naturally induced by the given ring homomorphism $R \to R'$. (For more details about this sequence, and a proof of its exactness, see [71, Chapter 15].)

It is well known that for any object $C^\bullet$ of $D^{\mathrm{p}}(R)$ one can define a canonical Euler characteristic $\chi_R(C^\bullet)$ in the group $K_0(R)$. We recall further that, for any such complex $C^\bullet$ and any (bounded) exact sequence of R'-modules of the form

$$\epsilon \colon 0 \longrightarrow \cdots \longrightarrow R' \otimes_R H^i(C^\bullet) \longrightarrow R' \otimes_R H^{i+1}(C^\bullet)$$
$$\longrightarrow R' \otimes_R H^{i+2}(C^\bullet) \longrightarrow \cdots \longrightarrow 0,$$

one can define a canonical pre-image $\chi_{R,R'}^{\mathrm{ref}}(C^\bullet, \epsilon)$ of $\chi_R(C^\bullet)$ under the connecting homomorphism $K_0(R, R') \to K_0(R)$ that occurs in the sequence (1.1). (For more details of this natural 'refined Euler characteristic' construction, see either [13] or [14].)

In particular, if $C^\bullet$ belongs to $D^{\mathrm{p}}(R)$ and the complex $R' \otimes_R^{\mathbb{L}} C^\bullet$ is acyclic, then we shall set

$$\chi_{R,R'}^{\mathrm{ref}}(C^\bullet) := \chi_{R,R'}^{\mathrm{ref}}(C^\bullet, \epsilon_{R'}),$$

where $\epsilon_{R'}$ denotes the exact sequence of zero R'-modules.

In the case that R' is the total quotient ring of R, we often abbreviate the connecting homomorphism $\partial_{R,R'}$ in the exact sequence (1.1) to ∂_R and the Euler characteristics $\chi^{\mathrm{ref}}_{R,R'}(-,-)$ and $\chi^{\mathrm{ref}}_{R,R'}(-)$ to $\chi^{\mathrm{ref}}_R(-,-)$ and $\chi^{\mathrm{ref}}_R(-)$ respectively. In those cases when R and R' are both clear from context, we also sometimes abbreviate the notation $\partial_{R,R'}$ and $\chi^{\mathrm{ref}}_{R,R'}(-,-)$ to ∂ and $\chi^{\mathrm{ref}}(-,-)$ respectively.

For any profinite group G and any finite extension $\mathcal{O}$ of $\mathbb{Z}_\ell$ (for any prime ℓ) we write $\Lambda_{\mathcal{O}}(\mathcal{G})$ for the $\mathcal{O}$-Iwasawa algebra $\varprojlim_U \mathcal{O}[G/U]$ of G, where U runs over the set of open normal subgroups of G (partially ordered by inclusion) and the limit is taken with respect to the obvious transition homomorphisms. We also write $Q_{\mathcal{O}}(\mathcal{G})$ for the total quotient ring of $\Lambda_{\mathcal{O}}(\mathcal{G})$, and if $\mathcal{O} = \mathbb{Z}_p$, then we omit the subscripts '$\mathcal{O}$' from both $\Lambda_{\mathcal{O}}(\mathcal{G})$ and $Q_{\mathcal{O}}(\mathcal{G})$.

Motivated by the approach of Coates, Fukaya, Kato, Sujatha and Venjakob in [29], we now assume that G lies in a group extension of the form

$$\{1\} \longrightarrow H \longrightarrow G \xrightarrow{\pi_G} \Gamma \longrightarrow \{1\} \tag{1.2}$$

in which Γ is (topologically) isomorphic to the additive group of p-adic integers $\mathbb{Z}_p$. We also fix an algebraic closure $\mathbb{Q}_p^c$ of $\mathbb{Q}_p$ and write $\mathcal{O}$ for the valuation ring of a finite extension of $\mathbb{Q}_p$ in $\mathbb{Q}_p^c$.

We write S for the subset of $\Lambda_{\mathcal{O}}(G)$ comprising elements f for which the quotient $\Lambda_{\mathcal{O}}(G)/\Lambda_{\mathcal{O}}(G)f$ is finitely generated as a module over the ring $\Lambda_{\mathcal{O}}(H)$ and we also set $S^* = \bigcup_{i \geq 0} p^{-i}S$.

We recall that in [29, §2] it is shown that S and S^* are both multiplicatively closed left and right Ore sets of non-zero divisors and so we can write $\Lambda_{\mathcal{O}}(G)_S$ and $\Lambda_{\mathcal{O}}(G)_{S^*} = \Lambda_{\mathcal{O}}(G)_S[\frac{1}{p}]$ for the corresponding localizations of $\Lambda_{\mathcal{O}}(G)$ (we however caution the reader that this notation does not explicitly indicate that S and S^* depend upon both $\mathcal{O}$ and the chosen extension (1.2)).

We often use the fact that the long exact sequence (1.1) is compatible with the formulation of scalar extensions in the sense that there exists a commutative diagram

$$\begin{array}{ccccc} K_1(\Lambda_{\mathcal{O}}(G)) & \longrightarrow & K_1(\Lambda_{\mathcal{O}}(G)_{S^*}) & \xrightarrow{\partial_{\mathcal{O},G,S^*}} & K_0(\Lambda_{\mathcal{O}}(G), \Lambda_{\mathcal{O}}(G)_{S^*}) \\ \Big\| & & \Big\uparrow & & \Big\uparrow \\ K_1(\Lambda_{\mathcal{O}}(G)) & \xrightarrow{\alpha_{\mathcal{O},G}} & K_1(\Lambda_{\mathcal{O}}(G)_S) & \xrightarrow{\partial_{\mathcal{O},G,S}} & K_0(\Lambda_{\mathcal{O}}(G), \Lambda_{\mathcal{O}}(G)_S) \end{array} \tag{1.3}$$

in which we set $\partial_{\mathcal{O},G,\Sigma} = \partial_{\Lambda_{\mathcal{O}}(G),\Lambda_{\mathcal{O}}(G)_\Sigma}$ for both $\Sigma = S$ and $\Sigma = S^*$.

If Σ denotes either of the Ore sets S or S^* defined above, then we shall write $D^{\mathrm{p}}_\Sigma(\Lambda_{\mathcal{O}}(G))$ for the full triangulated subcategory of $D^{\mathrm{p}}(\Lambda_{\mathcal{O}}(G))$ comprising those (perfect) complexes $C^\bullet$ for which $\Lambda_{\mathcal{O}}(G)_\Sigma \otimes_{\Lambda_{\mathcal{O}}(G)} C^\bullet$ is acyclic.

We note, in particular, that any object $C^\bullet$ of $D^{\mathrm{p}}_\Sigma(\Lambda_{\mathcal{O}}(G))$ gives rise to an Euler characteristic element

$$\chi^{\mathrm{ref}}(C^\bullet) = \chi^{\mathrm{ref}}_{\Lambda_{\mathcal{O}}(G),\Lambda_{\mathcal{O}}(G)_\Sigma}(C^\bullet, \epsilon_{\Lambda_{\mathcal{O}}(G)_\Sigma})$$

in $K_0(\Lambda_{\mathcal{O}}(G), \Lambda_{\mathcal{O}}(G)_\Sigma)$ that depends only upon $C^\bullet$.

We recall that, if G has no element of order p and Σ denotes either S or S^*, then the group $K_0(\Lambda_{\mathcal{O}}(G), \Lambda_{\mathcal{O}}(G)_\Sigma)$ is naturally isomorphic to the Grothendieck group of the category of finitely generated $\Lambda_{\mathcal{O}}(G)$-modules M with the property that $\Lambda_{\mathcal{O}}(G)_\Sigma \otimes_{\Lambda_{\mathcal{O}}(G)} M$ vanishes (for an explicit description of this isomorphism, see, for example, [24, § 1.2]).

In particular, if G is also abelian then the determinant functor induces a natural isomorphism between $K_0(\Lambda_{\mathcal{O}}(G), \Lambda_{\mathcal{O}}(G)_\Sigma)$ and the multiplicative group of invertible $\Lambda_{\mathcal{O}}(G)$-lattices in $\Lambda_{\mathcal{O}}(G)_\Sigma$. For any (finitely generated torsion) $\Lambda_{\mathcal{O}}(G)$-module M as above, this isomorphism sends the element $\chi^{\mathrm{ref}}(M[0])$ defined above to the (classical) characteristic ideal $\mathrm{ch}_{\Lambda_{\mathcal{O}}(G)}(M)$ of M.

In what follows, the Euler characteristic $\chi^{\mathrm{ref}}_{\Lambda_{\mathcal{O}}(G), Q_{\mathcal{O}}(G)}(-, -)$ will often be abbreviated to $\chi^{\mathrm{ref}}_{\mathcal{O}, G}(-, -)$.

We now fix a topological generator γ of the group Γ that occurs in the extension (1.2) and also an Ore set $\Sigma \in \{S, S^*\}$.

Then, for the valuation ring $\mathcal{O}'$ of any finite extension of $\mathbb{Q}_p$ in $\mathbb{Q}_p^c$ which contains $\mathcal{O}$ and any continuous homomorphism of the form

$$\rho: G \longrightarrow \mathrm{GL}_n(\mathcal{O}'), \tag{1.4}$$

there is an induced ring homomorphism

$$\Lambda_{\mathcal{O}}(G)_\Sigma \longrightarrow \mathrm{M}_n(\mathcal{O}') \otimes_{\mathcal{O}'} Q_{\mathcal{O}'}(\Gamma) \cong \mathrm{M}_n(Q_{\mathcal{O}'}(\Gamma))$$

that sends every element g of G to $\rho(g) \otimes \pi_G(g)$. This ring homomorphism induces in turn a homomorphism of abelian groups

$$\Phi_{\mathcal{O}, G, \Sigma, \rho}: K_1(\Lambda_{\mathcal{O}}(G)_\Sigma) \longrightarrow K_1(\mathrm{M}_n(Q_{\mathcal{O}'}(\Gamma))) \cong K_1(Q_{\mathcal{O}'}(\Gamma))$$
$$\cong Q_{\mathcal{O}'}(\Gamma)^\times \cong Q(\mathcal{O}'[[u]])^\times$$

where we write $\mathcal{O}'[[u]]$ for the ring of power series over $\mathcal{O}'$ in the formal variable u, the first isomorphism is induced by Morita equivalence, the second by taking determinants (over the ring $Q_{\mathcal{O}'}(\Gamma)$), and the last by sending $\gamma - 1$ to u. In the sequel we shall often not distinguish between the homomorphisms $\Phi_{\mathcal{O}, G, S, \rho}$ and $\Phi_{\mathcal{O}, G, S^*, \rho}$ and also abbreviate $\Phi_{\mathcal{O}, G, \Sigma, \rho}$ to $\Phi_{G, \rho}$ if we feel that $\mathcal{O}$ and Σ are both clear from context.

For any element ξ of the Whitehead group $K_1(\Lambda_{\mathcal{O}}(G)_\Sigma)$ and any representation ρ as in (1.4), one then defines the 'value' $\xi(\rho)$, resp. the 'leading term' $\xi^*(\rho)$, of ξ at ρ to be the value, resp. the leading term, at $u = 0$ of the series $\Phi_{\mathcal{O}, G, \Sigma, \rho}(\xi)$. In particular, one has $\xi^*(\rho) \in \mathbb{Q}_p^c \setminus \{0\}$ for all ρ and one regards the value $\xi(\rho)$ to be equal to '∞' if the algebraic order of $\Phi_{\mathcal{O}, G, \Sigma, \rho}(\xi)$ at $u = 0$ is strictly negative.

We finally recall that a continuous representation ρ as in (1.4) (with $\mathcal{O} = \mathbb{Z}_p$) is said to be an 'Artin representation' if its image $\rho(G)$ is finite.

1.2   Pro-coverings, pro-sheaves and perfect complexes

We quickly review some standard material concerning pro-coverings and perfect complexes of pro-sheaves. The interested reader will also find a nice (and much more detailed) treatment of all of the necessary material in Witte's articles [82] and [83].

For any prime power n we write $\mathbb{F}_n$ for the finite field of cardinality n. We also fix now a prime p and a strictly positive integral power q of p.

Let X be a connected scheme over $\mathbb{F}_q$ and write $\mathbf{FEt}/X$ for the category of X-schemes that are finite and étale over X.

Then, for any geometric point $\overline{x}$ of X, the functor that takes each scheme Y to the set $F_{\overline{x}}(Y) = \mathrm{Hom}_X(\overline{x}, Y)$ of geometric points of Y that lie over $\overline{x}$ gives an equivalence of categories between projective systems in $\mathbf{FEt}/X$ and the category of projective systems of finite sets upon which the group $\pi_1(X, \overline{x})$ acts continuously (on the left).

In particular, for any continuous quotient G of $\pi_1(X, \overline{x})$ there exists a corresponding pro-covering of X of group G which we denote by $(f\colon Y \to X, G)$, or more simply by $f\colon Y \to X$.

Fix a prime number ℓ, a finite extension $\mathcal{O}$ of $\mathbb{Z}_\ell$ and an étale sheaf of $\mathcal{O}$-modules $\mathcal{L}$ on X. Then for any morphism $h\colon Y \to X$ in $\mathbf{FEt}/X$ the sheaf $h_*h^*\mathcal{L}$ is the sheaf of $\mathcal{O}$-modules that is associated to the pre-sheaf $U \mapsto \bigoplus_{\mathrm{Hom}_X(U,Y)} \mathcal{L}(U)$ where the transition morphisms $\bigoplus_{\mathrm{Hom}_X(V,Y)} \mathcal{L}(V) \to \bigoplus_{\mathrm{Hom}_X(U,Y)} \mathcal{L}(U)$ for each $\alpha\colon U \to V$ are given by mapping each vector $(x_\psi)_\psi$ to $(\mathcal{L}(\alpha)(x_\psi))_{\psi \circ \alpha}$.

If the morphism h is Galois, then the group $\mathcal{G} = \mathrm{Aut}_X(Y)$ acts on the left on $h_*h^*\mathcal{L}$ and for the stalk at $\overline{x}$ there is an isomorphism of $\mathcal{O}[\mathcal{G}]$-modules

$$(h_*h^*\mathcal{L})_{\overline{x}} \cong \bigoplus_{F_{\overline{x}}(Y)} \mathcal{L}_{\overline{x}} \cong \mathcal{O}[\mathcal{G}] \otimes_{\mathcal{O}} \mathcal{L}_{\overline{x}}, \tag{1.5}$$

where the second map results from the fact that (since h is Galois) any choice of an element ξ of $F_{\overline{x}}(Y)$ induces an isomorphism $\sigma \mapsto \sigma \circ \xi$ of $\mathcal{G}$-sets $\mathcal{G} \cong F_{\overline{x}}(Y)$.

For any pro-covering $f\colon Y \to X$ of group G as above and any flat, smooth $\mathcal{O}$-sheaf $\mathcal{L}$ on X, the associated pro-sheaf $\mathcal{L}_G = f_*f^*\mathcal{L}$ has a natural action of the algebra $\Lambda_{\mathcal{O}}(G)$.

In any such case the isomorphism (1.5) implies that each stalk of $\mathcal{L}_G$ is a finitely generated projective $\Lambda_{\mathcal{O}}(G)$-module and this in turn implies that the étale cohomology complex $R\Gamma(X_{\text{ét}}, \mathcal{L}_G)$ is an object of $D^{\mathrm{p}}(\Lambda_{\mathcal{O}}(G))$.

2   Higher direct images

As above, we fix an integral power q of a given prime number p. In addition, we now fix a separated variety X that is of finite type over $\mathbb{F}_q$ and a geometric point $\overline{x}$ of X. We also fix a separable closure $\mathbb{F}_q^c$ of $\mathbb{F}_q$.

We then consider compact p-adic Lie groups G which lie in a commutative diagram of continuous homomorphisms of the form

$$\pi_1(X, \overline{x}) \xrightarrow{\pi_{X,\overline{x}}} \pi_1^p(\mathbb{F}_q, \mathbb{F}_q^c) \qquad (2.1)$$

$$\pi_G' \searrow \qquad \nearrow \pi_G$$

$$G$$

where $\pi_{X,\overline{x}}$ denotes the canonical homomorphism to the maximal pro-p quotient $\pi_1^p(\mathbb{F}_q, \mathbb{F}_q^c)$ of $\pi_1(\mathbb{F}_q, \mathbb{F}_q^c)$ and π_G' is surjective.

The composite homomorphism

$$\pi_1(\mathbb{F}_q, \mathbb{F}_q^c) \longrightarrow \mathrm{Gal}(\mathbb{F}_q^c/\mathbb{F}_q) \longrightarrow \mathbb{Z}_p,$$

where the first map is the canonical isomorphism and the second map sends the Frobenius automorphism $z \mapsto z^q$ in $\mathrm{Gal}(\mathbb{F}_q^c/\mathbb{F}_q)$ to 1, induces an identification of $\mathrm{im}(\pi_{X,\overline{x}}) = \mathrm{im}(\pi_G)$ with a subgroup of $\mathbb{Z}_p$ of finite index, d_X say. This identification gives rise in turn to a canonical group extension

$$\{1\} \longrightarrow \ker(\pi_G) \longrightarrow G \xrightarrow{\pi_G} \mathrm{im}(\pi_G) \longrightarrow \{1\}$$

of the form (1.2).

In the rest of this section we shall therefore set

$$H = \ker(\pi_G) \quad \text{and} \quad \Gamma = \mathrm{im}(\pi_{X,\overline{x}}) = \mathrm{im}(\pi_G),$$

and write γ for the topological generator of Γ that is given by the d_Xth power of the image of the Frobenius automorphism $z \mapsto z^q$ under the natural projection map $\mathrm{Gal}(\mathbb{F}_q^c/\mathbb{F}_q) \cong \pi_1(\mathbb{F}_q, \mathbb{F}_q^c) \to \pi_1^p(\mathbb{F}_q, \mathbb{F}_q^c)$.

In particular, we shall always use the K-theoretical constructions that are discussed in § 1 with respect to this choice of data.

For any finite extension $\mathcal{O}$ of $\mathbb{Z}_p$ and any flat, smooth $\mathcal{O}$-sheaf $\mathcal{L}$ on X, we shall write $Z(X, \mathcal{L}, t)$ for the associated Zeta function in a formal variable t. We also write $r_{\mathcal{L}}$ for the algebraic order of $Z(X, \mathcal{L}, t)$ at $t = 1$ and then define the (normalized) leading term of $Z(X, \mathcal{L}, t)$ at $t = 1$ to be the limit

$$Z^*(X, \mathcal{L}, 1) := d_X^{-r_{\mathcal{L}}} \lim_{t \to 1} (1 - t)^{-r_{\mathcal{L}}} Z(X, \mathcal{L}, t) \in \mathbb{Q}_p^{c \times}. \qquad (2.2)$$

Since the structure morphism $s \colon X \to \mathrm{Spec}(\mathbb{F}_q)$ is assumed to be separated, it factors as $s = s' \circ j$ with $j \colon X \to X'$ an open immersion and $s' \colon X' \to \mathrm{Spec}(\mathbb{F}_q)$ a proper morphism. Then the cohomology with compact support

$$R\Gamma_c(X_{\text{ét}}, \mathcal{L}_G) := R\Gamma(X'_{\text{ét}}, j_!\mathcal{L}_G)$$

and the higher direct image with proper support

$$R\Gamma(\mathbb{F}_q, Rs_{X,!}\mathcal{L}_G) := R\Gamma(\mathrm{Spec}(\mathbb{F}_q)_{\text{ét}}, Rs'_*j_!\mathcal{L}_G)$$

of $\mathcal{L}_G$ are naturally isomorphic as objects of $D(\Lambda_{\mathcal{O}}(G))$ and the results of Deligne in [31, p. 95, Théorème 4.9 and Arcata, IV, § 5] show further that these complexes belong to $D^{\mathrm{p}}(\Lambda_{\mathcal{O}}(G))$ and are also independent, up to natural isomorphism, of the chosen compactification $s = s' \circ j$.

We can now state the main result of this section. This result has been proved by the first-named author in [20] and is the natural analogue in our present setting of the main conjecture of non-commutative Iwasawa theory that is formulated by Coates et al. in [29]. Note that it also constitutes a natural non-abelian generalization of the result of Emerton and Kisin in [35, Corollary 1.8] and hence of the 'main conjecture' that was proved by Crew in [30].

The appropriate special case of this result can also presumably be explicitly related to the results of Vauclair and the second-named author that we will discuss later in § 4 (although, at this stage, we do not know of a precise such connection).

We recall that any Artin representation of the form $\rho \colon G \to \mathrm{GL}_m(\mathcal{O})$ gives rise in a natural way to a flat, smooth $\mathcal{O}$-sheaf $\mathcal{L}_\rho$ on X.

Theorem 2.1. *Let $s_X \colon X \to \mathrm{Spec}(\mathbb{F}_q)$ be a separated morphism of finite type. Let G be any compact p-adic Lie group as in (2.1) and write $f \colon Y \to X$ for the corresponding pro-covering of X of group G. Let $\mathcal{L}$ be a flat, smooth $\mathbb{Z}_p$-sheaf on X. Then the complex $R\Gamma(\mathbb{F}_q, Rs_{X,!}f_*f^*\mathcal{L})$ belongs to $D_S^{\mathrm{p}}(\Lambda(G))$ and there exists an element $\xi_{X,G,\mathcal{L}}$ of $K_1(\Lambda(G)_S)$ with*

$$\partial_G(\xi_{X,G,\mathcal{L}}) = -\chi^{\mathrm{ref}}(R\Gamma(\mathbb{F}_q, Rs_{X,!}f_*f^*\mathcal{L}))$$

and such that for all Artin representations ρ of G one has

$$\xi_{X,G,\mathcal{L}}^*(\rho) = Z^*(X, \mathcal{L} \otimes \mathcal{L}_\rho, 1).$$

To end this section we make some general remarks about this result.

The first thing to note is that the argument of [20] can be used to prove a slightly more general result in which $\mathcal{L}$ is replaced by any bounded complex of flat, smooth $\mathbb{Z}_p$-sheaves on X. However, since we have no use for this generalization, we give no further details here.

We next note that if X is an affine curve, G is abelian and $\mathcal{L}$ has abelian monodromy, then the proof of Theorem 2.1 can be reduced to the main result of Lai, Tan and the first-named author in [22] and hence relies on Weil's formula for the Zeta function in terms of ℓ-adic cohomology for any prime $\ell \neq p$ and thus avoids any use of crystalline cohomology.

However, the proof given in [20] of Theorem 2.1 in the general case makes pivotal use of a result of [35] which itself relies on important aspects of the theory of F-crystals. (Indeed, it is with the results of loc. cit. in mind that Theorem 2.1 is phrased in terms of higher direct images with proper support rather than cohomology with compact support.)

Finally we note that it is, of course, natural to ask for an analogue of Theorem 2.1 in which q is coprime to p whilst G is still assumed to be a p-adic Lie group (as in (2.1)) and $\mathcal{L}$ a flat, smooth $\mathbb{Z}_p$-sheaf.

In fact, in [82, 83] Witte has independently formulated and proved (a generalization of) the natural analogue of Theorem 2.1 in this setting. We further note that the argument given in loc. cit. is very different in style to the proof of Theorem 2.1 that we sketch below. Indeed, Witte is able to use a natural interpretation of the relevant relative algebraic K_0-groups in terms of Waldhausen K-theory (that is due essentially to Muro and Tonks [60]) to deduce the desired result in a rather elegant fashion from the aforementioned results of Weil. However, the more complicated nature of the results of Emerton and Kisin in [35], in comparison to the results of Weil, mean that such an approach does not seem to apply in the setting of Theorem 2.1. For more details in this context, see also Remarks 3.3 and 9.2.

3 Proof of Theorem 2.1

In this section we shall discuss the main points of the proof of Theorem 2.1 that is given in [20].

It is important to note at the outset that Theorem 2.1 does not assert the existence of a unique, or even a canonical, validating element $\xi_{X,G,\mathcal{L}}$. Reflecting this fact, the proof of Theorem 2.1 that is given in [20] is both rather involved and indirect in nature and, in particular, does not construct a validating element $\xi_{X,G,\mathcal{L}}$ explicitly (but see Remark 3.8 for details of a possible alternative approach).

3.1 Finite generation

The starting point of the proof of Theorem 2.1 is the following important result of Witte (from [83, Theorem 8.1]).

Theorem 3.1. *If G is any compact p-adic Lie group as in the diagram (2.1), then for any finite extension $\mathcal{O}$ of $\mathbb{Z}_p$ and any flat, smooth $\mathcal{O}$-sheaf $\mathcal{L}$ on X the complex $R\Gamma(\mathbb{F}_q, Rs_{X,!}\mathcal{L}_G)$ belongs to $D_S^p(\Lambda_{\mathcal{O}}(G))$.*

This result is equivalent to asserting that, under the given hypotheses, in each degree i the cohomology group $H^i(R\Gamma(\mathbb{F}_q, Rs_{X,!}\mathcal{L}_G))$ is finitely generated as a module over the Iwasawa algebra $\Lambda_{\mathcal{O}}(H)$. A slightly more general version of this fact is proved in [83] by a careful reduction to consideration of constant field extensions.

An alternative, and rather more elementary, proof of Theorem 3.1 in the special case that X is an affine curve is given in [17, Proposition 4.1(iii)].

3.2 The Hochschild–Serre exact triangle

We now write s_c for the morphism $\mathrm{Spec}(\mathbb{F}_q^c) \to \mathrm{Spec}(\mathbb{F}_q)$ that is induced by the inclusion $\mathbb{F}_q \subset \mathbb{F}_q^c$ and ϕ for the geometric Frobenius automorphism in $\mathrm{Gal}(\mathbb{F}_q^c/\mathbb{F}_q)$. Then the result of Deligne in [31, p. 95, Théorème 4.9] implies that both of the complexes $R\Gamma(\mathbb{F}_q, Rs_{X,!}\mathcal{L}_G)$ and $R\Gamma(\mathbb{F}_q^c, s_c^* Rs_{X,!}\mathcal{L}_G)$ belong to $D^p(\Lambda_{\mathcal{O}}(G))$ and, in addition, the Hochschild–Serre spectral sequence implies the existence of a canonical exact triangle in $D^p(\Lambda_{\mathcal{O}}(G))$ of the form

$$R\Gamma(\mathbb{F}_q^c, s_c^* Rs_{X,!}\mathcal{L}_G) \xrightarrow{1-\phi} R\Gamma(\mathbb{F}_q^c, s_c^* Rs_{X,!}\mathcal{L}_G) \longrightarrow R\Gamma(\mathbb{F}_q, Rs_{X,!}\mathcal{L}_G)[1]$$
$$\longrightarrow R\Gamma(\mathbb{F}_p^c, s_c^* Rs_{X,!}\mathcal{L}_G)[1].$$

$$(3.1)$$

The next result constructs a concrete realization of this triangle in the case that G has rank one as a p-adic Lie group. This construction is useful for the purposes of making certain explicit computations in algebraic K-theory.

We note that if G has rank one, then the total quotient ring $Q_{\mathcal{O}}(G)$ of $\Lambda_{\mathcal{O}}(G)$ is both semisimple and Artinian.

Lemma 3.2. *We assume that $\ell = p$ and that G is any group as in (2.1) that has rank one as a p-adic Lie group. Let $P^\bullet$ be a complex in $C^p(\Lambda_{\mathcal{O}}(G))$ that is isomorphic in $D^p(\Lambda_{\mathcal{O}}(G))$ to $R\Gamma(\mathbb{F}_q^c, s_c^* Rs_{X,!}\mathcal{L}_G)$.*

Then there exists an endomorphism $\hat{\phi}$ of the complex $P^\bullet$ which is such that $1 - \hat{\phi}^i$ is injective in each degree i and there exists a commutative diagram in $D^p(\Lambda_{\mathcal{O}}(G))$ of the form

$$\begin{array}{ccc} P^\bullet & \xrightarrow{\;\iota\;} & R\Gamma(\mathbb{F}_q^c, s_c^* Rs_{X,!}\mathcal{L}_G) \\ {\scriptstyle 1-\hat{\phi}}\downarrow & & \downarrow{\scriptstyle 1-\phi} \\ P^\bullet & \xrightarrow{\;\iota\;} & R\Gamma(\mathbb{F}_q^c, s_c^* Rs_{X,!}\mathcal{L}_G) \end{array}$$

$$(3.2)$$

in which ι is an isomorphism.

Further, in each degree i the homomorphism $1 - \hat{\phi}^i$ induces an automorphism of the module $Q_{\mathcal{O}}(G) \otimes_{\Lambda_{\mathcal{O}}(G)} P^i$, the product

$$Z_G(1 - \hat{\phi}) := \prod_{i \in \mathbb{Z}} \langle 1 - \hat{\phi}^i \mid Q_{\mathcal{O}}(G) \otimes_{\Lambda_{\mathcal{O}}(G)} P^i \rangle^{(-1)^{i+1}}$$

belongs to $\mathrm{im}(K_1(\Lambda_{\mathcal{O}}(G)_S) \to K_1(Q_{\mathcal{O}}(G)))$ and in the group $K_0(\Lambda_{\mathcal{O}}(G), \Lambda_{\mathcal{O}}(G)_S)$ one has an equality

$$\partial_{\mathcal{O},G}(Z_G(1 - \hat{\phi})) = -\chi_{\mathcal{O},G}^{\mathrm{ref}}(R\Gamma(\mathbb{F}_q, Rs_{X,!}\mathcal{L}_G)).$$

$$(3.3)$$

Remark 3.3. The final equality of Lemma 3.2 combines with the sequences in (1.3) to imply that the existence of a diagram (3.2) determines the element $Z_G(1 - \hat{\phi})$ up to multiplication by an element of $\mathrm{im}(\alpha_{\mathcal{O},G})$.

The alternative approach of Witte in [83] relies on Waldhausen K-theory and can actually be used to show that $Z_G(1 - \hat{\phi})$, and also a natural analogue of this element for higher rank G, is uniquely determined by a diagram of the form (3.2). However, since one only needs to use this fact (via the results of Emerton and Kisin that are applied in § 3.4) in the case that G is abelian and in this case it can be proved directly as in [31, p. 115, Corollaire 1.13], the concrete approach of Lemma 3.2 allows one to avoid any use of Waldhausen K-theory and the rather delicate constructions that it entails when proving results of the form Theorem 2.1.

3.3 An important reduction step

Given the result of Theorem 3.1, the good functorial behaviour of the Euler characteristics $\chi^{\mathrm{ref}}(R\Gamma(\mathbb{F}_q, Rs_{X,!}\mathcal{L}_G))$ under change of the Galois group G and of the leading terms $Z^*(X, \mathcal{L} \otimes \mathcal{L}_\rho, 1)$ under change of Artin representation ρ can be used to reduce the proof of Theorem 2.1 to consideration of the special class of pro-covers that is described in the following result.

Proposition 3.4. *It suffices to prove Theorem 2.1 for all pro-covers of the form $(f': Y' \to X', G')$ in which the group G' is a direct product $H \times P$ with H a finite abelian group of order prime to p and P a rank one pro-p p-adic Lie group.*

The reduction of Theorem 2.1 to consideration of rank one pro-covers follows directly from the general result of [19, Theorem 2.1] which itself relies on an important K-theoretical observation made by Fukaya and Kato in [38, Proposition 1.5.1].

The subsequent reduction from the case of general rank one pro-covers to pro-covers that are of the form described in Proposition 3.4 can then be proved by using certain standard (if rather involved) 'induction' techniques that involve either the very general results of Dress in [33] or the related results that are described in the context of classical Galois module theory by Fröhlich in [37].

3.4 The abelian case

In this subsection we assume that G is abelian and of rank one and discuss how this case of Theorem 2.1 can be deduced from the results of Emerton and Kisin in [35].

In this case G is isomorphic to a direct product $P' \times P$ with P' finite abelian of order prime to p and P a rank one abelian pro-p p-adic Lie group and an easy argument shows that, after changing $\mathcal{O}$ if necessary, we can assume that $G = P$ is pro-p.

The validity of Theorem 2.1 in this special case then follows from the following result.

Theorem 3.5. *Let G be a rank one abelian pro-p p-adic Lie group as in (2.1). Fix a finite unramified extension $\mathcal{O}$ of $\mathbb{Z}_p$ and a flat and smooth $\mathcal{O}$-sheaf $\mathcal{F}$ on X. Then the complex $R\Gamma(\mathbb{F}_p, Rs_{X,!}\mathcal{F}_G)$ belongs to $D_S^{\mathrm{p}}(\Lambda_\mathcal{O}(G))$ and there exists a unique element $\xi_\mathcal{F}$ of $K_1(\Lambda_\mathcal{O}(G)_S)$ with*

$$\partial_{\mathcal{O},G}(\xi_\mathcal{F}) = -\chi^{\mathrm{ref}}_{\mathcal{O},G}(R\Gamma(\mathbb{F}_q, Rs_{X,!}\mathcal{F}_G)) \tag{3.4}$$

and such that for all Artin representations ρ of G one has

$$\xi_{\mathcal{F}}^{*}(\rho) = Z_{\mathcal{O}}^{*}(X, \mathcal{F} \otimes \mathcal{L}_{\rho}, 1). \tag{3.5}$$

It is perhaps interesting to note that if X is an affine curve and $\mathcal{F}$ has abelian monodromy, then (after translating from the language of graded invertible modules to that of relative algebraic K-theory in the natural way) the argument of Lai, Tan and the first-named author in [17, § A.2] can be used to give a proof of Theorem 3.5 that relies solely on Weil's formula for the Zeta function in terms of ℓ-adic cohomology for any prime $\ell \neq p$ (and hence avoids any use of either crystalline cohomology or the theory of F-crystals). In general, however, the proof of Theorem 3.5 can be derived from the main result of Emerton and Kisin in [35] as follows.

Since G is pro-p the ring $A := \Lambda_{\mathcal{O}}(G)$ is local and we write $\mathfrak{m}$ for its maximal ideal and $A\langle t \rangle$ for the $\mathfrak{m}$-adic completion of the polynomial ring $A[t]$. We also write $\mathcal{F}_{G}^{\bullet}$ for the complex comprising $\mathcal{F}_{G}$ placed in degree 0 (and with the zero sheaf placed in all other degrees).

One makes an easy reduction to the case $q = p$ and in this case the function $L(X, \mathcal{F}_{G}^{\bullet}, t)$ defined in [35, (1.7)] is equal to the series

$$Z_{A}(X, \mathcal{F}^{\bullet}, t) := \prod_{x \in X^{0}} \det_{A}(1 - \phi^{\mathrm{d}(x)} t^{\mathrm{d}(x)} \mid \mathcal{F}_{G,\bar{x}})^{-1} \in 1 + tA[[t]],$$

where X^{0} is the set of closed points of X and for each x in X^{0} we write $\mathrm{d}(x)$ for the degree of x over $\mathbb{F}_{p}$.

Further, if $\bar{z}$ denotes the geometric point $\mathrm{Spec}(\mathbb{F}_{p}^{c})$ of $\mathrm{Spec}(\mathbb{F}_{p})$, then the result of [59, Chapter II, Theorem 3.2(a)] (with π equal to the morphism s_{c} that was defined in § 3.2) implies that the complex $(Rs_{X,!}\mathcal{F}_{G})_{\bar{z}}^{\bullet}$ identifies with $R\Gamma(\mathbb{F}_{p}^{c}, s_{c}^{*}Rs_{X,!}\mathcal{F}_{G}^{\bullet})$. From the result of [31, p. 115, Corollaire 1.13] it therefore follows that

$$L(\mathbb{F}_{p}, Rs_{X,!}\mathcal{F}_{G}^{\bullet}, t) = \prod_{i \in \mathbb{Z}} \det_{\Lambda_{\mathcal{O}}(G)}(1 - \hat{\phi}^{i} \cdot t \mid P^{i})^{(-1)^{i+1}} \in 1 + tA[[t]]$$

where the series $L(\mathbb{F}_{p}, Rs_{X,!}\mathcal{F}_{G}^{\bullet}, t)$ is defined as in [35] and the morphism $\hat{\phi}$ and bounded complex $P^{\bullet}$ are chosen as in Lemma 3.2.

In particular, Lemma 3.2 implies that the series $L(\mathbb{F}_{p}, Rs_{X,!}\mathcal{F}_{G}^{\bullet}, t)$ converges at $t = 1$ to give an element of $A_{S}^{\times} \cong \mathrm{im}(K_{1}(A_{S}) \to K_{1}(Q_{\mathcal{O}}(G)))$ and also satisfies

$$\partial_{\mathcal{O}, G}(L(\mathbb{F}_{p}, Rs_{X,!}\mathcal{F}_{G}^{\bullet}, 1)) = -\chi_{\mathcal{O}, G}^{\mathrm{ref}}(R\Gamma(\mathbb{F}_{p}, Rs_{X,!}\mathcal{F}_{G})). \tag{3.6}$$

In addition, the result of Emerton and Kisin in [35, Corollary 1.8] implies that

$$Z_{A}(X, \mathcal{F}_{G}^{\bullet}, t) = v(X, \mathcal{F}_{G}^{\bullet}, t)L(\mathbb{F}_{p}, Rs_{X,!}\mathcal{F}_{G}^{\bullet}, t) \in 1 + tA[[t]] \tag{3.7}$$

for an element $v(X, \mathcal{F}_{G}^{\bullet}, t)$ of $1 + \mathfrak{m}tA\langle t \rangle$.

Now, since any series in $1 + \mathfrak{m}tA\langle t\rangle$ converges at $t = 1$ to give an element of $A^\times$, the equality (3.7) implies that $Z_A(X, \mathcal{F}^\bullet_G, t)$ converges at $t = 1$ to the value

$$\xi_{\mathcal{F}} := Z_A(X, \mathcal{F}^\bullet_G, 1) = v(1)L(\mathbb{F}_p, Rs_{X,!}\mathcal{F}^\bullet_G, 1) \in A^\times_S,$$

where we have set $v(1) := v(X, \mathcal{F}^\bullet_G, 1)$.

Then the equality (3.6) implies that this element $\xi_{\mathcal{F}}$ satisfies (3.4) since $v(1)$ belongs to $A^\times$ and an explicit computation shows that it also satisfies (3.5), as required to prove Theorem 3.5.

3.5 Strategy of Burns and Kato

We next review a general proof strategy in non-commutative Iwasawa theory that was introduced by the first-named author and Kato. This strategy will then be used to deduce Theorem 2.1 from the results of Proposition 3.4 and Theorem 3.5.

To do this we write Σ_G for the set of rank one abelian subquotients $\mathcal{A} = U_{\mathcal{A}}/J_{\mathcal{A}}$ of G and let Σ^*_G be any subset of Σ_G which satisfies the following condition:

(∗) For each Artin representation ρ of G there is a finite subset $\{\mathcal{A}_i : i \in I_\rho\}$ of subquotients in Σ^*_G and for each index i an integer m_i and a degree one representation ρ_i of $\mathcal{A}_i$ for which there is an isomorphism of (virtual) representations of the form

$$\rho \cong \sum_{i \in I_\rho} m_i \cdot \mathrm{Ind}^G_{U_{\mathcal{A}_i}} \mathrm{Inf}^{U_{\mathcal{A}_i}}_{\mathcal{A}_i} \rho_i.$$

(We note that Brauer's Induction Theorem implies this condition is automatically satisfied by setting $\Sigma^*_G = \Sigma_G$, but that in practice it can be very convenient to choose a smaller set Σ^*_G.)

We fix a finite unramified extension $\mathcal{O}$ of $\mathbb{Z}_p$ and for each $\mathcal{A}$ in Σ_G consider the composite homomorphism

$$q_{\mathcal{O},\mathcal{A}}\colon K_1(\Lambda_{\mathcal{O}}(G)_{S^*}) \to K_1(\Lambda_{\mathcal{O}}(U_{\mathcal{A}})_{S^*}) \to K_1(\Lambda_{\mathcal{O}}(\mathcal{A})_{S^*}) \to Q_{\mathcal{O}}(\mathcal{A})^\times$$

where the first arrow is the natural restriction map, the second is the natural projection and the third is the isomorphism induced by taking determinants over $\Lambda_{\mathcal{O}}(\mathcal{A})_{S^*} = Q_{\mathcal{O}}(\mathcal{A})$. We regard $q_{\mathcal{O},\mathcal{A}}$ as a homomorphism on both $K_1(\Lambda_{\mathcal{O}}(G))$ and $K_1(\Lambda_{\mathcal{O}}(G)_S)$ in the obvious way and often abbreviate it to $q_{\mathcal{A}}$ (believing that $\mathcal{O}$ is clear from context).

We fix a flat, smooth $\mathcal{O}$-sheaf $\mathcal{L}$ on X and, following Theorem 3.5, for each $\mathcal{A}$ in Σ_G we write $\xi_{\mathcal{A},\mathcal{L}}$ for the unique element of the group $K_1(\Lambda_{\mathcal{O}}(\mathcal{A})_S) \cong \Lambda_{\mathcal{O}}(\mathcal{A})^\times_S$ that validates Theorem 2.1 with the scheme X, group G and sheaf $\mathcal{L}$ replaced by the scheme $Y_{U_{\mathcal{A}}}$ that corresponds via a morphism $f_{U_{\mathcal{A}}}\colon Y_{U_{\mathcal{A}}} \to X$ to the (open) subgroup $U_{\mathcal{A}}$ of G, the abelian group $\mathcal{A} = U_{\mathcal{A}}/J_{\mathcal{A}}$ and the sheaf $f^*_{U_{\mathcal{A}}}\mathcal{L}$ respectively.

For more general versions of the following result see [19, Theorem 2.2 and Remark 2.5] and [50, Proposition 2.5].

Proposition 3.6. *Let G be a pro-p compact p-adic Lie group and $\mathcal{L}$ a flat, smooth $\mathcal{O}$-sheaf on X.*

If there exists an element ξ of $K_1(\Lambda_\mathcal{O}(G)_S)$ and an element u of $K_1(\Lambda_\mathcal{O}(G))$ which together satisfy both

(i) $\partial_{\mathcal{O},G}(\xi) = -\chi^{\mathrm{ref}}(R\Gamma(\mathbb{F}_q, Rs_{X,!}\mathcal{L}_G))$, *and*

(ii) $q_{\mathcal{O},\mathcal{A}}(u) = q_{\mathcal{O},\mathcal{A}}(\xi)^{-1}\xi_{\mathcal{A},\mathcal{L}}$ *for all $\mathcal{A}$ in Σ_G^*,*

then the product $\xi_{X,G,\mathcal{L}} := u\xi$ validates Theorem 2.1.

3.6   Application to Theorem 2.1

Just as above, for each (abelian) subquotient $\mathcal{A} = U_\mathcal{A}/J_\mathcal{A}$ in Σ_G one can define a natural composite homomorphism

$$q_{\mathcal{O},\mathcal{A}}^t \colon K_1(\Lambda_\mathcal{O}(G)[[t]]) \longrightarrow K_1(\Lambda_\mathcal{O}(U_\mathcal{A})[[t]]) \longrightarrow \Lambda_\mathcal{O}(\mathcal{A})[[t]]^\times.$$

The key fact that we use concerning these homomorphisms is described in the following result (the proof of which will be discussed in § 3.7).

Theorem 3.7. *Let G be a rank one pro-p p-adic Lie group and $\mathcal{O}$ a finite unramified extension of $\mathbb{Z}_p$. Let u^t be an element of $K_1(\Lambda_\mathcal{O}(G)[[t]])$ such that for each subquotient $\mathcal{A}$ in Σ_G^* the series $q_{\mathcal{O},\mathcal{A}}^t(u^t)$ converges at $t = 1$ to give an element of $\Lambda_\mathcal{O}(\mathcal{A})^\times$.*

Then $q_{\mathcal{O},\mathcal{A}}^t(u^t)$ converges at $t = 1$ for all $\mathcal{A}$ in Σ_G and there exists an element u of $K_1(\Lambda_\mathcal{O}(G))$ with

$$q_{\mathcal{O},\mathcal{A}}(u) = q_{\mathcal{O},\mathcal{A}}^t(u^t)|_{t=1}$$

for all $\mathcal{A}$ in Σ_G.

To apply this result in the context of Theorem 2.1 we note that if η is any endomorphism of a finitely generated projective $\Lambda_\mathcal{O}(G)$-module M, then for any series in $\mathbb{Z}[\eta][[t]]$ of the form

$$f(\eta, t) := 1 + \sum_{j \geq 1} \sum_{i \geq 0} m_{ij}\eta^i t^j$$

one obtains a well-defined automorphism of the associated finitely generated projective $\Lambda_\mathcal{O}(G)[[t]]$-module $M_t := \mathcal{O}[[t]] \otimes_\mathcal{O} M$ by setting

$$f(\eta, t)(\lambda \otimes m) := \lambda \otimes m + \sum_{j \geq 1} \sum_{i \geq 0} m_{ij}t^j \lambda \otimes \eta^i(m)$$

for each $\lambda \in \mathcal{O}[[t]]$ and $m \in M$. In any such case we write $\langle f(\eta, t) \mid M_t \rangle$ for the corresponding element of $K_1(\Lambda_\mathcal{O}(G)[[t]])$.

In particular, for any flat, smooth $\mathcal{O}$-sheaf $\mathcal{L}$ on X one obtains in this way well-defined elements of $K_1(\Lambda_\mathcal{O}(G)[[t]])$ by setting

$$Z_G(X, \mathcal{L}_G, t) := \left\langle \prod_{x \in X^0} (1 - \phi^{d(x)} t^{d(x)})^{-1} \mid (\mathcal{L}_{G, \bar{x}})_t \right\rangle$$

and

$$Z_G(1 - \hat{\phi}, t) := \prod_{i \in \mathbb{Z}} \langle 1 - \hat{\phi}^i t \mid P_t^i \rangle^{(-1)^{i+1}}$$

where here the endomorphism $\hat{\phi}$ and (bounded) complex $P^\bullet$ in $C^p(\Lambda_\mathcal{O}(G))$ are chosen as in Lemma 3.2.

Now the equality (3.3) combines with the result of Theorem 3.1 and the exact commutative diagram (1.3) to imply that the element $Z_G(1 - \hat{\phi})$ belongs to the image of $K_1(\Lambda_\mathcal{O}(G)_S)$ in $K_1(Q_\mathcal{O}(G))$ and further that any fixed pre-image ξ of $Z_G(1 - \hat{\phi})$ in $K_1(\Lambda_\mathcal{O}(G)_S)$ satisfies the condition (i) of Proposition 3.6.

Theorem 2.1 is then proved by using Theorem 3.7 to construct an element of $K_1(\Lambda_\mathcal{O}(G))$ which together with ξ satisfies condition (ii) of Proposition 3.6.

To explain this, we set

$$u^t := Z_G(1 - \hat{\phi}, t)^{-1} Z_G(X, \mathcal{L}_G, t) \in K_1(\Lambda_\mathcal{O}(G)[[t]]).$$

Then the functorial properties of $Z_G(X, \mathcal{L}_G, t)$ and $Z_G(1 - \hat{\phi}, t)$ under change of G combine with the equality (3.7) to imply that for each subquotient $\mathcal{A}$ in Σ_G the series $q_\mathcal{A}^t(u^t)$ converges at $t = 1$ to give an element of $\Lambda_\mathcal{O}(\mathcal{A})^\times$.

Theorem 3.7 (with $\Sigma_G^* = \Sigma_G$) then combines with Theorem 3.5 to imply the existence of an element u in $K_1(\Lambda_\mathcal{O}(G))$ such that at each $\mathcal{A}$ in Σ_G one has an equality

$$q_\mathcal{A}(u) = q_\mathcal{A}^t(u^t)|_{t=1} = q_\mathcal{A}(\xi)^{-1} \xi_\mathcal{A},$$

where $\xi_\mathcal{A}$ is the unique element of $K_1(\Lambda_\mathcal{O}(\mathcal{A})) = \Lambda_\mathcal{O}(\mathcal{A})^\times$ which validates Theorem 2.1 with $G = \mathcal{A}$.

This shows that the elements ξ and u validate the conditions of Proposition 3.6 and hence proves that Theorem 2.1 is valid for the given group G and sheaf $\mathcal{L}$.

Remark 3.8. At each geometric point $\bar{x}$ of X the $\Lambda(G)$-module $\mathcal{L}_{G, \bar{x}}$ is isomorphic to $\Lambda(G) \otimes_{\mathbb{Z}_p} \mathcal{L}_{\bar{x}}$ and so is free of rank, d say, independent of $\bar{x}$. In particular, if we fix a $\Lambda(G)$-basis of $\mathcal{L}_{G, \bar{x}}$ and write $M(\phi^{d(x)})$ for the matrix that corresponds to the action of $\phi^{d(x)}$ on $\mathcal{L}_{G, \bar{x}}$, then after fixing an ordering of the set X^0 we obtain an element of $GL_d(\Lambda(G)[[t]])$ by setting $\hat{Z}_G(X, \mathcal{L}_G, t) := \prod_{x \in X^0} (\mathrm{Id} - M(\phi^{d(x)}) t^{d(x)})^{-1}$.

The natural homomorphism of groups $GL_d(\Lambda(G)[[t]]) \to K_1(\Lambda(G)[[t]])$ sends this series $\hat{Z}_G(X, \mathcal{L}_G, t)$ to the element $Z_G(X, \mathcal{L}_G, t)$ that plays a key role in the argument given above. However, the proof of Theorem 2.1 does not show that $\hat{Z}_G(X, \mathcal{L}_G, t)$ converges at $t = 1$. In fact, if this series does converge at $t = 1$ (for any choice of $\Lambda(G)$-bases of the stalks $\mathcal{L}_{G,\bar{x}}$ and any ordering of the set X^0), then one could use it to generalize the more elementary approach to Theorem 2.1 that is discussed just after the statement of Theorem 3.5.

3.7   Proof of Theorem 3.7

Theorem 3.7 is proved by adapting a very elegant approach that was introduced by Kakde in [49] (and which is itself a natural development of methods that were first used by Kato in [50] and is also closely related to methods used by Ritter and Weiss in [68]).

One first shows easily that Theorem 3.7 is valid if for any particular choice of set Σ_G^*, there exists an element u of $K_1(\Lambda_{\mathcal{O}}(G))$ with $q_{\mathcal{A}}(u) = q_{\mathcal{A}}^t(u^t)|_{t=1}$ for all $\mathcal{A}$ in Σ_G^*. Having shown this, one is then able to carefully choose a set Σ_G^* just as in Kakde [49].

To do this we fix a lift $\tilde{\Gamma}$ of Γ in G and thereby identify G with the semi-direct product $H \rtimes \Gamma$ with $H := \ker(\pi_G)$. We also fix a natural number e such that the group $Z := \tilde{\Gamma}^{p^e}$ is central in G and set $\overline{G} := G/Z$. For any subgroup P of $\overline{G}$ we write U_P for the inverse image of P in G and then set

$$\Sigma_G^* := \{U_P^{\mathrm{ab}} : P \leq \overline{G}\}$$

where $\mathcal{G}^{\mathrm{ab}}$ denotes the abelianization of a group $\mathcal{G}$. (Brauer's Induction Theorem implies that this set satisfies the necessary condition $(*)$.)

We then write

$$\Delta_{\mathcal{O},\Sigma_G^*}^t : K_1(\Lambda_{\mathcal{O}}(G)[[t]]) \longrightarrow \prod_{P \leq \overline{G}} \Lambda_{\mathcal{O}}(U_P^{\mathrm{ab}}[[t]])^\times$$

for the group homomorphism which has each P-component equal to $q_{U_P^{\mathrm{ab}}}^t$.

The key step now is to combine the argument of Kakde in [49] together with a natural variant of the classical theory of integral group logarithms (as originally developed by R. Oliver [63, 64] and M. Taylor [79]) in the setting of the non-commutative power series ring $\Lambda_{\mathcal{O}}(G)[[t]]$ in order to prove the existence of an explicit subgroup $\Phi_{\mathcal{O}}^{G,t}$ of $\prod_{P \leq \overline{G}} \Lambda_{\mathcal{O}}(U_P^{\mathrm{ab}}[[t]])^\times$ which has the following two properties:

(P1) $\mathrm{im}(\Delta_{\mathcal{O},\Sigma_G^*}^t) \subseteq \Phi_{\mathcal{O}}^{G,t}$;

(P2) if $(\xi_P(t))_{P \leq \overline{G}}$ is any element of $\Phi_{\mathcal{O}}^{G,t}$ with the property that at each P the series $\xi_P(t)$ converges at $t = 1$ to give an element of $\Lambda_{\mathcal{O}}(U_P^{\mathrm{ab}})^\times$, then the element $(\xi_P(1))_{P \leq \overline{G}}$ belongs to the group $\Phi_{\mathcal{O}}^G$ that is defined by Kakde in [49].

Note that to prove the existence of an appropriate subgroup $\Phi_{\mathcal{O}}^{G,t}$ one follows the approach of [49] to describe it concretely in terms of congruence relations between the individual components of elements of $\prod_{P \leq \overline{G}} \Lambda_{\mathcal{O}}(U_P^{\mathrm{ab}}[[t]])^\times$. However, to prove Theorem 3.7 the properties (P1) and (P2) are all that one actually requires of $\Phi_{\mathcal{O}}^{G,t}$.

Indeed, if u^t is any element of $K_1(\Lambda_{\mathcal{O}}(G)[[t]])$ that satisfies the assumptions of Theorem 3.7, then the properties (P1) and (P2) combine to imply that the vector $(q_{U_P^{\mathrm{ab}}}^t(u^t)|_{t=1})_{P \leq \overline{G}}$ belongs to the group $\Phi_{\mathcal{O}}^G$. The main algebraic result of Kakde in [49] then implies the existence of an element u of $K_1(\Lambda_{\mathcal{O}}(G))$ with $q_{U_P^{\mathrm{ab}}}(u) = q_{U_P^{\mathrm{ab}}}^t(u^t)|_{t=1}$ for all $P \leq \overline{G}$, as required.

This completes our discussion of the proof of Theorem 3.7, and hence also of the proof of Theorem 2.1.

The reader will find a discussion of several interesting, and comparatively explicit, consequences of Theorem 2.1 in § 8.

4 Semistable abelian varieties over unramified extensions

There is an extensive existing literature concerning the Iwasawa theory of abelian varieties over compact ℓ-adic Lie extensions of global function fields of characteristic p, with $p \neq \ell$. In this regard we mention, in particular, the interesting work of Ellenberg [34], Sechi [70], Bandini and Longhi [8], and Pacheco [65].

In the next four sections, however, we shall focus on results that concern the Iwasawa theory of abelian varieties over compact p-adic Lie extensions of such function fields (of characteristic p).

In this first section we state (in Theorem 4.1) a natural analogue of Theorem 2.1 that was recently proved by Vauclair and the second-named author in [80] and concerns the Selmer modules of semistable abelian varieties over certain unramified compact p-adic Lie-extensions. We note that, in contrast to the very general result of Theorem 2.1, in this case Vauclair and the second-named author are able to construct a canonical element of the Whitehead group $K_1(\Lambda(G)_{S^*})$ that validates the relevant main conjecture.

4.1 Hypotheses and notations

We assume to be given a global function field K of characteristic p and field of constants $\mathbb{F}_q$. We then write K_{ar} for the constant $\mathbb{Z}_p$-extension of K.

We also fix an abelian variety A over K that has semistable reduction and write Z for the set of places at which A has bad reduction. We write C for the proper smooth curve over $\mathbb{F}_q$ which has field of functions K and $\mathcal{A}$ for the Néron model over C of A over K. We denote by $D^{\log}(A)$ the Dieudonné crystal defined by Kato and the second-named author [51] on the log-scheme $C^{\#}$ which has underlying scheme C and log-structure defined by the smooth divisor Z.

We recall that the p-Selmer group of A over any algebraic field extension F of K is defined to be

$$\mathrm{Sel}_{p^\infty}(A/F) := \ker\left(\mathrm{H}^1_{\mathrm{fl}}(F, A_{p^\infty}) \longrightarrow \prod_v \mathrm{H}^1_{\mathrm{fl}}(F_v, A) \right),$$

where A_{p^∞} is the p-divisible group associated to A, $\mathrm{H}^1_{\mathrm{fl}}(-, -)$ denotes flat cohomology and in the product v runs over all places of F.

In the sequel we shall study the Pontrjagin dual

$$X_p(A/F) := \mathrm{Hom}_{\mathbb{Z}_p}(\mathrm{Sel}_{p^\infty}(A/F), \mathbb{Q}_p/\mathbb{Z}_p)$$

of $\mathrm{Sel}_{p^\infty}(A/F)$. If F/K is Galois, then we endow $X_p(A/F)$ with the natural contragredient action of $\mathrm{Gal}(F/K)$.

In the rest of this section we always assume that A satisfies the following hypothesis which, for convenience, we refer to as '$\mu_A \sim 0$':

- A is isogeneous to an abelian variety A' such that $X_p(A'/K_{\mathrm{ar}})$ has trivial μ-invariant as a $\Lambda(\Gamma)$-module.

We fix a compact p-adic Lie extension K_∞ of K that is unramified everywhere and contains K_{ar} and then set $G := \mathrm{Gal}(K_\infty/K)$. This gives a canonical group extension

$$\{1\} \longrightarrow \mathrm{Gal}(K_\infty/K_{\mathrm{ar}}) \longrightarrow G \xrightarrow{\pi_G} \mathrm{Gal}(K_{\mathrm{ar}}/K) \longrightarrow \{1\}$$

of the form (1.2) in which π_G is the natural restriction map.

In the sequel we therefore set

$$H := \mathrm{Gal}(K_\infty/K_{\mathrm{ar}}) \quad \text{and} \quad \Gamma := \mathrm{Gal}(K_{\mathrm{ar}}/K)$$

and use the canonical Ore sets S and S^* in $\Lambda(G)$ that are discussed in §1 (with respect to this choice of data).

The extension K_∞/K corresponds to a pro-étale covering $C_\infty \to C$ (in the sense of §1.2). For convenience we also fix a cofinal system G_n of finite quotients of G, identify G with the inverse limit $\varprojlim_n G_n$ and for each n write $C_n \to C$ for the finite layer of C_∞ that corresponds to the group G_n.

4.2   Statement of the main results

In this section we write $\Gamma(-, -)$ for the global section functor for the flat (rather than étale) topology.

In [80] the authors first define a canonical element of the relative algebraic K_0-group $K_0(\Lambda(G), \Lambda(G)_{S^*})$ that plays the role of a (non-commutative) characteristic ideal in the formulation of the main conjecture.

To do this they define complexes of $\Lambda(G)$-modules by setting

- $N_\infty = N_{\infty,C,Z} := R\varprojlim_{n,k} R\Gamma^Z(C_n, \mathcal{A}_{p^k})$, where Γ^Z is the functor of global sections (for the flat topology) that vanish at Z;
- $L_\infty := R\varprojlim_n R\Gamma(C_n, \mathrm{Lie}(\mathcal{A})(-Z))$.

Under the hypothesis that $\mu_A \sim 0$, it can then be shown that these complexes belong to the category $D^p_{S^*}(\Lambda(G))$ that was introduced in § 1 (see, for example, Proposition 5.9).

One thereby obtains a well-defined element of $K_0(\Lambda(G), \Lambda(G)_{S^*})$ by setting

$$\chi(A/K_\infty) := \chi^{\mathrm{ref}}(N_\infty) + \chi^{\mathrm{ref}}(L_\infty),$$

where $\chi^{\mathrm{ref}}(-)$ is the Euler characteristic that was discussed in § 1.

The authors of [80] then associate a canonical p-adic L-function to the arithmetic of A over K.

To recall their construction we write $D^{\log,0}$ for the crystalline sheaf that is obtained as the kernel of the canonical map $D^{\log}(A) \to \mathrm{Lie}(\mathcal{A})$ that is constructed in [51] and note that one obtains well-defined objects of $D^p(\Lambda(G))$ by setting

- $P_\infty := R\varprojlim_n R\Gamma(C_n^\#/\mathbb{Z}_p, D^{\log}(A)(-Z))$,
- $I_\infty := R\varprojlim_n R\Gamma(C_n^\#/\mathbb{Z}_p, D^{\log,0}(A)(-Z))$.

For convenience, for each Y in $\{I, L, N, P\}$ we denote by Y_0 and Y_{ar} the complex Y_∞ when $K_\infty = K$ and $K_\infty = K_{\mathrm{ar}}$ respectively.

Now, under the stated hypotheses on A, there are morphisms in $D^p(\Lambda(G))$

$$\mathbf{1} \colon I_\infty \longrightarrow P_\infty$$
$$\varphi \colon I_\infty \longrightarrow P_\infty$$

which have the following property: at the base level and after inverting p, the map $\mathbf{1}$ coincides with the identity on $I_0[1/p] = P_0[1/p]$ and the map φ coincides with p^{-1} times the endomorphism on $P_0[1/p]$ that is induced by the Frobenius operator of $D^{\log}(A)$.

Next write $S_\infty = S_{\infty,C^\#}$ for a complex that lies in an exact triangle in $D^p(\Lambda(G))$ of the form

$$S_\infty \longrightarrow I_\infty \xrightarrow{\;\mathbf{1}-\varphi\;} P_\infty \longrightarrow S_\infty[1]$$

and then define the complexes S_0 and S_{ar} in just the same way as above. We recall in particular that the complex S_0 was first defined in [51] where it was referred to as the 'syntomic complex' of A over K.

It is clear that the scalar extension $\mathbf{1}_{S^*}$ of the morphism $\mathbf{1}$ is an isomorphism in $D^p(\Lambda(G)_{S^*})$ and this is in fact also true of the morphism $\mathbf{1} - \varphi$ (see Corollary

5.12). Given this observation, the article [80] defines a p-adic L function for A over K_∞ by setting

$$\mathcal{L}_{A/K_\infty} := \mathrm{Det}_{\Lambda(G)_{S^*}}((\mathbf{1} - \varphi)_{S^*} \circ (\mathbf{1}_{S^*})^{-1}),$$

where $\mathrm{Det}_{\Lambda(G)_{S^*}}$ is the universal determinant functor of [52] (see also [38]) which, in particular, induces a multiplicative map from the set of automorphisms of $D^{\mathrm{p}}(\Lambda(G)_{S^*})$ to the Whitehead group $K_1(\Lambda(G)_{S^*})$.

We can now collect together the main results of [80] in the statement of the following result.

This result verifies a natural analogue for the abelian variety A and the compact p-adic Lie extension K_∞ of K of the main conjecture formulated by Coates et al. in [29]. It is thus reasonable to expect that there should be a direct connection between this result and the appropriate special case of Theorem 2.1 (but, at this stage, we are not aware that the precise such connection has been worked out).

Theorem 4.1. *Fix a semistable abelian variety A over K and an unramified p-adic Lie extension K_∞ of K as above.*

(i) *Then for every Artin representation ρ of G whose values generate a (finite) totally ramified extension of $\mathbb{Q}_p$, one has*

$$\rho(\mathcal{L}_{A/K_\infty}) = L_Z(A, \rho, 1),$$

where the value of $\mathcal{L}_{A/K_\infty}$ at ρ is as defined in §1.

(ii) *In $K_0(\Lambda(G), \Lambda(G)_{S^*})$ one has an equality*

$$\partial_{G,S^*}(\mathcal{L}_{A/K_\infty}) = \chi(A/K_\infty).$$

For a brief discussion of the key ideas which underlie the proof of the interpolation formula given in Theorem 4.1(i), see §5.5. The proof of Theorem 4.1(ii) relies, on the other hand, mainly on the existence of certain natural exact triangles in $D^{\mathrm{p}}(\Lambda(G))$ of the form

$$I_\infty \xrightarrow{\ \mathbf{1}\ } P_\infty \longrightarrow L_\infty \longrightarrow I_\infty[1] \tag{4.1}$$

and

$$N_\infty \longrightarrow I_\infty \xrightarrow{\ \mathbf{1}-\varphi\ } P_\infty \longrightarrow N_\infty[1]. \tag{4.2}$$

In fact, whilst the existence of the triangle (4.1) follows essentially directly from the definitions of the complexes I_∞, P_∞ and L_∞, the existence of (4.2) is equivalent to the existence of an isomorphism in $D^{\mathrm{p}}(\Lambda(G))$ between S_∞ and N_∞ and proving that such an isomorphism exists is a difficult task. It was first proved by Kato and the second-named author in [51] when G is the trivial group. However, constructing such an isomorphism more generally requires a close re-examination of the constructions made in Chapters 4 and 5 of [51] in order to extend them into the necessary Iwasawa-theoretic setting, and much of [80] is taken up with this rather detailed work.

5   Proof of Theorem 4.1

In this section we explain in greater detail some of the key steps that are involved in the proof of Theorem 4.1.

As in § 4.2, we shall continue to write $\Gamma(-,-)$ for the global section functor for the flat topology.

5.1   The complexes N_0 and S_0

In both this subsection and in § 5.2, we shall discuss the construction of an isomorphism in $D(\Lambda(G))$ between the complexes S_∞ and N_∞ (or equivalently, of an exact triangle of the form (4.2)).

We start by explaining how to show that the complexes N_0 and S_0 are isomorphic by means of a careful dévissage argument.

We first discuss the appropriate dévissage for N_0.

At each place v in Z at which A has semistable reduction we use the natural 'Raynaud extensions'

$$0 \longrightarrow T_v \longrightarrow G_v \longrightarrow B_v \longrightarrow 0$$
$$0 \longrightarrow T_v^* \longrightarrow G_v^* \longrightarrow B_v^* \longrightarrow 0.$$

Here T_v and T_v^* are tori and B_v and B_v^* are abelian varieties over the ring of integers $\mathcal{O}_v$ of the completion K_v of K at v. At each such v we also write Γ_v for the Cartier dual of T_v^*.

Theorem 5.1. *There exists a natural exact triangle in $D^+(\mathbb{Z}_p)$ of the form*

$$N_{0,C,Z} \to N_{0,U,\emptyset} \oplus \prod_{v \in Z} N_{0,\mathcal{O}_v,k(v)} \to \prod_{v \in Z} N_{0,K_v,\emptyset} \to N_{0,C,Z}[1].$$

For each place v in Z there is also a natural isomorphism

$$N_{0,\mathcal{O}_v,k(v)} \simeq R\varprojlim_k R\Gamma^{k(v)}(\mathcal{O}_v, G_{v,p^k})$$

and an exact triangle in $D^+(\mathbb{Z}_p)$

$$N_{0,\mathcal{O}_v,k(v)} \to R\varprojlim_k R\Gamma(\mathcal{O}_v, \mathcal{A}_{v,p^k}) \to R\varprojlim_k R\Gamma(k(v), \overline{A}_{k(v),p^k}) \to N_{0,\mathcal{O}_v,k(v)}[1].$$

Note that, modulo proving the acyclicity of $R\Gamma^{k(v)}(\mathcal{O}_v, \Gamma_v \otimes \mathbb{Q}_p/\mathbb{Z}_p)$, the main point of the proof of Theorem 5.1 is the construction of a natural Mayer–Vietoris triangle.

We next discuss the dévissage that is necessary to deal with S_0.

Let H be a p-divisible group on $X \in \{C, \mathcal{O}_v\}$ that is endowed with the log structure induced by either $T = Z$ or $T = k(v)$ or even the trivial log-structure.

We then write $S_{0,X^\#,H}$ for a complex that is defined by means of an exact triangle in $D^+(\mathbb{Z}_p)$ of the form

$$S_{0,X^\#,H} \longrightarrow R\Gamma(X^\#/\mathbb{Z}_p, D^{\log,0}(H)(-T))$$

$$\xrightarrow{1-\varphi} R\Gamma(X^\#/\mathbb{Z}_p, D^{\log}(H)(-T)) \longrightarrow S_{0,X^\#,H}[1],$$

where $D^{\log}(H)$ denotes the inverse image of the crystalline Dieudonné crystal of H/X under the natural morphism of topoi $(X^\#/\mathbb{Z}_p)_{\text{Crys}} \to (X/\mathbb{Z}_p)_{\text{Crys}}$.

Theorem 5.2. *There exists a natural exact triangle in $D^+(\mathbb{Z}_p)$*

$$S_{0,C^\#} \longrightarrow S_{0,U} \oplus \prod_{v \in Z} S_{0,\mathcal{O}_v^\#} \longrightarrow \prod_{v \in Z} S_{0,K_v} \longrightarrow S_{0,C^\#}[1].$$

For each place v in Z there are also exact triangles in $D^+(\mathbb{Z}_p)$

$$S_{0,\mathcal{O}_v^\#} \longrightarrow S_{0,\mathcal{O}_v^\#,\Gamma_{v,p\infty}} \longrightarrow S_{0,\mathcal{O}_v^\#,G_{v,p\infty}}[1] \longrightarrow S_{0,C^\#}[1]$$

and for $H \in \{\Gamma_{v,p\infty}, G_{v,p\infty}\}$ also

$$S_{0,\mathcal{O}_v^\#,H} \longrightarrow S_{0,\mathcal{O}_v,H} \longrightarrow S_{0,k(v),\bar{H}} \longrightarrow S_{0,\mathcal{O}_v^\#,H}[1],$$

where $\bar{H}$ denotes the reduction of H modulo p.

The proof of this result relies on the construction of the Dieudonné crystal $D^{\log}(A)$, an explicit computation on de Rham complexes and a natural adaptation of the Hyodo–Kato construction to the setting of the diagram of schemes that underlies the associated Mayer–Vietoris triangle.

We can now deduce the existence of the required isomorphism between the complexes N_0 and S_0.

Theorem 5.3. *The complexes N_0 and S_0 are isomorphic.*

Proof. After taking account of the results of Theorems 5.1 and 5.2, this is reduced to proving the following claims:

- $S_{0,\mathcal{O}_v^\#,\Gamma_{v,p\infty}}$ is acyclic;
- $S_{0,\mathcal{O}_v^\#,G_{v,p\infty}}$ is naturally isomorphic to the complex $\ker(\mathcal{A}(\mathcal{O}_v) \to \mathcal{A}(k(v)))[1]$;
- If H is a p-divisible group on a scheme X having finite p-bases, then $S_{0,X,H}$ is naturally isomorphic to $R\varprojlim_k R\Gamma(X, H_{p^k})$.

The first two claims here are verified by means of an explicit computation of the syntomic complexes. The final claim is proved by applying the syntomic topology (as in [36] and [11]) to each of the p-divisible groups $G_{v,p\infty}/\mathcal{O}_v$, $\overline{G}_{v,p\infty}/k(v)$, $A_{v,p\infty}/K_v$ and $\mathcal{A}_{U,p\infty}/U$. $\square$

Remark 5.4. There is a rather delicate technical difficulty that we have, for simplicity, ignored in the above reasoning: the mapping cone construction is not functorial in the derived category. To overcome this difficulty the authors of [80] are forced to work with an appropriate derived category of diagrams.

5.2 Extending to the complexes S_∞ and N_∞

It is shown in [80] that the constructions described above are functorial both in the category of semistable abelian varieties as well as with respect to étale base change. This fact then allows the authors to construct objects $\mathcal{N}$ and $\mathcal{S}$ in $D^+(C_{\text{ét}})$ that are respectively associated to the cohomology theory of N and S.

For example, the complex $\mathcal{N}$ is defined to be $R \varprojlim_k R\epsilon_*^Z \mathcal{A}_{p^k}$ where the functor $\epsilon^Z : C_{\text{FL}} \to C_{\text{ét}}$ is defined as $\epsilon^Z(F) := \ker(\epsilon_* F \to \epsilon_* i_* i^* F)$ with i the closed immersion $Z \subset C$ and ϵ is the natural morphism $C_{\text{FL}} \to C_{\text{ét}}$ of change of topologies.

Having constructed $\mathcal{N}$ and $\mathcal{S}$ in this way, the authors of loc. cit. are then able to prove the following result.

Theorem 5.5. *There exists an isomorphism in $D^+(C_{\text{ét}})$ of the form*

$$\mathcal{N} \simeq \mathcal{S}$$

which induces, upon applying the global section functor over C, the isomorphism

$$N_0 \simeq S_0$$

that is constructed in Theorem 5.3.

Deducing the existence of an isomorphism of complexes $N_\infty \cong S_\infty$ from the above isomorphism $\mathcal{N} \simeq \mathcal{S}$ is now a rather formal process involving the theory of normic systems.

Recall that a 'normic system' for a group G is a collection (M_n), with each M_n a $\mathbb{Z}_p[G_n]$-module, together with transition maps $M_n \to M_{n+1}$ and $M_{n+1} \to M_n$ that satisfy certain natural compatibilities. In particular, one can define a 'normic section functor' from the derived category of étale sheaves of $\mathbb{Z}_p$-modules on C to the category of normic systems for G by associating to each abelian sheaf F of $\mathbb{Z}_p$-modules on the small étale site of C the collection $(F(C_n))$ together with its natural restriction and corestriction maps (in this regard note that C_n is an étale C-scheme because the extension K_n/K is unramified and that the action of G_n on C_n endows $F(C_n)$ with a natural structure as $\mathbb{Z}_p[G_n]$-module).

For the present purposes it is actually sufficient to consider the projective system of $\Lambda(G)$-modules that underlies this normic system and in this way one constructs a well-defined exact functor

$$D^+(C_{\text{ét}}) \longrightarrow D^+(\mathcal{C}_{\Lambda(G)}^{\mathbb{N}}) \longrightarrow D^+(\Lambda(G))$$

where $D^+(\mathcal{C}_{\Lambda(G)}^{\mathbb{N}})$ denotes the derived category of projective systems of $\Lambda(G)$-modules and the second arrow is the natural 'passage to inverse limit' functor.

The required isomorphism $N_\infty \cong S_\infty$ is then obtained by applying this functor to the isomorphism $\mathcal{N} \simeq \mathcal{S}$ in Theorem 5.5.

5.3   The complex N_{ar}

We now write k_∞ for the $\mathbb{Z}_p$-extension $\bigcup_{n\geq 0} k_n$ of $\mathbb{F}_q$ (and note that $K_{\mathrm{ar}} = Kk_\infty$).

Proposition 5.6.

  (i) *For any complex $Y \in \{I, P, L\}$, there is a canonical isomorphism in $D(\Lambda(\Gamma))$
  of the form $W(k_\infty) \otimes^{\mathbb{L}}_{\mathbb{Z}_p} Y_0 \simeq Y_{\mathrm{ar}}$.*

  (ii) *The induced morphism $\mathbf{1} - \varphi : I_{\mathrm{ar}} \to P_{\mathrm{ar}}$ is $W(k_\infty)$-σ-linear.*

Proof. Claim (i) results from the base change theorem for log crystalline cohomology and claim (ii) is straightforward to verify. □

We recall that $Q(\Gamma)$ denotes the total quotient ring of $\Lambda(\Gamma)$.

Corollary 5.7. *The complex $Q(\Gamma) \otimes^{\mathbb{L}}_{\Lambda(\Gamma)} N_{\mathrm{ar}}$ is acyclic.*

Proof. The result of Proposition 5.6(i) combines with the exact triangle (4.2) to imply that it is enough to prove that the morphism

$$Q(\Gamma) \otimes^{\mathbb{L}}_{\Lambda(\Gamma)} I_{\mathrm{ar}} \longrightarrow Q(\Gamma) \otimes^{\mathbb{L}}_{\Lambda(\Gamma)} P_{\mathrm{ar}}$$

in $D^{\mathrm{p}}(Q(\Gamma))$ that is induced by $\mathbf{1} - \varphi$ is an isomorphism. Since $\mathrm{Lie}(D)$ is a finite-dimensional $\mathbb{F}_p$-vector space, $\mathbf{1}$ gives an isomorphism $\mathbb{Q}_p \otimes^{\mathbb{L}}_{\mathbb{Z}_p} I_{\mathrm{ar}} \to \mathbb{Q}_p \otimes^{\mathbb{L}}_{\mathbb{Z}_p} P_{\mathrm{ar}}$. The required result can thus be proved by using the fact that crystalline cohomology over a proper log-smooth base is finite dimensional and applying the following easy fact from σ-linear algebra: if ψ is a linear endomorphism of a finite-dimensional $\mathbb{Q}_p$-vector space M then any map of the form $\mathrm{id} - \sigma \otimes \psi$ on $W(k_\infty) \otimes_{\mathbb{Z}_p} M$ is surjective and has kernel a finite-dimensional $\mathbb{Q}_p$-vector space. □

The connection between N_{ar} and the arithmetic invariants of the abelian variety A is described in the next result.

Proposition 5.8.

  (i) *The complex N_{ar} is concentrated in degrees 1, 2 and 3. There is also a canonical exact sequence*

$$0 \longrightarrow H^1(N_{\mathrm{ar}}) \longrightarrow \varprojlim_{k,n} \mathrm{Sel}_{p^k}(A/Kk_n) \longrightarrow \varprojlim_{k,n} \prod_{v \in Z} \mathcal{A}(k(v)k_n)/p^k \longrightarrow 0$$

  and canonical isomorphisms

$$H^2(N_{\mathrm{ar}}) \cong X_p(A^t/K_{\mathrm{ar}}) \quad \text{and} \quad H^3(N_{\mathrm{ar}}) \cong A^t_{p^\infty}(K_{\mathrm{ar}}).$$

  (ii) *The following conditions are equivalent:*

   (a) *The complex N_{ar} belongs to $D^{\mathrm{p}}(\mathbb{Z}_p)$.*

   (b) *The module $X_p(A^t/K_{\mathrm{ar}})$ is finitely generated over $\mathbb{Z}_p$.*

   (c) *The module $X_p(A/K_{\mathrm{ar}})$ is finitely generated over $\mathbb{Z}_p$.*

Proof. Claim (i) is proved by an explicit computation using the definition of N_{ar}. Claim (ii) then follows easily from the descriptions in claim (i) and the natural isogeny between A and A^t. $\qquad\square$

5.4 The complex N_∞

Proposition 5.9.

(i) *The descriptions of Proposition 5.8(i) remain valid if one replaces the complex N_{ar} by N_∞.*

(ii) *Under the hypothesis $\mu_A \sim 0$ the complex N_∞ belongs to $D^{\mathrm{p}}_{S*}(\Lambda(G))$.*

Proof. For each natural number m we choose a finite extension K'_m of K inside K_∞ such that $K_\infty = \bigcup_{m,i} K'_m k_i$. Then, since each of the modules $H^q(N_{K'_m k_i})$ is compact, the projective system $H^q(N_{K'_m k_i})$ is $\varprojlim$-acyclic and so there are natural isomorphisms

$$H^q(N_\infty) \cong \varprojlim_{m,i} H^q(N_{K'_m k_i}) \cong \varprojlim_{m} H^q(N_{K'_m k_\infty}).$$

The descriptions in claim (i) can therefore be obtained by passing to the inverse limit over m of the descriptions in Proposition 5.8(i) with N_{ar} replaced by $N_{K'_m k_\infty} = N_{K'_m,\mathrm{ar}}$

Next we combine the hypothesis $\mu_A \sim 0$ with the result of Proposition 5.8(ii) to deduce that the complex $\Lambda(\Gamma) \otimes^{\mathbb{L}}_{\Lambda(G)} N_\infty \simeq N_{\mathrm{ar}}$ belongs to $D^p(\mathbb{Z}_p)$.

Claim (ii) therefore follows from the easy algebraic fact that any complex M in $D^{\mathrm{p}}(\Lambda(G))$ belongs to $D^{\mathrm{p}}_S(\Lambda(G))$, and hence also to $D^{\mathrm{p}}_{S*}(\Lambda(G))$, if the complex $\Lambda(\Gamma) \otimes^{\mathbb{L}}_{\Lambda(G)} M$ belongs to $D^p(\mathbb{Z}_p)$. $\qquad\square$

Remark 5.10. Propositions 5.8(ii) and 5.9(ii) combine to imply that the module $X_p(A/K_\infty)$ is $\Lambda(G)$-torsion. We are aware of two situations in which this observation has been either strengthened or generalized.

(i) Let L be any $\mathbb{Z}_p$-power extension of K that is unramified outside a finite set Σ of places of K at each of which A has ordinary reduction, and set $G :=$ $\mathrm{Gal}(L/K)$. Then in this case Tan [74] has proved the following strengthening of Proposition 5.9(ii):

- $X_p(A/L)$ is finitely generated as a $\Lambda(G)$-module if and only if the group $H^1(G_v, A(L))$ is cofinitely generated as a $\mathbb{Z}_p$-module for all v in Σ, where G_v denotes the decomposition group of G at v.

- If $X_p(A/L)$ is finitely generated as a $\Lambda(G)$-module and $K_{\mathrm{ar}} \subseteq L$, then $X_p(A/L)$ is a torsion $\Lambda(G)$-module.

- If $X_p(A/L)$ is a torsion $\Lambda(G)$-module, then there exists a finite set T of proper intermediate $\mathbb{Z}_p$-power extensions of L/K with the following property: for each $\mathbb{Z}_p$-power extension M of K inside L the

$\Lambda(\mathrm{Gal}(M/K))$-module $X_p(A/M)$ is torsion unless M is a subfield of some field in T.

(ii) Let now L be any compact p-adic Lie extension of K which contains K_{ar}, is unramified outside a finite set of places Σ and is such that $G := \mathrm{Gal}(L/K)$ contains no element of order p. Then in [62] Ochiai and the second-named author prove that $X_p(A/L)$ is a finitely generated $\Lambda(\mathrm{Gal}(L/K_{\mathrm{ar}}))$-module for any abelian variety A over K which satisfies both of the following conditions:

- A has good reduction at all places outside Σ and ordinary reduction at all places in Σ.

- The (classical) μ-invariant of the module $X_p(A/K_{\mathrm{ar}})$ vanishes.

Note also that this latter μ-invariant always vanishes if A is constant ordinary.

Finally we note that the recent preprint [10] of Bandini and Valentino uses a natural generalization of Mazur's Control Theorem to prove similar results concerning the structure of the module $X_p(A/L)$.

Remark 5.11. If one does not assume that L contains K_{ar}, then the $\Lambda(G)$-module $X_p(A/L)$ need not be torsion. For an explicit example (taken from the Appendix of [55]), fix a quadratic extension K/k and a non-isotrivial semistable elliptic curve A over k that has analytic rank zero and split multiplicative reduction at a given place v_0. Then $X_p(A/L)$ fails to be a torsion $\Lambda(\mathrm{Gal}(L/K))$-module whenever L is a $\mathbb{Z}_p$-extension of K that is dihedral over k, totally ramified above v_0 and unramified at all other places.

5.5   Main Conjecture for A over K_∞

We first record an easy consequence of Proposition 5.9.

Corollary 5.12. *Under the hypothesis that $\mu_A \sim 0$, the morphisms in $D^p(\Lambda(G)_{S^*})$*

$$\mathbf{1}_{S^*} : (I_\infty)_{S^*} \longrightarrow (P_\infty)_{S^*} \quad and \quad (\mathbf{1} - \varphi)_{S^*} : (I_\infty)_{S^*} \longrightarrow (P_\infty)_{S^*}$$

that are induced by the respective scalar extensions of $\mathbf{1}$ and $\mathbf{1} - \varphi$ are isomorphisms.

Proof. For the morphism $\mathbf{1}_{S^*}$ this claim follows directly from the scalar extension of the triangle (4.1) and the fact that L_∞ is annihilated by p. For the morphism $(\mathbf{1} - \varphi)_{S^*}$, the claim is a direct consequence of Proposition 5.9(ii) and scalar extension of the triangle (4.2). $\qquad\square$

This result leads naturally to the definition of the "p-adic L-function" which occurs in the statement of Theorem 4.1.

Definition 5.13. Under the hypothesis $\mu_A \sim 0$, the p-adic L-function $\mathcal{L}_{A/K_\infty}$ for A over K_∞ is the element of $K_1(\Lambda(G)_{S^*})$ that is defined by setting

$$\mathcal{L}_{A/K_\infty} := \mathrm{Det}_{\Lambda(G)_{S^*}}((\mathbf{1} - \varphi)_{S^*} \circ (\mathbf{1}_{S^*})^{-1}).$$

Given this definition of the p-adic L-function, the interpolation property stated in Theorem 4.1(i) is proved by comparing log-crystalline cohomology to rigid cohomology and then using the base change theorems and Künneth formula in rigid cohomology together with a description of the Hasse–Weil L function of A twisted by an Artin representation in terms of rigid cohomology. Note that the additional hypothesis that is made on the Artin representation (that is, that its values should generate a totally ramified extension of $\mathbb{Q}_p$) is used to identify the representation with a unit F-crystal on the curve in order to obtain an explicit p-adic description of the twisted Hasse–Weil L-function.

In view of Corollary 5.12, and the explicit definition of $\mathcal{L}_{A/K_\infty}$ given above, the equality of Theorem 4.1(ii) is then obtained by applying the following (straightforward) algebraic observation to the exact triangles of (4.1) and (4.2) (so that one has $R = \Lambda(G)$ and $\Sigma = S^*$).

Lemma 5.14. *Let R be an associative unital left noetherian ring and Σ a left Ore set of non-zero divisors of R. Let*

$$C \xrightarrow{a} C' \longrightarrow C(a) \longrightarrow C[1]$$

and

$$C \xrightarrow{b} C' \longrightarrow C(b) \longrightarrow C[1]$$

be exact triangles in $D^{\mathrm{p}}(R)$ which have the property that the complexes $R_\Sigma \otimes_R C(a)$ and $R_\Sigma \otimes_R C(b)$ are acyclic.

Then in $K_0(R, R_\Sigma)$ one has an equality

$$\partial_{R,R_\Sigma}(\mathrm{Det}_{R_\Sigma}(a_\Sigma \circ b_\Sigma^{-1})) = \chi_{R,R_\Sigma}^{\mathrm{ref}}(C(a)) - \chi_{R,R_\Sigma}^{\mathrm{ref}}(C(b))$$

with $a_\Sigma := R_\Sigma \otimes_R a$ and $b_\Sigma := R_\Sigma \otimes_R b$.

6 Constant ordinary abelian varieties over abelian extensions

In this section we discuss the main results, and key ideas, of the recent article [55] of Lai, Longhi, Tan and the second-named author concerning the Iwasawa theory of Selmer modules of constant ordinary abelian varieties over certain $\mathbb{Z}_p$-power extensions of global function fields.

Throughout, we fix a function field in one variable K with field of constants a finite field $\mathbb{F}_q$ of characteristic $p > 0$ and an abelian variety A over K of dimension g. We always assume, unless explicitly stated otherwise, that the following hypothesis is satisfied.

- A/K is constant ordinary and L is an infinite $\mathbb{Z}_p$-power extension of K that is unramified outside a finite set of places S and contains K_{ar}.

For this fixed data we set $G := \mathrm{Gal}(L/K)$.

The involution $x \mapsto x^{\#}$ that we defined previously on $\mathbb{Z}[\mathcal{G}]$ for each finite quotient $\mathcal{G}$ of G extends by continuity to give an involution $\iota^{\#}$ of the ring $\Lambda(G)$. For each f in $\Lambda(G)$ we set $f^{\#} := \iota^{\#}(f)$ and for each $\Lambda(G)$-module M we also write $M^{\#}$ for the module $\Lambda(G) \otimes_{\Lambda(G), \iota^{\#}} M$ (this makes sense since G is abelian).

6.1   The p-adic L-function of A/L

We next recall the definition of the p-adic L-function that the authors of [55] associated to the abelian variety A over L. This function is derived in a fairly straightforward way from the Stickelberger element that is naturally related to the Iwasawa theory of divisor class groups.

Let T be a finite non-empty set of places of K that is disjoint from S and set $\delta_S := 0$ if $S \neq \emptyset$ with $\delta_{\emptyset} := 1$.

Then, with Fr_q denoting the arithmetic Frobenius in $\mathrm{Gal}(K_{\mathrm{ar}}/K)$, one can show that the 'Stickelberger function'

$$\theta_{L,S,T}(s) := (1 - \mathrm{Fr}_q \cdot q^{-s})^{\delta_S} \prod_{v \in T} (1 - [v] \cdot q_v^{1-s}) \prod_{v \notin S} (1 - [v] \cdot q_v^{-s})^{-1}$$

gives a well-defined element of $\Lambda(G)[[q^{-s}]]$, where we write $[v]$ for the Frobenius element in G that is associated to any place v of K that is unramified in L/K. One finds that the coefficients of this power series tend to zero in $\Lambda(G)$ and so the 'Stickelberger element'

$$\theta_{L,S} := \theta_{L,S,T}(0) \in \Lambda(G)$$

is well defined. One also checks easily that this element is indeed independent of the choice of the auxiliary set T, at least up to multiplication by units of $\Lambda(G)$.

For any finite extension F of K inside L we write $\mathfrak{W}_F$ for the p-adic completion of its group of divisor classes. We then set

$$\mathfrak{W}_L := \varprojlim_F \mathfrak{W}_F,$$

with the limit taken with respect to the natural norm maps.

The following important result was first proved by Crew in [30] and a simplified proof has recently been given by Lai, Tan and the first-named author in [22] (and, in this regard, see also Kueh, Lai and Tan [54]).

Theorem 6.1 (Main Conjecture for $\mathfrak{W}_L$). *One has* $\mathrm{ch}_{\Lambda(G)}(\mathfrak{W}_L) = \Lambda(G) \cdot \theta_{L,S}^{\#}$.

We write $\{\alpha_i : 1 \leq i \leq g\}$ for the eigenvalues of the Frobenius endomorphism acting on the module $A[p^{\infty}]$. For each index i we then define $\theta_{A,L,S,T,i}^{+}(s)$ to be the series that is obtained from the Stickelberger function $\theta_{L,S,T}(s)^{\#}$ by making the change of variable

$$q^{-s} \longmapsto \alpha_i^{-1} q^{-s}. \tag{6.1}$$

For each place v we also set $\alpha_{i,v} := \alpha_i^{\deg(v)}$ and define

$$\theta^+_{A,L,S,i}(s) := \theta^+_{A,L,S,T,i}(s) \cdot \prod_{v \in T} (1 - [v]^{-1} \cdot \alpha_{i,v}^{-1} \cdot q_v^{1-s})^{-1}, \tag{6.2}$$

which we consider as an element of $\Lambda(G)[[q^{-s}]]$.

One then defines a series

$$\theta^+_{A,L,S}(s) := \prod_{i=1}^{g} \theta^+_{A,L,S,i}(s). \tag{6.3}$$

In particular, since each series $\theta^+_{A,L,S,i}(s)$ converges at $s = 0$, one can set

$$\theta^+_{A,L,S} := \theta^+_{A,L,S}(0) \in \Lambda(G)$$

and then finally define $\theta_{A,L,S}$ to be the product

$$\theta_{A,L,S} := \theta^+_{A,L,S}(\theta^+_{A,L,S})^{\#}.$$

6.2 The interpolation formula

We now describe the key interpolation property of the element $\theta_{A,L,S}$ defined above. To do this, we fix a continuous character $\omega \colon G \longrightarrow \mu_{p^\infty}$ and write S_ω for the subset of S comprising places at which ω ramifies. We then set

$$\Xi_S(\omega, s) := \prod_{i=1}^{g} \prod_{v \in S - S_\omega} \frac{1 - \omega([v])^{-1} \alpha_{i,v}^{-1} q_v^s}{1 - \omega([v]) \beta_{i,v}^{-1} q_v^{-s}}$$

with $\beta_{i,v} := q_v \alpha_{i,v}^{-1}$ for each index i. We also set

$$\Delta_S(s) := \prod_{i=1}^{g} (1 - \mathrm{Frob}_q^{-1} \alpha_i^{-1} q^s)^{\delta_S} (1 - \mathrm{Frob}_q \alpha_i^{-1} q^{-s})^{\delta_S}.$$

We denote by κ the genus of the smooth projective curve C over $\mathbb{F}_q$ that corresponds to K and by d_ω the degree of the conductor of ω. Fix an additive character $\Psi \colon \mathbf{A}_K / K \to \mu_{p^\infty}$ on the adele classes of K and a differential idele $b = (b_v)$ attached to Ψ and then, for every place v, let α_v be the self-dual Haar measure on K_v with reference to Ψ_v. One then sets

$$\tau(\omega) := \begin{cases} \omega(\mathrm{Frob}_q)^{2-2\kappa} & \text{if } \omega \text{ has trivial conductor,} \\ \dfrac{1}{\omega(b)} \displaystyle\prod_{v \in S_\omega} \dfrac{1}{|b_v|_v^{1/2}} \int_{O_v^\times} \omega_v(x) \Psi_v(b_v^{-1} x) \, d\alpha_v(x) & \text{otherwise,} \end{cases}$$

where O_v is the ring of integers of K_v.

The following result explains precisely how the element $\theta_{A,L,S}$ interpolates the values of the L-functions of twists of A.

Theorem 6.2. *For any continuous character* $\omega\colon G \longrightarrow \mu_{p^\infty}$ *as above one has an equality*

$$\omega(\theta_{A,L,S}) = \tau(\omega)^g \left(q^{g/2} \prod_{i=1}^g \alpha_i^{-1} \right)^{2\kappa-2+d_\omega} \omega(\Delta_S(0))\,\Xi_S(\omega,0)\,L_S(A,\omega,1),$$

where $L_S(A,\omega,1)$ *denotes the value at* $s = 1$ *of the* S-*truncated Hasse–Weil* L *function of* A/K *twisted by* ω.

The proof of this interpolation property is actually straightforward, the key point (of course) being that A is constant. Indeed, since the Stickelberger element interpolates, essentially by its very construction, all twists of the S-truncated Dedekind zeta function of K by Dirichlet characters that are unramified outside of S, one has both

$$\omega\big(\theta_{A,L,S,i}^+(0)\big) = (1 - \omega^{-1}(\mathrm{Frob}_q)\alpha_i^{-1})^{\delta_S} L_S(\omega^{-1},s_i)$$

and

$$\omega\big(\theta_{A,L,S,i}^+(0)^{\#}\big) = (1 - \omega(\mathrm{Frob}_q)\alpha_i^{-1})^{\delta_S} L_S(\omega,s_i),$$

where the complex number s_i is such that $\alpha_i = q^{s_i}$.

On the other hand, the fact that A is constant implies that the Frobenius eigenvalues at any place v are equal to $\{\alpha_{i,v}, \beta_{i,v} : 1 \le i \le g\}$ and so one has

$$L_S(A,\omega,1) = \prod_{i=1}^g L_S(\omega,s_i)L_S(\omega,1-s_i),$$

since $\beta_{i,v} = q_v^{1-s_i}$ for each index i.

Given these facts, Theorem 6.2 follows directly from the functional equation that relates $L_S(\omega^{-1},s_i)$ and $L_S(\omega,1-s_i)$.

6.3   The Main Conjecture for A over L

By applying the results of Ochiai and the second-named author that were recalled in Remark 5.10(ii) one can show that the Pontrjagin dual $X_p(A/F)$ of the p-Selmer group $\mathrm{Sel}_{p^\infty}(A/L)$ of A over L is both finitely generated and torsion over $\Lambda(G)$. We may therefore in the sequel write $\mathrm{ch}_{\Lambda(G)}(X_p(A/L))$ for the (classical) characteristic ideal of this module.

We can now state the main result of Lai et al. in [55]. (Note that it is also shown in loc. cit. that if $L = K_{\mathrm{ar}}$, then this result is equivalent to the relevant special case of Theorem 4.1(ii).)

Theorem 6.3 (Main Conjecture for A over L). *In* $\Lambda(G)$ *one has*

$$\mathrm{ch}_\Lambda(X_p(A/L)) = \Lambda(G) \cdot \theta_{A,L,S}.$$

A very brief summary of the proof of Theorem 6.3 is as follows. The authors of [55] first make a canonical decomposition of $X_p(A/L)$ into a 'Frobenius part' and a 'Verschiebung part'. The Frobenius part can then be related to the result of Theorem 6.1 concerning divisor class groups and hence linked to a Stickelberger element that can be shown to divide $\theta_{A,L,S}$. To deduce the equality of Theorem 6.3 from this divisibility relation one then uses a functional equation between the Frobenius and Verschiebung parts of $X_p(A/L)$. (This functional equation is in fact a specialization of a general result that is valid for any abelian variety over a global field.)

An interested reader will find more details of this argument in the next section.

7   Proof of Theorem 6.3

We now discuss the main ideas that lie behind the proof of Theorem 6.3 that is given in [55].

7.1   Frobenius-Verschiebung decomposition

Let A/K be a constant ordinary abelian variety and N/K be any Galois extension. We denote F_q the Frobenius endomorphism and V_q the Verschiebung of A and recall that the composition of these two maps is the endomorphism given by multiplication by q. For any finitely generated $\mathrm{End}(A) \otimes \mathbb{Z}_p$-module M we define the 'Frobenius part' and 'Verschiebung part' of M by setting

$$F(M) = \bigcap_{m \geq 0} F_q^m(M),$$

$$V(M) = \bigcap_{m \geq 0} F_q^m(M).$$

In particular, the decomposition

$$\mathrm{End}(A) \otimes \mathbb{Z}_p = (F(\mathrm{End}(A)) \otimes \mathbb{Z}_p) \oplus (V(\mathrm{End}(A)) \otimes \mathbb{Z}_p)$$

induces a natural direct sum decomposition of M as

$$M = F(M) \oplus V(M). \tag{7.1}$$

By applying this decomposition with M equal to each of $A(N) \otimes \mathbb{Q}_p/\mathbb{Z}_p$, the Selmer module $\mathrm{Sel}_{p^\infty}(A/N)$ and the p-primary part of the Tate–Shafarevich group $\mathrm{III}_{p^\infty}(A/N)$ of A/N, one obtains short exact sequences

$$0 \longrightarrow F(A(N)) \otimes \mathbb{Q}_p/\mathbb{Z}_p \longrightarrow F(\mathrm{Sel}_{p^\infty}(A/N)) \longrightarrow F(\mathrm{III}_{p^\infty}(A/N)) \longrightarrow 0$$

and

$$0 \longrightarrow V(A(N)) \otimes \mathbb{Q}_p/\mathbb{Z}_p \longrightarrow V(\mathrm{Sel}_{p^\infty}(A/N)) \longrightarrow V(\mathrm{III}_{p^\infty}(A/N)) \longrightarrow 0.$$

The decomposition (7.1) also respects the Néron–Tate height pairing and the Cassels–Tate pairing in the sense that they induce a natural duality between the Frobenius and Verschiebung parts of the Mordell–Weil groups and Tate–Shafarevich groups respectively.

7.2   Selmer modules and class groups

We relate the module $\mathfrak{W}_L$ defined in §6.1 to the Pontrjagin dual $F(X_p(A/L))$ of the Frobenius part of $\mathrm{Sel}_{p^\infty}(A/L)$.

This connection relies upon the following general result, which allows one to pass between flat and étale cohomology groups.

Lemma 7.1. *Let B/K be an abelian variety and for any non-negative integer m let $F_q^{(m)}: B \to B^{(q^m)}$ denote the relative Frobenius that is induced by $x \mapsto x^{q^m}$. Then for any natural number n there exists a non-negative integer m for which $F_{q,*}^{(m)}(\mathrm{Sel}_{p^n}(B/K)) \subset \mathrm{H}^1_{\mathrm{fl}}(K, (B^{\text{ét}}_{p^n})^{(q^m)})$, where $B^{\text{ét}}_{p^n}$ is the maximal étale quotient of the group scheme B_{p^n}.*

In our setting, this result implies that for any finite extension F of K that contains $\mathbb{F}_q(A[p^n])K$ there is an identification of the form

$$F_q^m(\mathrm{Sel}_{p^n}(A/F)) = \mathrm{Hom}(\mathfrak{W}_F, A[p^n]).$$

By taking inductive limits of this identification over m and n and then dualizing one obtains a canonical isomorphism of $\Lambda(G)$-modules of the form

$$F(X_p(A/L)) \cong \mathfrak{W}_L \otimes_{\mathbb{Z}_p} \mathrm{Hom}(A(L)[p^\infty], \mathbb{Q}_p/\mathbb{Z}_p). \tag{7.2}$$

By combining this isomorphism with the result of Theorem 6.1 one can then prove the following important result.

Theorem 7.2. *One has $\mathrm{ch}_{\Lambda(G)}(F(X_p(A/L))) = \Lambda(G) \cdot \theta^+_{A,L,S}$.*

7.3   Functional equations

We next recall a very general source of arithmetic 'functional equations'.

To do this, we first recall some basic facts about finitely generated torsion $\Lambda(G)$-modules. By the general theory of modules over a Krull domain, one knows that for each such module M there exists a finite set of irreducible elements $\{\xi_i : 1 \le i \le m\}$ of $\Lambda(G)$ and a 'pseudo-isomorphism'

$$\Phi: \bigoplus_{i=1}^m \Lambda(G)/\Lambda(G)\xi_i^{r_i} \longrightarrow M.$$

This means that Φ is a homomorphism of $\Lambda(G)$-modules for which both $\ker(\Phi)$ and $\operatorname{coker}(\Phi)$ are 'pseudo-null' (that is, that they vanish after localization at each height one prime ideal of $\Lambda(G)$). We use the fact that the associated 'elementary module'

$$[M] := \bigoplus_{i=1}^{m} \Lambda(G)/\Lambda(G)\xi_i^{r_i}$$

depends only upon (the isomorphism class of) M.

We now consider a collection of data

$$\mathfrak{A} = \left\{ \mathfrak{a}_n, \mathfrak{b}_n, \langle \ , \ \rangle_n, \mathfrak{r}_m^n, \mathfrak{k}_m^n \ \mid \ n, m \in \mathbb{N}, \ n \geq m \right\}$$

of the following sort:

- $\mathfrak{a}_n, \mathfrak{b}_n$ are finite abelian p-groups, each with an action of $\Lambda(G)$ that factors through the natural projection homomorphism $\Lambda(G) \to \mathbb{Z}_p[G_n]$ with $G_n \cong (\mathbb{Z}/p^n)^d$.

- For each pair of integers n and m with $n \geq m$ there are homomorphisms of $\mathbb{Z}_p[G]$-modules

$$\mathfrak{r}_m^n \colon \mathfrak{a}_m \times \mathfrak{b}_m \longrightarrow \mathfrak{a}_n \times \mathfrak{b}_n,$$

$$\mathfrak{k}_m^n \colon \mathfrak{a}_n \times \mathfrak{b}_n \longrightarrow \mathfrak{a}_m \times \mathfrak{b}_m,$$

which satisfy $\mathfrak{r}_m^n(\mathfrak{a}_m) \subset \mathfrak{a}_n$, $\mathfrak{r}_m^n(\mathfrak{b}_m) \subset \mathfrak{b}_n$, $\mathfrak{k}_m^n(\mathfrak{a}_n) \subset \mathfrak{a}_m$, $\mathfrak{k}_m^n(\mathfrak{b}_n) \subset \mathfrak{b}_m$ and $\mathfrak{r}_n^n = \mathfrak{k}_n^n = \mathrm{id}$. In addition, the sets $\{\mathfrak{a}_n \times \mathfrak{b}_n, \mathfrak{r}_m^n\}_n$ and $\{\mathfrak{a}_n \times \mathfrak{b}_n, \mathfrak{k}_m^n\}_n$ form an inductive and projective system respectively.

- For each pair of integers n and m with $n \geq m$ one has both

$$\mathfrak{r}_m^n \circ \mathfrak{k}_m^n = \mathrm{N}_{G_n/G_m} \colon \mathfrak{a}_n \times \mathfrak{b}_n \longrightarrow \mathfrak{a}_n \times \mathfrak{b}_n$$

and

$$\mathfrak{k}_m^n \circ \mathfrak{r}_m^n = p^{d(n-m)} \cdot \mathrm{id} \colon \mathfrak{a}_m \times \mathfrak{b}_m \longrightarrow \mathfrak{a}_m \times \mathfrak{b}_m,$$

where N_{G_n/G_m} denotes the norm map $\sum_{\sigma \in \ker(G_n \to G_m)} \sigma$ associated with the natural projection $G_n \twoheadrightarrow G_m$.

- For each natural number n,

$$\langle \ , \ \rangle_n \colon \mathfrak{a}_n \times \mathfrak{b}_n \longrightarrow \mathbb{Q}_p/\mathbb{Z}_p$$

is a perfect pairing (and hence $\mathfrak{a}_n$ and $\mathfrak{b}_n$ are dual p-groups) that respects both G-action and the morphisms $\mathfrak{r}_m^n$ and $\mathfrak{k}_m^n$ in the sense that one has

$$\langle \gamma \cdot a, \gamma \cdot b \rangle_n = \langle a, b \rangle_n \quad \text{for all } \gamma \in G,$$
$$\langle a, \mathfrak{r}_m^n(b) \rangle_n = \langle \mathfrak{k}_m^n(a), b \rangle_m$$

and

$$\langle \mathfrak{r}_m^n(a), b \rangle_n = \langle a, \mathfrak{k}_m^n(b) \rangle_m.$$

Definition 7.3. In [55] a collection of data $\mathfrak{A}$ as above is said to be a 'Γ-system' if both of the limits $\mathfrak{a} := \varprojlim_n \mathfrak{a}_n$ and $\mathfrak{b} := \varprojlim_n \mathfrak{b}_n$ are finitely generated torsion $\Lambda(G)$-modules.

   A Γ-system $\mathfrak{A}$ is said to be 'strongly controlled' if for every natural number n the $\Lambda(G)$-module $\mathfrak{a}_n^0 \times \mathfrak{b}_n^0 := \bigcup_{n' \geq n} \ker(\mathfrak{r}_n^{n'})$ is pseudo-null.

   A Γ-system $\mathfrak{A}$ is called 'pseudo-controlled' if the limit $\mathfrak{a}^0 \times \mathfrak{b}^0 := \varprojlim_n \mathfrak{a}_n^0 \times \mathfrak{b}_n^0$ is a pseudo-null $\Lambda(G)$-module.

   In order to state the main property of such Γ-systems we further recall that in [55] an element f of $\Lambda(G)$ is said to be 'simple' if there exists an element ξ of $\boldsymbol{\mu}_{p^\infty}$ and an element γ of $G \setminus G^p$ with $f = \prod_{\sigma \in \mathrm{Gal}(\mathbb{Q}_p(\xi)/\mathbb{Q}_p)}(\gamma - \sigma(\xi))$.

Theorem 7.4. *Let $\mathfrak{A} = \{\mathfrak{a}_n, \mathfrak{b}_n, \langle\ ,\ \rangle_n, \mathfrak{r}_m^n, \mathfrak{t}_m^n \mid n, m \in \mathbb{N},\ n \geq m\}$ be a pseudo-controlled Γ-system for which the associated module $\mathfrak{b}$ is annihilated by an element of $\Lambda(G)$ that is not divisible by any simple element.*
   Then there is a pseudo-isomorphism of $\Lambda(G)$-modules of the form $\mathfrak{a}^\# \sim \mathfrak{b}$.

Proof. It suffices, by symmetry (between the roles of $\mathfrak{a}$ and $\mathfrak{b}$), to construct a homomorphism of $\Lambda(G)$-modules $\mathfrak{a}^\# \to \mathfrak{b}$ the kernel of which is pseudo-null.

   The authors of [55] show that this is true for a composite homomorphism of the form

$$\mathfrak{a}^\# \longrightarrow \mathrm{Hom}_{\Lambda(G)}\big(\mathfrak{b}, (1/y)\Lambda(G)/\Lambda(G)\big) = \mathrm{Hom}_{\Lambda(G)}\big(\mathfrak{b}, Q(G)/\Lambda(G)\big) \longrightarrow \mathfrak{b}.$$

Here y is a choice of element of $\Lambda(G)$ that annihilates $\mathfrak{b}$ and is not divisible by any simple element and the first arrow is induced by the given duality between $\mathfrak{a}_n^\#$ and $\mathfrak{b}_n$ for all n and so has kernel equal to the pseudo-null module $(\mathfrak{a}^0)^\#$; the second arrow is the pseudo-isomorphism that is induced by the obvious identification

$$\mathrm{Hom}_{\Lambda(G)}\big(\Lambda(G)/z\Lambda(G), Q(G)/\Lambda(G)\big) = z^{-1}\Lambda(G)/\Lambda(G) \xrightarrow{\times z} \Lambda(G)/\Lambda(G)z$$

that one has for any non-zero element z of $\Lambda(G)$. $\qquad\square$

   For each finitely generated torsion $\Lambda(G)$-module M we now write

$$\mathfrak{S}(M) := \big\{\Lambda(G) \cdot f \mid \mathrm{ch}_{\Lambda(G)}(M) \subset \Lambda(G) \cdot f \text{ and } f \text{ is simple}\big\}$$

for the set of principal ideals of $\Lambda(G)$ which both divide $\mathrm{ch}_{\Lambda(G)}(M)$ and are generated by a simple element. The authors of [55] define the 'simple part' of M to be

$$[M]_{\mathrm{si}} := \bigoplus_{\Lambda(G)\xi_i \in \mathfrak{S}(M)} \Lambda(G)/\Lambda(G)\xi_i^{r_i} \subset [M]$$

and also write $[M]_{\mathrm{ns}}$ for its complement (the 'non-simple part' of M), so that there is a direct sum decomposition

$$[M] = [M]_{\mathrm{si}} \oplus [M]_{\mathrm{ns}}.$$

   We now recall the following important (and very general) result of Tan [74].

Theorem 7.5. *Let A be an abelian variety that is defined over a global field K. Let L be an infinite $\mathbb{Z}_p$-power extension of K that is unramified outside a finite set of places at each of which A has ordinary reduction and is also such that the $\Lambda(G)$-module $X_p(A/L)$ is torsion.*

For each finite extension F of K inside L write $Z_p(A/F)$ the Pontrjagin dual of the quotient of $\text{III}_{p^\infty}(A/F)$ by its maximal divisible subgroup and $Z_p(A/L)$ for the projective limit $\varprojlim_F Z_p(A/F)$ over all such subfields of L.

Then one has $[X_p(A/L)]_{\mathrm{ns}} = [Z_p(A/L)]_{\mathrm{ns}}$.

By combining this result with Theorem 7.4 one can then prove the following key 'functional equation' for Selmer modules.

Theorem 7.6. *Let A be an abelian variety that is defined over a global field K and suppose that the $\Lambda(G)$-module $X_p(A/L)$ is torsion. Then there are pseudo-isomorphisms of $\Lambda(G)$-modules*

$$X_p(A/L) \sim X_p(A/L)^\# \sim X_p(A^t/L),$$

where A^t is the dual abelian variety of A.

Note that it is clearly enough to prove this result for both of the modules $[X_p(A/L)]_{\mathrm{si}}$ and $[X_p(A/L)]_{\mathrm{ns}}$. In addition, the proof for the simple part $[X_p(A/L)]_{\mathrm{si}}$ is comparatively straightforward because $[M]_{\mathrm{si}} = [M]_{\mathrm{si}}^\#$ for any finitely generated torsion $\Lambda(G)$-module M.

However, when proving Theorem 7.6 for the module $[X_p(A/L)]_{\mathrm{ns}}$ one first uses the result of Theorem 7.5 to replace $X_p(A/L)$ by $Z_p(A/L)$ and then the Cassels–Tate pairings on each of the groups $Z_p(A/F)$ to obtain from $Z_p(A/L)$ a Γ-system to which one can apply Theorem 7.4.

7.4 Completion of the proof

We are now ready to sketch the completion of the proof of Theorem 6.1.

From the Frobenius-Verschiebung decomposition of $X_p(A/L)$ and the multiplicativity of characteristic ideals, one immediately obtains equalities

$$\mathrm{ch}_{\Lambda(G)}(X_p(A/L)) = \mathrm{ch}_{\Lambda(G)}(F(X_p(A/L)) \oplus V(X_p(A/L)))$$

$$= \mathrm{ch}_{\Lambda(G)}(F(X_p(A/L)))\mathrm{ch}_{\Lambda(G)}(V(X_p(A/L))).$$

Further, under the hypotheses discussed at the beginning of §6, Theorem 7.6 specializes to give pseudo-isomorphisms of $\Lambda(G)$-modules of the form

$$F(X_p(A^t/L)) \sim F(X_p(A^t/L))^\# \sim V(X_p(A/L)).$$

Substituting both this fact and the result of Theorem 7.2 into the last displayed equality one deduces that

$$\mathrm{ch}_{\Lambda(G)}(X_p(A/L)) = \mathrm{ch}_{\Lambda(G)}(F(X_p(A^t/L)^\#)) \cdot \theta^+_{A,L,S}.$$

Now the abelian varieties A and A^t are isogenous, whilst the isomorphism (7.2) implies that the module $F(X_p(A/L))$ has no p-torsion, and these facts combine to imply that

$$\mathrm{ch}_{\Lambda(G)}(F(X_p(A^t/L)^\#)) = \mathrm{ch}_{\Lambda(G)}(F(X_p(A/L)^\#)).$$

Given this, the above displayed equality implies that

$$\mathrm{ch}_{\Lambda(G)}(X_p(A/L)) = \mathrm{ch}_{\Lambda(G)}(F(X_p(A/L)^\#)) \cdot \theta^+_{A,L,S}$$

$$= \Lambda(G) \cdot (\theta^+_{A,L,S}(\theta^+_{A,L,S})^\#) = \Lambda(G) \cdot \theta_{A,L,S},$$

as claimed in Theorem 6.1.

8   Explicit consequences

In the next three sections we take some time to try to indicate the interest of the sort of main conjectures in non-commutative Iwasawa theory that were discussed in § 2 and § 4.

To do this we shall first recall in this section some of the more explicit consequences of Theorem 2.1 that are proved in [20] and then in § 9 and § 10 we shall quickly review the methods that are used to deduce these sorts of consequences.

At the outset we note that the general formalism of descent in non-commutative Iwasawa theory that was recently developed by Venjakob and the first-named author in [24] plays a key role in the derivation of all such results and that it seems reasonable to believe this formalism will in time allow similar deductions to be made from the results of Theorems 4.1 and 6.3.

8.1   Weil-étale cohomology

For any scheme Y' of finite type over $\mathbb{F}_q$ we write $Y'_{W\text{ét}}$ for the Weil-étale site on Y' that is defined by Lichtenbaum in [58, § 2].

We also write ϕ for the geometric Frobenius automorphism $z \mapsto z^{\frac{1}{q}}$ in $\mathrm{Gal}(\mathbb{F}_p^c/\mathbb{F}_q)$, θ for the element of $H^1(\mathrm{Spec}(\mathbb{F}_q)_{W\text{ét}}, \mathbb{Z}) = \mathrm{Hom}(\langle\phi\rangle, \mathbb{Z})$ that sends the (arithmetic) Frobenius automorphism to 1 (and hence sends ϕ to -1) and $\theta_{Y'}$ for the pullback of θ to $H^1(Y'_{W\text{ét}}, \mathbb{Z})$.

We recall that for any sheaf $\mathcal{F}$ on $Y'_{W\text{ét}}$, taking cup product with $\theta_{Y'}$ gives a complex of the form

$$0 \longrightarrow H^0(Y'_{W\text{ét}}, \mathcal{F}) \xrightarrow{\cup\theta_{Y'}} H^1(Y'_{W\text{ét}}, \mathcal{F}) \xrightarrow{\cup\theta_{Y'}} H^2(Y'_{W\text{ét}}, \mathcal{F}) \xrightarrow{\cup\theta_{Y'}} \cdots . \quad (8.1)$$

For any finite Galois covering $f \colon Y \to X$ of group $\mathcal{G}$ we set

$$Z(f,t) := \sum_{\chi \in \mathrm{Ir}(\mathcal{G})} L^{\mathrm{Artin}}(Y, \chi, t) e_\chi.$$

Here $\mathrm{Ir}(\mathcal{G})$ denotes the set of irreducible complex characters of $\mathcal{G}$ and for each such χ we write $L^{\mathrm{Artin}}(Y, \chi, t)$ for the Artin L-function that is defined by Milne in [59, Chapter VI, Example 13.6(b)] and e_χ for the primitive central idempotent of $\mathbb{C}[\mathcal{G}]$ that is obtained by setting

$$e_\chi := \chi(1)|\mathcal{G}|^{-1} \sum_{g \in \mathcal{G}} \chi(g^{-1})g.$$

Each of the functions $L^{\mathrm{Artin}}(Y, \chi, t)$ is known to be a rational function of t and we write $r_{f,\chi}$ for its order of vanishing at $t = 1$. We further recall that the leading term

$$Z^*(f, 1) := \sum_{\chi \in \mathrm{Ir}(\mathcal{G})} \lim_{t \to 1} (1-t)^{-r_{f,\chi}} L^{\mathrm{Artin}}(Y, \chi, t) e_\chi$$

of $Z(f, t)$ at $t = 1$ belongs to $\zeta(\mathbb{Q}[\mathcal{G}])^\times$, where we write $\zeta(A)$ for the centre of any ring A.

We next write

$$\delta_\mathcal{G} : \zeta(\mathbb{Q}[\mathcal{G}])^\times \longrightarrow K_0(\mathbb{Z}[\mathcal{G}], \mathbb{Q}[\mathcal{G}])$$

for the canonical 'extended boundary homomorphism' that is defined by Flach and the first-named author in [21, Lemma 9]. We recall that this homomorphism lies in a commutative diagram (of abelian groups) of the form

$$
\begin{array}{ccc}
K_1(\mathbb{Q}[\mathcal{G}]) & \xrightarrow{\partial_{\mathbb{Z}[\mathcal{G}], \mathbb{Q}[\mathcal{G}]}} & K_0(\mathbb{Z}[\mathcal{G}], \mathbb{Q}[\mathcal{G}]) \\
{\scriptstyle \mathrm{Nrd}_{\mathbb{Q}[\mathcal{G}]}} \downarrow & \nearrow{\scriptstyle \delta_\mathcal{G}} & \\
\zeta(\mathbb{Q}[\mathcal{G}])^\times & &
\end{array}
$$

where the canonical boundary homomorphism $\partial_{\mathbb{Z}[\mathcal{G}], \mathbb{Q}[\mathcal{G}]}$ occurs in the appropriate case of the exact sequence (1.1) and $\mathrm{Nrd}_{\mathbb{Q}[\mathcal{G}]}$ denotes the injective homomorphism that is induced by taking reduced norms with respect to the semisimple algebra $\mathbb{Q}[\mathcal{G}]$. (We recall further that, in general, the latter homomorphism is not surjective.)

By the general philosophy of 'equivariant Tamagawa number conjectures' that was introduced in [21], an explicit formula for the image of $Z^*(f, 1)$ under the homomorphism $\delta_\mathcal{G}$ constitutes a natural leading term formula for the $\zeta(\mathbb{C}[\mathcal{G}])$-valued function $Z(f, t)$. In this regard, the formula stated in the following result is proved in [20] by combining a special case of Theorem 2.1 with the explicit descent formalism in non-commutative Iwasawa theory that is developed by Venjakob and the first-named author in [24].

In this result we write $\chi_\mathcal{G}^{\mathrm{ref}}(-, -)$ for the refined Euler characteristic construction $\chi_{\mathbb{Z}[\mathcal{G}], \mathbb{Q}[\mathcal{G}]}^{\mathrm{ref}}(-, -)$ that was discussed in § 1.

Theorem 8.1. *We assume to be given a Cartesian diagram*

$$
\begin{array}{ccc}
Y & \xrightarrow{\ j_Y\ } & Y' \\
{\scriptstyle f}\downarrow & & \downarrow{\scriptstyle f'} \\
X & \xrightarrow{\ j\ } & X'
\end{array}
\qquad (8.2)
$$

in which f' is a finite Galois covering of group $\mathcal{G}$, X is geometrically connected, X' and Y' are proper and j and j_Y are open immersions. We also assume that both

(i) *in each degree i the (abelian) group $H^i(Y'_{W\text{ét}}, j_{Y,!}\mathbb{Z})$ is finitely generated, and*

(ii) *the complex (8.1) with $\mathcal{F} = j_{Y,!}\mathbb{Z}$ has finite cohomology groups.*

Then the complex $R\Gamma(Y'_{W\text{ét}}, j_{Y,!}\mathbb{Z})$ belongs to $D^{\mathrm{p}}(\mathbb{Z}[\mathcal{G}])$ and in $K_0(\mathbb{Z}[\mathcal{G}], \mathbb{Q}[\mathcal{G}])$ one has an equality

$$
\delta_{\mathcal{G}}(Z^*(f,1)) = -\chi_{\mathcal{G}}^{\mathrm{ref}}(R\Gamma(Y'_{W\text{ét}}, j_{Y,!}\mathbb{Z}), \epsilon_{f,j}),
$$

where $\epsilon_{f,j}$ is the exact sequence of $\mathbb{Q}[\mathcal{G}]$-modules that is induced by (8.1) with $\mathcal{F} = j_{Y,!}\mathbb{Z}$.

Note that if $Y' = X'$ (so $Y = X$ and $\mathcal{G}$ is trivial), then the equality of Theorem 8.1 coincides with the leading term formula that is proved by Lichtenbaum in [58, Theorem 8.2]. In fact the latter result also shows that (for any group $\mathcal{G}$) the conditions (i) and (ii) in Theorem 8.1 are valid if $X = X'$ (so j is the identity) and Y' is both smooth and projective, or if Y' is a curve, or if both Y and Y' are smooth surfaces.

In general, however, if Y' is either non-smooth or non-proper, then condition (i) in Theorem 8.1 is not always satisfied (and the validity of condition (ii) is related to Tate's conjecture on the bijectivity of the cycle-class map).

In particular, relaxing these hypotheses, it is natural to ask if the same methods as used in [20] can prove a result similar to Theorem 8.1 concerning the theory of 'arithmetic cohomology' that is introduced by Geisser in [39].

8.2   Leading term formulas for affine curves

In the special case that X is an affine curve and $\mathbb{F}_q$ is isomorphic to the field of constants of the function field of X Theorem 8.1 implies the following unconditional result.

For an explicit statement of this result, and a discussion of its proof, see §9.

Corollary 8.2. *The conjectural leading term formula of [15] is valid.*

This result cannot, as far as we are aware, be proved by using only classical (commutative) Iwasawa-theoretic techniques and is of interest for several reasons.

Firstly, it is both a non-abelian generalization of the main result of [17] and also validates a natural function field analogue of an important special case of the 'equivariant Tamagawa number conjecture' that was formulated by Flach and the first-named author in [21] as a natural refinement of the seminal conjecture of Bloch and Kato from [12] (for more details of this connection see [15, Remark 2 and Remark 3]).

In addition, the approach that is discussed in [17, Remark 3.3] gives an interpretation of Corollary 8.2 in terms of certain explicit integral congruence relations between (suitably normalized) leading terms at $s = 0$ of the Artin L-series of families of complex representations.

We further recall that the central conjecture of [15] was also in effect a universal refined Stark Conjecture and, as a result, Corollary 8.2 has a range of more explicit consequences. Before stating the first such consequence we recall that the so-called 'Ω-Conjectures' were formulated by Chinburg in [25, §4.2] as natural analogues for global function fields of the central conjectures of Galois module theory that he had earlier formulated in [26, 27]. In particular, the $\Omega(3)$-Conjecture predicts a precise connection between the signs of the Artin root numbers of the irreducible complex symplectic characters of a finite Galois extension of global function fields and the global canonical class of class field theory that is defined by Tate in [75].

The following result follows directly by combining Corollary 8.2 with the result of [15, Theorem 4.1].

Corollary 8.3. *The $\Omega(3)$-Conjecture of Chinburg is valid for all global function fields.*

Note that Corollaries 8.2 and 8.3 represented strong improvements of previous results in this area since (as far as the present authors are aware) neither the central conjecture of [15] or Chinburg's $\Omega(3)$-Conjecture had hitherto been proved for any non-abelian Galois extension of degree divisible by p.

In addition, by combining Corollary 8.3 with results of Chinburg in [25] and [28] one can also deduce that Chinburg's $\Omega(1)$-Conjecture is valid for all tamely ramified Galois extensions of global function fields and obtain an explicit algebraic formula for the epsilon constants associated to tamely ramified Galois covers of curves over finite fields (for more details, see Ward's recent thesis [81]). Making progress on the $\Omega(2)$-Conjecture in the case of wildly Galois ramified extensions still seems however to be a very difficult problem.

Let now F/k be a finite Galois extension of global function fields, set $\mathcal{G} := \mathrm{Gal}(F/k)$ and fix a finite non-empty set of places Σ of k that contains all places which ramify in F/k. For any intermediate field E of F/k we write w_E for the order of the (finite) torsion subgroup of $E^\times$, $\Sigma(E)$ for the set of places of E above those in Σ, $\mathcal{O}_{E,\Sigma}$ for the subring of E comprising all elements that are integral at all places outside $\Sigma(E)$, C_E^Σ for the affine curve $\mathrm{Spec}(\mathcal{O}_{E,\Sigma})$, and $f_{E/k}^\Sigma \colon C_E^\Sigma \to C_k^\Sigma$ for the morphism that is induced by the inclusion $\mathcal{O}_{k,\Sigma} \subseteq \mathcal{O}_{E,\Sigma}$. We also fix q so that $\mathbb{F}_q$ identifies with the constant field of k, regard each $f_{E/k}^{\Sigma}$ as a morphism of $\mathbb{F}_q$-schemes and note that C_k^{Σ} is geometrically connected as a scheme over $\mathbb{F}_q$. Just as before, we write $x \mapsto x^{\#}$ for the $\mathbb{Q}$-linear involution of $\zeta(\mathbb{Q}[\mathcal{G}])$ that is induced by setting $g^{\#} := g^{-1}$ for each element g of $\mathcal{G}$.

Deligne has famously proved that if $\mathcal{G}$ is abelian, then $w_F Z(f_{E/k}^{\Sigma}, 1)^{\#}$ belongs to $\mathbb{Z}[\mathcal{G}]$ and annihilates $\mathrm{Pic}^0(F)$ (cf. [76, Chapter V]), and the next consequence of Corollary 8.2 that we record is a natural generalization of this theorem.

To state this result we note that for every natural number m and matrix H in $\mathrm{M}_m(\mathbb{Z}[\mathcal{G}])$ there is a unique matrix H^* in $\mathrm{M}_m(\mathbb{Q}[\mathcal{G}])$ with

$$HH^* = H^*H = \mathrm{Nrd}_{\mathbb{Q}[\mathcal{G}]}(H) \cdot I_m$$

and such that for every primitive idempotent e of $\zeta(\mathbb{Q}[\mathcal{G}])$ the matrix H^*e is invertible if and only if the product $\mathrm{Nrd}_{\mathbb{Q}[\mathcal{G}]}(H)e$ is non-zero.

We then obtain an ideal of $\zeta(\mathbb{Z}[\mathcal{G}])$ by setting

$$\mathcal{A}(\mathbb{Z}[\mathcal{G}]) := \big\{ x \in \zeta(\mathbb{Q}[\mathcal{G}]) \mid \text{if } d > 0 \text{ and } H \in \mathrm{M}_d(\mathbb{Z}[\mathcal{G}]) \text{ then } xH^* \in \mathrm{M}_d(\mathbb{Z}[\mathcal{G}]) \big\}.$$

(This ideal was first introduced by Nickel and has recently been described explicitly in many cases by Johnston and Nickel in [48]: in particular, we recall from loc. cit. that $\mathbb{Z}_\ell \otimes_{\mathbb{Z}} \mathcal{A}(\mathbb{Z}[\mathcal{G}])$ is equal to $\mathbb{Z}_\ell \otimes_{\mathbb{Z}} \zeta(\mathbb{Z}[\mathcal{G}]) = \zeta(\mathbb{Z}_\ell[\mathcal{G}])$ whenever the prime ℓ does not divide the order of the commutator subgroup of $\mathcal{G}$.)

We also set $w_{F/k} := \sum_{\chi \in \mathrm{Ir}(\mathcal{G})} w_F^{\chi(1)} e_\chi \in \zeta(\mathbb{Q}[\mathcal{G}])^{\times}$.

Corollary 8.4. *For every a in $\mathcal{A}(\mathbb{Z}[\mathcal{G}])$ the element $a w_{F/k} Z(f_{F/k}^{\Sigma}, 1)^{\#}$ belongs to $\mathbb{Z}[\mathcal{G}]$ and annihilates the module $\mathrm{Pic}^0(F)$.*

Note that this result recovers Deligne's theorem because if $\mathcal{G}$ is abelian, then $w_{F/k} = w_F$ and $\mathcal{A}(\mathbb{Z}[\mathcal{G}]) = \mathbb{Z}[\mathcal{G}]$ so that one can take $a = 1$ in the above.

The proof of Corollary 8.4 that is given in [20] proceeds via an explicit description of the non-commutative Fitting invariants of certain natural Weil-étale cohomology groups (see Theorem 10.2 below) and a more general annihilation result in which the term $Z(f_{F/k}^{\Sigma}, 1)$ is replaced by the value at $t = 1$ of suitable higher-order derivatives of $Z(f_{F/k}^{\Sigma}, t)$.

The final consequence of Corollary 8.2 that we record here concerns a range of very explicit integral refinements of Stark's Conjecture.

This result is proved in [20] by applying the general approach of [18].

Corollary 8.5. *Natural non-abelian generalizations of all of the following refinements of Stark's conjecture are valid for all global function fields.*

- *The Rubin–Stark Conjecture of [69, Conjecture B′].*
- *The 'guess' formulated by Gross in [44, top of p. 195].*
- *The 'refined class number formula' of Gross [44, Conjecture 4.1].*
- *The 'refined class number formula' of Tate [77] (see also [78, (∗)]).*

- *The 'refined class number formula' of Aoki, Lee and Tan* [2, Conjecture 1.1].
- *The 'refined $\mathfrak{p}$-adic abelian Stark Conjecture' of Gross* [44, Conjecture 7.6].

This result shows, in particular, that the equality of Corollary 8.2 also incorporates aspects of the celebrated leading term formula proved by Hayes in [47].

For a detailed discussion of the rather extensive theory of integral refinements of Stark's Conjecture and its various connections to other more general conjectures, an interested reader can consult either Tate's classic book [76], the volume of conference proceedings [23] or the more recent article [16].

9 Proofs of Theorem 8.1 and Corollary 8.2

In this section we fix a Cartesian diagram as in (8.2) and assume both that X is geometrically connected and that the conditions (i) and (ii) in Theorem 8.1 are satisfied.

We note in particular that in this case the result of [46, Exposé V, Proposition 6.9] implies that the homomorphism π_G that occurs in the diagram (2.1) is surjective, and hence that the index d_X that occurs in the definition (2.2) of normalized leading terms is equal to 1.

In the sequel we shall abbreviate the Euler characteristics $\chi^{\mathrm{ref}}_{\mathbb{Z}[\mathcal{G}],\mathbb{Q}[\mathcal{G}]}(-,-)$ and $\chi^{\mathrm{ref}}_{\mathbb{Z}_\ell[\mathcal{G}],\mathbb{Q}_\ell[\mathcal{G}]}(-,-)$ for each prime ℓ that are discussed in §1 to $\chi^{\mathrm{ref}}_{\mathcal{G}}(-,-)$ and $\chi^{\mathrm{ref}}_{\mathcal{G},\ell}(-,-)$ respectively.

For each prime ℓ we consider the complex of $\mathbb{Z}_\ell[\mathcal{G}]$-modules

$$0 \longrightarrow H^0_c(X_{\mathrm{ét}}, f_*f^*\mathbb{Z}_\ell) \xrightarrow{\beta^0_{f,\ell}} H^1_c(X_{\mathrm{ét}}, f_*f^*\mathbb{Z}_\ell) \xrightarrow{\beta^1_{f,\ell}} H^2_c(X_{\mathrm{ét}}, f_*f^*\mathbb{Z}_\ell) \xrightarrow{\beta^2_{f,\ell}} \cdots$$
(9.1)

where each $\beta^i_{f,\ell}$ denotes the composite homomorphism

$$H^i_c(X_{\mathrm{ét}}, f_*f^*\mathbb{Z}_\ell) \cong H^i(\mathbb{F}_q, Rs_{X,!}f_*f^*\mathbb{Z}_\ell) \longrightarrow H^i(\mathbb{F}^c_q, s^*_c Rs_{X,!}f_*f^*\mathbb{Z}_\ell)$$
$$\longrightarrow H^{i+1}(\mathbb{F}_q, Rs_{X,!}f_*f^*\mathbb{Z})_\ell) \cong H^{i+1}_c(X_{\mathrm{ét}}, f_*f^*\mathbb{Z}_\ell)$$

where the arrows denote the maps that occur in the cohomology sequence of the exact triangle (3.1) with G and $\mathcal{L}$ now replaced by $\mathcal{G}$ and $\mathbb{Z}_\ell$ respectively.

We also write $\delta_{\mathcal{G},\ell}$ for the composite homomorphism of abelian groups

$$\zeta(\mathbb{Q}_\ell[\mathcal{G}])^\times \longrightarrow K_1(\mathbb{Q}_\ell[\mathcal{G}]) \longrightarrow K_0(\mathbb{Z}_\ell[\mathcal{G}], \mathbb{Q}_\ell[\mathcal{G}])$$

where the first map is induced by the inverse of the (bijective) reduced norm map $\mathrm{Nrd}_{\mathbb{Q}_\ell[\mathcal{G}]}$ and the second map is the standard connecting homomorphism of relative K-theory.

We fix an isomorphism of fields $\iota\colon \mathbb{C} \cong \mathbb{C}_p$ and for each χ in $\mathrm{Ir}(\mathcal{G})$ we use this isomorphism to identify the composite character $\iota \circ \chi$ with any fixed choice of Artin representation of $\mathcal{G}$ that has character $\iota \circ \chi$.

For every prime ℓ there is a natural isomorphism

$$\mathbb{Z}_\ell[\mathcal{G}] \otimes^{\mathbb{L}}_{\mathbb{Z}[\mathcal{G}]} R\Gamma(Y'_{W\text{ét}}, j_{Y,!}\mathbb{Z}) \cong R\Gamma_c(X_{\text{ét}}, f_* f^* \mathbb{Z}_\ell) \tag{9.2}$$

in $D^-(\mathbb{Z}_\ell[\mathcal{G}])$, and by using this one we can prove the following reduction result.

Lemma 9.1. *The complex $R\Gamma(X'_{W\text{ét}}, f'_* f'^* j_! \mathbb{Z})$ belongs to $D^{\mathrm{p}}(\mathbb{Z}[\mathcal{G}])$ and for each prime ℓ the complex $R\Gamma_c(X_{\text{ét}}, f_* f^* \mathbb{Z}_\ell)$ belongs to $D^{\mathrm{p}}(\mathbb{Z}_\ell[\mathcal{G}])$.*

Furthermore, the cohomology groups of the complex (9.1) are finite and Theorem 8.1 is valid if and only if for every prime ℓ there is in $K_0(\mathbb{Z}_p[\mathcal{G}], \mathbb{Q}_p[\mathcal{G}])$ an equality

$$\delta_{\mathcal{G},\ell}(Z_\mathcal{G}^*(X,1)) = -\chi_{\mathcal{G},\ell}^{\mathrm{ref}}(R\Gamma_c(X_{\text{ét}}, f_* f^* \mathbb{Z}_\ell), \beta_{f,\ell}). \tag{9.3}$$

Here we set

$$Z_\mathcal{G}^*(X,1) := \sum_{\chi \in \mathrm{Ir}(\mathcal{G})} Z^*(X, \mathcal{L}_{\iota \circ \chi}, 1) e_{\iota \circ \chi} \in \zeta(\mathbb{Q}_\ell^c[\mathcal{G}])^\times$$

and write $\beta_{f,\ell}$ for the exact sequence of $\mathbb{Q}_\ell[\mathcal{G}]$-modules that is obtained by scalar extension of the complex (9.1).

For each prime $\ell \neq p$ the equality (9.3) can be proved by simply combining a general K-theoretical observation from [24, Proposition 5.10] with the celebrated description of the Zeta functions of ℓ-adic sheaves that is given by Grothendieck in [45] and described by Milne in [59, Chapter VI, Theorem 13.3 and Example 13.6(b)]. (This case of (9.3) can in fact also be proved by combining the main result of Witte in [83] together with the descent formalism developed in [24].)

In the case that $\ell = p$ one can derive the equality (9.3) from Theorem 2.1 by means of a rather delicate application of the explicit descent result of [24, Theorem 2.2]. Note that when applying this result the assumption that is made in Theorem 8.1(ii) ensures that the necessary 'semisimplicity' hypotheses of [24, Theorem 2.2] are satisfied and then to show that the exact sequences $\mathbb{Z}_p \otimes_{\mathbb{Z}} \epsilon_{f,j}$ and $\beta_{f,p}$ correspond under the isomorphism (9.2) (with $\ell = p$) it is convenient to use the explicit computation of Bockstein homomorphisms that is given by Rapoport and Zink in [67, 1.2].

By such an explicit analysis one then completes the proof of Theorem 8.1.

Remark 9.2. The occurrence of the unit series $v(X, \mathcal{F}_G^\bullet, t)$ in the formula (3.7) of Emerton and Kisin means that there is no direct analogue in the case that $\ell = p$ of Grothendieck's elegant formula for the Zeta functions of ℓ-adic sheaves in [45].

The inexplicit nature of the series $v(X, \mathcal{F}_G^\bullet, t)$ then prevents one either from using the K-theoretical result of [24, Proposition 5.10] to give a more direct proof of the equality (9.3) or from applying the general approach of Witte in [83] to prove the result of Theorem 2.1.

To discuss the proof of Corollary 8.2 we first quickly review the notation introduced just after Corollary 8.3.

For any global function field E we write C_E for the unique geometrically irreducible smooth projective curve with function field E. We fix a finite Galois extension of such fields F/k and set $\mathcal{G} := \mathrm{Gal}(F/k)$. We also fix a finite non-empty set of places Σ of k that contains all places which ramify in F/k and, with E denoting either k or F, we write C_E^Σ for the affine curve $\mathrm{Spec}(\mathcal{O}_{E,\Sigma})$, $j_E^\Sigma \colon C_E^\Sigma \to C_E$ for the corresponding open immersion and $f \colon C_F^\Sigma \to C_k^\Sigma$ and $f' \colon C_F \to C_k$ for the morphisms that are induced by the inclusion $k \subseteq F$. We fix q so that $\mathbb{F}_q$ identifies with the constant field of k and regard all of the above morphisms as morphisms of $\mathbb{F}_q$-schemes.

We also write $\mathcal{O}_{F,\Sigma}^\times$ for the unit group of $\mathcal{O}_{F,\Sigma}$, set $B_{F,\Sigma} := \bigoplus_w \mathbb{Z}$ where w runs over all places in $\Sigma(F)$ and write $B_{F,\Sigma}^0$ for the kernel of the homomorphism $B_{F,\Sigma} \to \mathbb{Z}$ that sends each element $(n_w)_w$ to the sum $\sum_w n_w$. We note that each of the groups $\mathcal{O}_{F,\Sigma}$, $\mathcal{O}_{F,\Sigma}^\times$, $B_{F,\Sigma}$ and $B_{F,\Sigma}^0$ has a natural action of $\mathcal{G}$.

It is well known that the complex $K_{F,\Sigma}^\bullet := R\Gamma(C_{F,\mathrm{W\acute et}}^\Sigma, \mathbb{G}_m)$ belongs to $D^{\mathrm{P}}(\mathbb{Z}[\mathcal{G}])$, is acyclic outside degrees zero and one and is such that $H^0(K_{F,\Sigma}^\bullet)$ can be identified with $\mathcal{O}_{F,\Sigma}^\times$, the torsion subgroup $H^1(K_{F,\Sigma}^\bullet)_{\mathrm{tor}}$ of $H^1(K_{F,\Sigma}^\bullet)$ with the ideal class group $\mathrm{Cl}(\mathcal{O}_{F,\Sigma})$ and the quotient $H^1(K_{F,\Sigma}^\bullet)/H^1(K_{F,\Sigma}^\bullet)_{\mathrm{tor}}$ with the lattice $B_{F,\Sigma}^0$ (see, for example, [15, Lemma 1]).

We identify each place w of F with the corresponding (closed) point of C_F and write val_w for the associated valuation on F and $\mathrm{d}(w)$ for the degree of w over $\mathbb{F}_q$. We then write

$$\mathrm{D}_{F,\Sigma}^q : \mathcal{O}_{F,\Sigma}^\times \longrightarrow B_{F,\Sigma}^0$$

for the homomorphism that sends each element u of $\mathcal{O}_{F,\Sigma}^\times$ to $(\mathrm{val}_w(u)\mathrm{d}(w))_{w \in \Sigma(F)}$. The induced map $\mathbb{Q} \otimes_\mathbb{Z} \mathrm{D}_{F,\Sigma}^q$ is bijective and therefore gives rise to an exact sequence of $\mathbb{Q}[\mathcal{G}]$-modules

$$\epsilon_{F,\Sigma} \colon 0 \longrightarrow \mathbb{Q} \otimes_\mathbb{Z} H^0(K_{F,\Sigma}^\bullet) \xrightarrow{\mathbb{Q} \otimes_\mathbb{Z} \mathrm{D}_{F,\Sigma}^q} \mathbb{Q} \otimes_\mathbb{Z} H^1(K_{F,\Sigma}^\bullet) \xrightarrow{0} \mathbb{Q} \otimes_\mathbb{Z} H^2(K_{F,\Sigma}^\bullet) \xrightarrow{0} \cdots$$

We can now give a more explicit statement of Corollary 8.2. This result constitutes the natural leading term formula at $t = 1$ for the $\zeta(\mathbb{C}[\mathcal{G}])$-valued Zeta-function $Z(f, t)$.

Recall that $x \mapsto x^\#$ denotes the $\mathbb{Q}$-linear involution of $\zeta(\mathbb{Q}[\mathcal{G}])$ that is induced by setting $g^\# := g^{-1}$ for each g in $\mathcal{G}$.

Theorem 9.3. $\delta_\mathcal{G}(Z^*(f, 1)^\#) = \chi_\mathcal{G}^{\mathrm{ref}}(R\Gamma(C_{F,\mathrm{W\acute et}}^\Sigma, \mathbb{G}_m), \epsilon_{F,\Sigma}).$

Remark 9.4. The result of Theorem 9.3 is known to have connections to several conjectures in the literature.

For example, it is easily shown that the equality in Theorem 9.3 is equivalent to that of [15, (3)] and hence implies the central conjecture (Conjecture C(F/k)) of [15]. From the computation of [15, Proposition 4.1] it thus follows that Theorem 9.3 is also equivalent to the function field case of [18, Conjecture LTC(F/k)].

In addition, one knows that the Artin–Verdier Duality Theorem induces (via the argument of Lichtenbaum in [58, proof of Theorem 6.5]) a canonical isomorphism in $D^{\mathrm{p}}(\mathbb{Z}[\mathcal{G}])$ of the form

$$\Delta_{\Sigma} : R\Gamma(C^{\Sigma}_{F,\mathrm{W\acute{e}t}}, \mathbb{G}_m) \cong R\operatorname{Hom}_{\mathbb{Z}}(R\Gamma(C_{F,\mathrm{W\acute{e}t}}, j^{\Sigma}_{F,!}\mathbb{Z}), \mathbb{Z}[-2]), \qquad (9.4)$$

where the linear dual complex is endowed with the natural contragredient action of $\mathcal{G}$, and this isomorphism implies that Theorem 9.3 generalizes the main result (Theorem 3.1) of [17].

Finally, the computations of Bae in [3] can be used to show that the equality of Theorem 9.3 implies a refinement of the leading term conjecture (for global function fields) that is formulated by Lichtenbaum in [57].

For each element x of $\zeta(\mathbb{Q}[\mathcal{G}])^{\times}$ one can show that

$$\delta_{\mathcal{G}}(x^{\#}) = -\psi^*(\delta_{\mathcal{G}}(x)). \qquad (9.5)$$

Here we write ψ^* for the involution of $K_0(\mathbb{Z}[\mathcal{G}], \mathbb{Q}[\mathcal{G}])$ which, for each pair of finitely generated projective $\mathbb{Z}[\mathcal{G}]$-modules P and Q and each isomorphism of $\mathbb{Q}[\mathcal{G}]$-modules $\mu \colon \mathbb{Q} \otimes_{\mathbb{Z}} P \to \mathbb{Q} \otimes_{\mathbb{Z}} Q$, satisfies

$$\psi^*((P, \mu, Q)) = (\operatorname{Hom}_{\mathbb{Z}}(P, \mathbb{Z}), \operatorname{Hom}_{\mathbb{Q}}(\mu, \mathbb{Q})^{-1}, \operatorname{Hom}_{\mathbb{Z}}(Q, \mathbb{Z})),$$

where the $\mathbb{Z}$-linear duals are endowed with the contragredient action of $\mathcal{G}$ and so are also projective $\mathbb{Z}[\mathcal{G}]$-modules.

In view of (9.5) the isomorphism $R\operatorname{Hom}_{\mathbb{Z}}(\Delta_{\Sigma}, \mathbb{Z}[-2])$ implies that Theorem 9.3 is valid if and only if one has an equality

$$\delta_{\mathcal{G}}(Z^*(f, 1)) = -\chi^{\mathrm{ref}}_{\mathcal{G}}(R\Gamma(C_{F,\mathrm{W\acute{e}t}}, j^{\Sigma}_{F,!}\mathbb{Z}), \epsilon^*_{F,\Sigma}). \qquad (9.6)$$

Here we write $\epsilon^*_{F,\Sigma}$ for the exact sequence of $\mathbb{Q}[\mathcal{G}]$-modules that is obtained by first taking the $\mathbb{Q}$-linear dual of $\epsilon_{F,\Sigma}$ and then using $\operatorname{Hom}_{\mathbb{Q}}(\mathbb{Q} \otimes_{\mathbb{Z}} H^i(\Delta_{\Sigma}), \mathbb{Q})$ in each degree i in order to identify $\operatorname{Hom}_{\mathbb{Q}}(\mathbb{Q} \otimes_{\mathbb{Z}} H^i(C^{\Sigma}_{F,\mathrm{W\acute{e}t}}, \mathbb{G}_m), \mathbb{Q})$ with $\mathbb{Q} \otimes_{\mathbb{Z}} H^{2-i}(C_{F,\mathrm{W\acute{e}t}}, j^{\Sigma}_{F,!}\mathbb{Z})$.

On the other hand, the equality (9.6) can be deduced from the result of Theorem 8.1 with our present choice of f (so that $Y = C^{\Sigma}_F$, $Y' = C_F$, $X = C^{\Sigma}_k$, $X' = C_k$, $j = j^{\Sigma}_k$ and $j_Y = j^{\Sigma}_F$). Indeed, since X is geometrically connected and Y' is a curve, the conditions (i) and (ii) of Theorem 8.1 are satisfied (by [58, Theorem 8.2]) and so that result implies that

$$\delta_{\mathcal{G}}(Z^*(f, 1)) = -\chi^{\mathrm{ref}}_{\mathcal{G}}(R\Gamma(C_{F,\mathrm{W\acute{e}t}}, j^{\Sigma}_{F,!}\mathbb{Z}), \epsilon_{f,j_k}).$$

To deduce (9.6) it is thus enough to show that the exact sequences $\epsilon^*_{F,\Sigma}$ and ϵ_{f,j_k} coincide and to prove this one verifies (by an explicit computation) that the following diagram commutes:

$$
\begin{array}{ccc}
\mathbb{Q} \otimes_{\mathbb{Z}} H^1(C_{F,\mathrm{W\acute{e}t}}, j_{F,!}\mathbb{Z}) & \xrightarrow{\mathrm{Hom}_{\mathbb{Z}}(H^1(\Delta_{\Sigma}),\mathbb{Q})} & \mathrm{Hom}_{\mathbb{Z}}(B^0_{F,\Sigma}, \mathbb{Q}) \\
{\scriptstyle \mathbb{Q}\otimes_{\mathbb{Z}}(\cup\theta_Y)}\downarrow & & \downarrow{\scriptstyle \mathrm{Hom}_{\mathbb{Z}}(\mathrm{D}^q_{F,\Sigma},\mathbb{Q})} \\
\mathbb{Q} \otimes_{\mathbb{Z}} H^2(C_{F,\mathrm{W\acute{e}t}}, j_{F,!}\mathbb{Z}) & \xrightarrow{\mathrm{Hom}_{\mathbb{Z}}(H^0(\Delta_{\Sigma}),\mathbb{Q})} & \mathrm{Hom}_{\mathbb{Z}}(\mathcal{O}^{\times}_{F,\Sigma}, \mathbb{Q}),
\end{array}
$$

where Δ_{Σ} is the duality isomorphism (9.4).

10   Fitting invariants and annihilation results

Ever since the proof (in 1870) of Stickelberger's classical theorem concerning the explicit structure of ideal class groups of abelian number fields, there has been much interest in proving theorems and formulating conjectures that use the values at integer arguments of Dirichlet L-functions to describe the annihilators and Fitting ideals of natural arithmetic Galois modules (see, for example, the nice survey of this area given by Greither in [42]).

In this section we shall discuss some consequences of Theorem 9.3 regarding the explicit computation of the (non-commutative) Fitting invariants of certain natural Weil-étale cohomology groups and their connection to the proof of Corollary 8.4. Throughout we fix a finite Galois cover $f\colon C^{\Sigma}_F \to C^{\Sigma}_k$ of affine curves over $\mathbb{F}_q$, as in Corollary 8.4 (where f is denoted by $f^{\Sigma}_{F/k}$).

We first introduce a convenient modification of the relevant cohomology complexes. Indeed, the natural complexes described below have the property that the first non-zero cohomology group is torsion-free and, perhaps surprisingly, this greatly simplifies the task of deriving explicit consequences from the equality in Theorem 9.3.

For any $\mathcal{G}$-module M we write $M^{\vee}$ for its Pontrjagin dual $\mathrm{Hom}(M,\mathbb{Q}/\mathbb{Z})$, endowed with the natural contragredient action of $\mathcal{G}$.

We also fix a finite non-empty set of closed points T of C^{Σ}_k.

Lemma 10.1. *We write $\mathbb{F}^{\times}_{T(F)}$ for the direct sum of the multiplicative groups of the residue fields of all places in $T(F)$. Then in $D^{\mathrm{p}}(\mathbb{Z}[\mathcal{G}])$ there are exact triangles of the form*

$$
R\Gamma_T(C^{\Sigma}_{F,\mathrm{W\acute{e}t}}, \mathbb{G}_m) \longrightarrow R\Gamma(C^{\Sigma}_{F,\mathrm{W\acute{e}t}}, \mathbb{G}_m) \longrightarrow \mathbb{F}^{\times}_{T(F)}[0] \longrightarrow R\Gamma_T(C^{\Sigma}_{F,\mathrm{W\acute{e}t}}, \mathbb{G}_m)[1]
\tag{10.1}
$$

and

$$
R\Gamma_T(C_{F,\mathrm{W\acute{e}t}}, j^{\Sigma}_{F,!}\mathbb{Z})[-1] \longrightarrow (F^{\times}_{T(F)})^{\vee}[-3]
$$
$$
\longrightarrow R\Gamma(C_{F,\mathrm{W\acute{e}t}}, j^{\Sigma}_{F,!}\mathbb{Z}) \longrightarrow R\Gamma_T(C_{F,\mathrm{W\acute{e}t}}, j^{\Sigma}_{F,!}\mathbb{Z})
\tag{10.2}
$$

as well as an isomorphism

$$
R\Gamma_T(C^{\Sigma}_{F,\mathrm{W\acute{e}t}}, \mathbb{G}_m) \cong R\mathrm{Hom}_{\mathbb{Z}}(R\Gamma_T(C_{F,\mathrm{W\acute{e}t}}, j^{\Sigma}_{F,!}\mathbb{Z}), \mathbb{Z}[-2]).
\tag{10.3}
$$

In view of this result, in each degree m one obtains natural 'T-modified Weil-étale cohomology groups' (that are defined up to canonical isomorphism) by setting

$$H_T^m(C_{F,\text{Wét}}^\Sigma, \mathbb{G}_m) := H^m(R\Gamma_T(C_{F,\text{Wét}}^\Sigma, \mathbb{G}_m))$$

and

$$H_T^m(C_{F,\text{Wét}}, j_{F,!}^\Sigma \mathbb{Z}) := H^m(R\Gamma_T(C_{F,\text{Wét}}, j_{F,!}^\Sigma \mathbb{Z})).$$

We next recall the very useful notion of non-commutative Fitting invariants that was first introduced by Parker in his PhD thesis [66] and then extensively developed and generalized by Nickel in [61].

One says that a finitely generated $\mathcal{G}$-module M has a 'locally-quadratic presentation' if for each prime ℓ there exists a natural number d_ℓ and an exact sequence of $\mathbb{Z}_\ell[\mathcal{G}]$-modules of the form

$$\mathbb{Z}_\ell[\mathcal{G}]^{d_\ell} \xrightarrow{\theta_\ell} \mathbb{Z}_\ell[\mathcal{G}]^{d_\ell} \longrightarrow \mathbb{Z}_\ell \otimes_\mathbb{Z} M \longrightarrow 0. \tag{10.4}$$

For any such M the (non-commutative) Fitting invariant $\text{Fit}_{\mathbb{Z}[\mathcal{G}]}(M)$ is then defined to be the unique full $\zeta(\mathbb{Z}[\mathcal{G}])$-sublattice of $\zeta(\mathbb{Q}[\mathcal{G}])$ with the property that at each prime ℓ one has

$$\mathbb{Z}_\ell \otimes_\mathbb{Z} \text{Fit}_{\mathbb{Z}[G]}(M) = \zeta(\mathbb{Z}_\ell[\mathcal{G}]) \cdot \left\{ \text{nr}_{\mathbb{Q}_\ell[\mathcal{G}]}(\theta_\ell) u : u \in \text{nr}_{\mathbb{Q}_\ell[\mathcal{G}]}(K_1(\mathbb{Z}_\ell[\mathcal{G}])) \right\}.$$

Note that, whilst it is straightforward that such a sublattice $\text{Fit}_{\mathbb{Z}[\mathcal{G}]}(M)$ exists, is independent of the choice of presentations (10.4) and agrees with the classical notion of (initial) Fitting ideal in the case that $\mathcal{G}$ is abelian, such invariants are usually very difficult to compute explicitly. For example, the trivial module $\{0\}$ has a locally-quadratic presentation but, whilst it is clear that $\text{Fit}_{\mathbb{Z}[\mathcal{G}]}(\{0\})$ both contains and is integral over $\zeta(\mathbb{Z}[\mathcal{G}])$, a complete description of it would require an explicit computation of the groups $\text{nr}_{\mathbb{Q}_\ell[\mathcal{G}]}(K_1(\mathbb{Z}_\ell[\mathcal{G}]))$ and this is in general very difficult.

We recall that $\mathbb{F}_q$ identifies with the constant field of k and, with T as above, we define a T-modified Zeta-function of the morphism $f \colon C_F^\Sigma \to C_k^\Sigma$ by setting

$$Z_T(f,t) = Z(f,t) \prod_{v \in T} \text{Nrd}_{\mathbb{Q}[\mathcal{G}]}([1 - \text{N}v \cdot \text{Fr}_w]_r).$$

Here for any x in $\mathbb{Q}[\mathcal{G}]$ we write $[x]_r$ for the endomorphism $y \mapsto yx$ of $\mathbb{Q}[\mathcal{G}]$, regarded as a left $\mathbb{Q}[\mathcal{G}]$-module in the obvious way.

By an easy computation using the exact triangle (10.2) (and the additivity criterion for refined Euler characteristics that is proved by Breuning and the first-named author in [13]) one finds that Theorem 9.3 is equivalent to the equality

$$\delta_{\mathcal{G}}(Z_T^*(f,1)^\#) = \chi_{\mathcal{G}}^{\text{ref}}(R\Gamma_T(C_{F,\text{Wét}}, j_{F,!}^\Sigma \mathbb{Z}), \epsilon_{F,\Sigma}).$$

Further, since the group $H^1_T(C_{F,\mathrm{W\acute{e}t}}, j^{\Sigma}_{F,!}\mathbb{Z})$ is torsion-free, the approach of [18, Lemma A.1(iii)] allows one to make explicit the latter refined Euler characteristic and hence to interpret it in terms of non-commutative Fitting invariants. In this way the following result is proved in [20].

We note that interesting related results have also recently been proved by Greither and Popescu in [43].

Theorem 10.2. *The $\mathcal{G}$-modules $H^1_T(C^{\Sigma}_{F,\mathrm{W\acute{e}t}}, \mathbb{G}_m)$ and $H^2_T(C_{F,\mathrm{W\acute{e}t}}, j^{\Sigma}_{F,!}\mathbb{Z})$ have locally-quadratic presentations and Fitting invariants equal to*

$$\mathrm{Fit}_{\mathbb{Z}[\mathcal{G}]}(H^1_T(C^{\Sigma}_{F,\mathrm{W\acute{e}t}}, \mathbb{G}_m)) = Z_T(f,1)^{\#} \cdot \mathrm{Fit}_{\mathbb{Z}[\mathcal{G}]}(\{0\})$$

and

$$\mathrm{Fit}_{\mathbb{Z}[\mathcal{G}]}(H^2_T(C_{F,\mathrm{W\acute{e}t}}, j^{\Sigma}_{F,!}\mathbb{Z})) = Z_T(f,1) \cdot \mathrm{Fit}_{\mathbb{Z}[\mathcal{G}]}(\{0\}).$$

Note that the last equality here implies the result of Corollary 8.4 because it can be shown that for each prime ℓ the $\mathbb{Z}_\ell[\mathcal{G}]$-module $\mathrm{Pic}^0(F)^{\vee}_\ell$ is isomorphic to a subquotient of $H^2_T(C_{F,\mathrm{W\acute{e}t}}, j^{\Sigma}_{F,!}\mathbb{Z})_\ell$. Indeed, this follows by means of an explicit computation of cohomology groups that uses the classical Artin–Verdier Duality Theorem if $\ell \neq p$ and uses Deninger's generalization [32] of this result for the logarithmic de Rham–Witt sheaf of Milne and Illusie if $\ell = p$. For more details of this computation, see [20, Lemma 11.3].

This completes our discussion of a selection of consequences of Theorem 2.1.

Remark 10.3. Whilst it is not actually needed for the proof of Corollary 8.4 that is sketched above, it would seem perfectly reasonable to the first-named author to expect that the $\mathcal{G}$-module $\mathrm{Pic}^0(F)^{\vee}$ is itself canonically isomorphic to a subquotient of $H^2_T(C_{F,\mathrm{W\acute{e}t}}, j^{\Sigma}_{F,!}\mathbb{Z})$. However, as far as we are aware, such a global isomorphism (whether canonical or not) is still not known to exist.

Appendix
On Non-Noetherian Iwasawa Theory

Francesc Bars

In this appendix we review some recent results of Bandini, Longhi and the present author in [4, 5, 6] which indicate possible directions in which several of the results (for commutative extensions) that are discussed in the main body of this article could be extended to the setting of Iwasawa algebras that are not noetherian.

To do this we fix a global function field K of characteristic p and assume that the field of constants is $\mathbb{F}_q$ with q a power of p. We also fix a place ∞ of K of degree d_∞ and write A for the subring of K comprising functions that are regular away from ∞. We then fix a prime ideal $\mathfrak{p}$ of A and write $\mathbb{F}_\mathfrak{p}$ for its residue field $A/\mathfrak{p}$ and q^d for the cardinality of $\mathbb{F}_\mathfrak{p}$.

We write H for the Hilbert class field of K that is totally split at ∞ and recall that the group $\mathrm{Gal}(H/K)$ is isomorphic to $\mathrm{Pic}(A)$. We also write H^+ for the normalizing field of K, and we recall that this field is an abelian extension of K which contains H, is unramified (over K) at every finite place of A and is such that the extension H^+/H is totally and tamely ramified at all places of H lying above ∞ and the group $\mathrm{Gal}(H^+/H)$ is naturally isomorphic to the quotient $\mathbb{F}_\infty^\times/\mathbb{F}_q^\times$.

Following Hayes' approach to explicit class field theory over K, we now fix a sgn-normalized rank one Drinfeld module Φ corresponding to a non-trivial embedding of $\mathbb{F}_q$-algebras of the form

$$\Phi \colon A \longrightarrow \mathrm{End}(\mathbb{G}_a/H^+) = H^+\{\tau\},$$

where τ satisfies $\tau a = a^q \tau$ for all $a \in H^+$ and $\tau^0 = \mathrm{id}$. (As an explicit example, we note that if $K = \mathbb{F}_q(T)$ and the distinguished place ∞ is chosen so that $A = \mathbb{F}_q[T]$, then $H^+ = H = K$ and the unique sgn-normalized rank one Drinfeld module that corresponds to sending T to $T \cdot \mathrm{id} + \tau$ is known as the 'Carlitz module'.)

For each natural number n we write $\Phi[\mathfrak{p}^n]$ for the set of zeroes of $\Phi(x)$, with $x \in \mathfrak{p}^n$, and then define K_n to be the field $K(\Phi[\mathfrak{p}^n])$. With respect to our fixed choice of place ∞, the '$\Phi[\mathfrak{p}^\infty]$-torsion extension' of K is then defined by setting

$$\mathcal{K}_\infty := \bigcup_{n \geq 1} K_n.$$

This field is an abelian extension of K that is unramified at any place of A that does not divide $\mathfrak{p}$, is tamely ramified at ∞ and is such that the (common) inertia and decomposition subgroup I_∞ of ∞ in $\mathrm{Gal}(\mathcal{K}_\infty/K)$ is isomorphic to $\mathbb{F}_\infty^\times$. In particular, therefore, the extension of K that is given by the field

$$\mathcal{K}_\infty^+ := \mathcal{K}_\infty^{I_\infty}$$

is totally split at ∞ and ramified only above the prime $\mathfrak{p}$.

The group $\mathrm{Gal}(\mathcal{K}_\infty/H^+)$ is a direct product of the form $\Delta \times \Gamma$ where Δ corresponds to the finite group $\mathbb{F}_\mathfrak{p}^\times$ (which has order prime to p) and Γ is isomorphic to the group $\mathbb{Z}_p^\mathbb{N}$. One also knows that the extension $\mathcal{K}_\infty/H^+$ is totally ramified at all primes of H^+ that lie above $\mathfrak{p}$.

We now assume that p does not divide the order of $\mathrm{Gal}(H^+/K)$. Then the above observations imply that $\mathrm{Gal}(\mathcal{K}_\infty/K)$ is a direct product $\Delta' \times \Gamma$ with Δ' a finite group of order prime to p, and we finally set

$$\mathcal{K}^{\mathrm{cyc}} := \mathcal{K}_\infty^{\Delta'}.$$

This field is a $\mathbb{Z}_p^\mathbb{N}$-extension of K that is totally ramified at $\mathfrak{p}$ and unramified at all other places. (In fact, for our purposes we could replace $\mathcal{K}^{\mathrm{cyc}}$ by any $\mathbb{Z}_p^\mathbb{N}$-extension of K in $\mathcal{K}_\infty$ which is totally ramified at $\mathfrak{p}$ and unramified at all other places.)

As yet, none of the above extensions $\mathcal{K}_\infty/K$, $\mathcal{K}_\infty^+/K$ or $\mathcal{K}^{\mathrm{cyc}}/K$ has been thoroughly investigated from an Iwasawa-theoretic standpoint, since the associated p-adic Iwasawa algebras are not noetherian and so there is no general structure theory in this setting. Notwithstanding this fact, in this appendix we shall describe some recent results of [4, 5, 6] which indicate how the results that are described in the main body of this article (relating to divisor class groups and to the Selmer groups of abelian varieties) could be extended to the setting of the $\mathbb{Z}_p^\mathbb{N}$-extension $\mathcal{K}^{\mathrm{cyc}}/K$.

The reader will find further details, and full proofs, of all the results that we discuss here in the given references.

A.1   The general setting

It is convenient for us to work in the following slightly more general setting:

- $\mathcal{K}$ is a fixed $\mathbb{Z}_p^\mathbb{N}$-extension of K that is unramified outside a finite set of places S of K.

In this setting we write Γ for the group $\mathrm{Gal}(\mathcal{K}/K)$ and Λ for the (non-noetherian) p-adic Iwasawa algebra $\Lambda(\Gamma)$. For any non-negative integer d we also fix a $\mathbb{Z}_p^d$-extension $\mathcal{K}_d$ of K that is contained in $\mathcal{K}$ and is such that both $\mathcal{K}_d \subset \mathcal{K}_{d+1}$ and $\mathcal{K} = \bigcup_{d\geq 0} \mathcal{K}_d$. For each index d we set $\Gamma_d := \mathrm{Gal}(\mathcal{K}_d/K)$ and write Λ_d for the p-adic Iwasawa algebra $\Lambda(\Gamma_d)$. In addition, we fix a commuting set of variables

$\{t_i : i \geq 0\}$ and then for each non-negative integer d we fix (as we may) an isomorphism of Λ_d with the power series ring $\mathbb{Z}_p[[t_1, \ldots, t_d]]$ in d variables in such a way that there are commuting diagrams

$$
\begin{array}{ccc}
\Lambda_{d+1} & \xrightarrow{\cong} & \mathbb{Z}_p[[t_1, \ldots, t_{d+1}]] \\
\pi_d^{d+1} \downarrow & & \downarrow \\
\Lambda_d & \xrightarrow{\cong} & \mathbb{Z}_p[[t_1, \ldots, t_d]]
\end{array}
$$

in which π_d^{d+1} is the natural projection map and the right-hand vertical map sends t_{d+1} to 0 and t_i to t_i if $1 \leq i \leq d$. For each index d and each Λ_d-module M we write M_{t_d} and M/t_d for the kernel and cokernel of the endomorphism of M given by multiplication by t_d.

We make much use of the following result.

Lemma A.1. *Let M be a finitely generated torsion Λ_d-module for which both M_{t_d} and M/t_dM are finitely generated torsion Λ_{d-1}-modules. Then one has*

$$
\mathrm{ch}_{\Lambda_{d-1}}(M_{t_d})\pi_{d-1}^d(\mathrm{ch}_{\Lambda_d}(M)) = \mathrm{ch}_{\Lambda_{d-1}}(M/t_d).
$$

This result follows by combining the (well-known) structure theory of Λ_d-modules with the fact that if N is any pseudo-null Λ_d-module, then $\mathrm{ch}_{\Lambda_{d-1}}(N_{t_d}) = \mathrm{ch}_{\Lambda_{d-1}}(N/t_d)$. (This latter observation itself refines a result of Greenberg from [41, after Lemma 2]; see also [74] or [5] in this regard.)

A.2 Divisor class groups

A.2.1 General observations

For any finite extension F of K we write $\mathfrak{W}_F$ and $\mathfrak{W}_F^0$ for the p-adic completion of its groups of divisor classes and degree zero divisor classes respectively. For an infinite extension $\mathcal{F}$ of K we then set $\mathfrak{W}_{\mathcal{F}} := \varprojlim_F \mathfrak{W}_F$ and $\mathfrak{W}_{\mathcal{F}}^0 := \varprojlim_F \mathfrak{W}_F^0$, where F runs over all finite extensions of K inside $\mathcal{F}$ and the limits are taken with respect to the natural norm maps. For finite extensions F and F' of K with $F \subseteq F'$, there is an exact commutative diagram

$$
\begin{array}{ccccc}
\mathfrak{W}_{F'}^0 & \hookrightarrow & \mathfrak{W}_{F'} & \xrightarrow{\deg_{F'}} & \mathbb{Z}_p \\
N_F^{F'} \downarrow & & N_F^{F'} \downarrow & & \downarrow {\times c_{F',F}} \\
\mathfrak{W}_F^0 & \hookrightarrow & \mathfrak{W}_F & \xrightarrow{\deg_F} & \mathbb{Z}_p \;,
\end{array}
$$

in which $\deg_{F'}$ and $\deg_F$ are the natural degree maps, $N_F^{F'}$ is the natural norm map, and $c_{F',F}$ is the relative degree of the fields of constants of F' and F. In particular, by passing to the limit as F runs over the finite extensions of K inside a given infinite extension $\mathcal{F}$ one obtains an exact sequence of $\mathbb{Z}_p$-modules of the form

$$\mathfrak{W}_{\mathcal{F}}^0 \lhook\joinrel\longrightarrow \mathfrak{W}_{\mathcal{F}} \xrightarrow{\ \deg_{\mathcal{F}}\ } \mathbb{Z}_{p,\mathcal{F}}$$

in which the module $\mathbb{Z}_{p,\mathcal{F}}$ vanishes if $K_{\mathrm{ar}} \subseteq \mathcal{F}$ and is otherwise isomorphic to $\mathbb{Z}_p$.

A.2.2   Pro-characteristic ideals

We first record some essentially classical facts.

Lemma A.2. *Assume the notation and hypotheses of* § A.1.

(i) *Then for each non-negative integer d the Λ_d-modules $\mathfrak{W}_{\mathcal{K}_d}^0$ and $\mathfrak{W}_{\mathcal{K}_d}$ are both finitely generated and torsion.*

(ii) *Assume that there are only finitely many places in $\mathcal{K}_1$ that lie above those in S. Then for each integer d with $d > 1$ the Λ_{d-1}-modules $\mathfrak{W}_{\mathcal{K}_d}^0/t_d$ and $\mathfrak{W}_{\mathcal{K}_d}/t_d$ are both finitely generated and torsion.*

Note that claim (ii) of this result applies, in particular, to the cyclotomic extension $\mathcal{K}^{\mathrm{cyc}}$ defined above.

In [5] this observation is combined with Lemma A.1 and the fact that for each integer d with $d > 1$ the Λ_{d-1}-module $(\mathfrak{W}_{\mathcal{K}_d}^0)_{t_d}$ is pseudo-null (as is proved in [4, Corollary 5.8] for $\mathcal{K} = \mathcal{K}^{\mathrm{cyc}}$, and in a more general setting in [53]) to deduce the following result.

Proposition A.3. *Assume that the field $\mathcal{K}$ in § A.1 is equal to $\mathcal{K}^{\mathrm{cyc}}$. Then for each strictly positive integer d one has $\pi_d^{d+1}(\mathrm{ch}_{\Lambda_{d+1}}(\mathfrak{W}_{\mathcal{K}_{d+1}}^0)) = \mathrm{ch}_{\Lambda_d}(\mathfrak{W}_{\mathcal{K}_d}^0)$.*

This result in turn allows one to define the *pro-characteristic ideal* of the Λ-module $\mathfrak{W}_{\mathcal{K}^{\mathrm{cyc}}}^0$ by setting

$$\widetilde{\mathrm{ch}}_\Lambda(\mathfrak{W}_{\mathcal{K}^{\mathrm{cyc}}}^0) := \bigcap_{d \geq k} \pi_d^{-1}(\mathrm{ch}_{\Lambda_d}(\mathfrak{W}_{\mathcal{K}_d}^0))$$

for any large enough integer k, where each π_d denotes the canonical projection map $\Lambda \to \Lambda_d$.

This ideal is easily seen to be independent of the choice of filtration $\mathcal{K}_d$ and plays the role of the characteristic ideal in the formulation of a main conjecture of Iwasawa theory for the module $\mathfrak{W}_{\mathcal{K}^{\mathrm{cyc}}}^0$.

A.2.3   The main conjecture

We fix a field $\mathcal{K}$ as in § A.1, set $\Gamma := \mathrm{Gal}(\mathcal{K}/K)$, and assume that the set of places S is non-empty. Then, just as in § 6.1, the formal product

$$\theta_{\mathcal{K},S}(u) := \prod_{v \notin S} (1 - [v]u^{\deg(v)})^{-1}$$

(where $[v]$ is the Frobenius automorphism of v in Γ) gives a well-defined series in $\mathbb{Z}[\Gamma][[u]]$. One can also show that this series evaluates at $u = 1$ to give a well-defined 'Stickelberger element' $\theta_{\mathcal{K},S} := \theta_{\mathcal{K},S}(1)$ in Λ.

For each strictly positive integer d the image of $\theta_{\mathcal{K},S}$ in Λ_d is equal to the element $\theta_{\mathcal{K}_d,S}$ that occurs in Theorem 6.1. In particular, by combining the latter result (for each d) with the functorial behaviour described in Lemma A.2 and the definition of the pro-characteristic ideal $\widetilde{\mathrm{ch}}_\Lambda(\mathfrak{W}^0_{\mathcal{K}^{\mathrm{cyc}}})$ given above, one verifies the following main conjecture for the module $\mathfrak{W}^0_{\mathcal{K}^{\mathrm{cyc}}}$.

Theorem A.4. *In the (non-noetherian algebra)* $\Lambda := \Lambda(\mathrm{Gal}(\mathcal{K}^{\mathrm{cyc}}/K))$*, one has*

$$\widetilde{\mathrm{ch}}_\Lambda(\mathfrak{W}^0_{\mathcal{K}^{\mathrm{cyc}}}) = \Lambda \cdot \theta^\#_{\mathcal{K}^{\mathrm{cyc}},\{\mathfrak{p}\}}. \qquad ^1$$

A.3 Selmer groups in the p-cyclotomic extension

In this section we fix an abelian variety A over K which has either good ordinary or split multiplicative reduction at all of the places in S. Under this hypothesis one knows that for each natural number d, the Λ_d-module $X_p(A_d) := X_p(A/\mathcal{K}_d)$ is finitely generated.

In addition, if either $K_{\mathrm{ar}} \subseteq \mathcal{K}_d$ or A has good ordinary reduction at all of the places which ramify in $\mathcal{K}$ and the group $\mathrm{Sel}_p(A/K)$ is finite, then one knows that the Λ_d-module $X_p(A_d) := X_p(A/\mathcal{K}_d)$ is both finitely generated and torsion.

A.3.1 Pro-Fitting ideals

From [7] (see also [4]) one has the following result.

Proposition A.5. *Fix A as above. Also fix a natural number d and assume that either $A(\mathcal{K})[p^\infty]$ vanishes or that the Fitting ideal $\mathrm{Fit}_{\Lambda_d}(X_p(A_d))$ is principal. Then for any integer e with $e > d$ one has $\pi^e_d(\mathrm{Fit}_{\Lambda_e}(X_p(A_e))) \subseteq \mathrm{Fit}_{\Lambda_d}(X_p(A_d))$.*

Remark A.6. We note that if A is any (non-isotrivial) elliptic curve whose j-invariant is not a pth power in $K^\times$, then the group $A(\mathcal{K})[p^\infty]$ vanishes. Note also that the Fitting ideal of a finitely generated torsion Λ_d-module is principal if it is an elementary module or, more generally, has a presentation with the same number of generators as relations.

The above result motivates the following definition: if either $A(\mathcal{K})[p^\infty]$ vanishes or the Λ_d-ideal $\mathrm{Fit}_{\Lambda_d}(X_p(A_d))$ is principal for all sufficiently large integers d,

---

[1]Very recently Anglés, Bandini, Longhi and the author proved a non-noetherian Iwasawa main conjecture for the χ-component of $\mathfrak{W}^0_{\mathcal{K}_\infty}$ when $K = \mathbb{F}_q(T)$ and χ is a non-trivial character of $\mathrm{Gal}(K1/K)$, the result is obtained via the usual definition of Fitting ideals in the algebraic side. The analytic side is related also to Stickelberger elements and have links to some sort of p-adic L-function which is related with Bernoulli–Goss polynomials. See details in [1].

then the *pro-Fitting ideal* of the Λ-module $X_p(A/\mathcal{K})$ is the Λ-ideal that is defined by setting

$$\widetilde{\mathrm{Fit}}_\Lambda(X_p(A/\mathcal{K})) := \bigcap_{d \geq k} \pi_d^{-1}(\mathrm{Fit}_{\Lambda_d}(X_p(A_d)))$$

for any large enough integer k.

This definition provides a natural algebraic element for which one could hope to formulate a conjectural analogue of Theorem A.4.

At this stage, however, there are still technical difficulties on the analytic side. In fact, the only case in which a natural p-adic L-function $\mathcal{L}_{A,\mathcal{K}}$ has been attached to this data is the one in which A is an elliptic curve which has split multiplicative reduction at all places which ramify in $\mathcal{K}$. (See the upcoming survey article of Bandini and Longhi [9] for details of the various possible constructions of p-adic L-functions.) In this context, therefore, one can at least formulate the following main conjecture.

Conjecture A.7. Fix an extension $\mathcal{K}$ of K as in § A.1 and also an elliptic curve A over K that has split multiplicative reduction at all places which ramify in $\mathcal{K}$. If either $A(\mathcal{K})[p^\infty]$ vanishes or the Λ_d-ideal $\mathrm{Fit}_{\Lambda_d}(X_p(A_d))$ is principal for all natural numbers d, then in $\Lambda = \Lambda(\mathrm{Gal}(\mathcal{K}/K))$ one has an equality

$$\widetilde{\mathrm{Fit}}_\Lambda(X_p(A/\mathcal{K})) = \Lambda \cdot \mathcal{L}_{A,\mathcal{K}}.$$

A.3.2 Pro-characteristic ideals

In another direction, the following result shows that, under certain hypotheses, one can also use Lemma A.1 to define a natural pro-characteristic ideal in this setting.

Proposition A.8. *Assume that A has good ordinary reduction at all places of K which ramify in $\mathcal{K}$ and that the group $\mathrm{Sel}_{p^\infty}(A/K)$ is finite. Then for any $d > 1$ the Λ_{d-1}-module $X_p(A_d)/t_d$ is both finitely generated and torsion. Moreover, if $d > 2$, then one has*

$$\mathrm{ch}_{\Lambda_{d-1}}(X_p(A_d)/t_d) = \mathrm{ch}_{\Lambda_{d-1}}(X_p(A_{d-1}))\mathrm{ch}_{\Lambda_{d-1}}(\mathrm{coker}(a_{d-1}^d)^\vee),$$

where a_{d-1}^d is the natural restriction map $\mathrm{Sel}_{p^\infty}(A_{d-1}) \to \mathrm{Sel}_{p^\infty}(A_d)^{\mathrm{Gal}(\mathcal{K}^d/\mathcal{K}^{d-1})}$.

In [74] Tan describes the precise relation between the ideals $\mathrm{ch}_{\Lambda_{d-1}}(X_p(A_{d-1}))$ and $\pi_{d-1}^d(\mathrm{ch}_{\Lambda_d}(X_p(A_d)))$ under the assumption that A has either good ordinary or split multiplicative reduction at all places in S. In [74] it is also shown that if $\mathcal{K} = \mathcal{K}^{\mathrm{cyc}}$, then $X_p(A_d)_{t_d}$ is a pseudo-null Λ_{d-1}-module and from this one easily deduces the following result (taken from [6]). In this result, we write $\mathcal{A}_{d-1,v}^0$ for the closed fiber of the Néron model of A over the field $\mathcal{K}_{d-1,v}$.

Corollary A.9. *Assume that $\mathcal{K} = \mathcal{K}^{\mathrm{cyc}}$. Fix an abelian variety A over K which has good ordinary reduction at $\mathfrak{p}$ and is such that the group $\mathrm{Sel}_{p^\infty}(A/K)$ is finite. If d is any integer with $d > 2$ and such that for any place v of $\mathcal{K}_{d-1}$ at which A has bad reduction the order of $\pi_0(\mathcal{A}_{d-1,v}^0)$ is not divisible by p, then*

$$\pi_{d-1}^d(\mathrm{ch}_{\Lambda_d}(X_p(A_d))) = \mathrm{ch}_{\Lambda_{d-1}}(X_p(A_{d-1})).$$

The above equality is always true for $d \gg 0$ if no unramified places $v \in S$ splits completely in $\mathcal{K}$.

In particular, if one assumes that the hypotheses of Corollary A.9 are satisfied for all sufficiently large integers d, then one can define the *pro-characteristic ideal* of $X_p(A/\mathcal{K}^{\mathrm{cyc}})$ to be the Λ-ideal obtained by setting

$$\widetilde{\mathrm{ch}}_\Lambda(X_p(A/\mathcal{K}^{\mathrm{cyc}})) := \bigcap_{d \geq k} \pi_d^{-1}(\mathrm{ch}_{\Lambda_d}(X_p(A_d)))$$

for any large enough integer k.

At this time, though, it remains an outstanding problem to formulate (for any abelian variety A for which it is defined) a conjectural description of this ideal in terms of natural p-adic L-functions.

Acknowledgement

The author is very grateful to David Burns and Fabien Trihan for helpful discussions and for a complete rewriting of a first preliminar version of this appendix. He also thanks Ignazio Longhi and Andrea Bandini for very useful comments. Support from the project MTM2009-10359 is acknowledged.

Bibliography

[1] B. Anglés, A. Bandini, F. Bars, I. Longhi, Iwasawa main conjecture for the Carlitz cyclotomic extension and applications, in preparation.

[2] N. Aoki, J. Lee, K-S. Tan, A refinement for a conjecture of Gross, preprint 2005.

[3] S. Bae, On the conjectures of Lichtenbaum and Chinburg over function fields, Math. Ann. **285** (1989), 417–445.

[4] A. Bandini, F. Bars, I. Longhi, Aspects of Iwasawa theory over function fields, preprint 2011, `arXiv:1005.2289v2[math.NT]`, to appear in the proceedings of a Banff workshop to be published by EMS.

[5] A. Bandini, F. Bars, I. Longhi, Characteristic ideals in Iwasawa theory, New York J. Math. **20** (2014), 759–778.

[6] A. Bandini, F. Bars, I. Longhi, Characteristic ideals and Selmer groups, preprint 2014, arXiv:1404.2788v1[math.NT].

[7] A. Bandini, I. Longhi, Control theorems for elliptic curves over function fields, Int. J. Number Theory **5** (2009), 229–256.

[8] A. Bandini, I. Longhi, Selmer groups for elliptic curves in $\mathbb{Z}_\ell^d$-extensions of function fields of characteristic p, Ann. Inst. Fourier **59** (2009), 2301–2327.

[9] A. Bandini, I. Longhi, p-adic L-functions for elliptic curves over function fields, manuscript in preparation.

[10] A. Bandini, M. Valentino, Control theorems for ℓ-adic Lie extensions of global function fields, to appear in Ann. Sc. Norm. Super. Pisa Cl. Sci.

[11] W. Bauer, On the conjecture of Birch and Swinnerton-Dyer for abelian varieties over function fields in characteristic $p > 0$, Invent. Math. **108** (1992), 263–287.

[12] S. Bloch, K. Kato, L-functions and Tamagawa numbers of motives, in: The Grothendieck Festschrift, vol. 1, Progress in Math. vol. 86, Birkhäuser, Boston, 1990, 333–400.

[13] M. Breuning, D. Burns, Additivity of Euler characteristics in relative algebraic K-theory, Homol. Homotopy Appl. **7** (2005), 11–36.

[14] D. Burns, Equivariant Whitehead torsion and refined Euler characteristics, CRM Proc. Lecture Notes vol. 36 (2004), 35–59.

[15] D. Burns, On the values of equivariant Zeta functions of curves over finite fields, Doc. Math. **9** (2004), 357–399.

[16] D. Burns, Leading terms and values of equivariant motivic *L*-functions, Pure Appl. Math. Q. **6** (2010), 83–172.

[17] D. Burns, Congruences between derivatives of geometric *L*-functions (with an appendix by D. Burns, K.F. Lai and K.-S. Tan), Invent. Math. **184** (2011), 221–256.

[18] D. Burns, On derivatives of Artin *L*-series, Invent. Math. **186** (2011), 291–371.

[19] D. Burns, On main conjectures in non-commutative Iwasawa theory and related conjectures, to appear in J. reine angew. Math.

[20] D. Burns, On main conjectures in geometric Iwasawa theory and related conjectures, submitted for publication.

[21] D. Burns, M. Flach, Tamagawa numbers for motives with (non-commutative) coefficients, Doc. Math. **6** (2001), 501–570.

[22] D. Burns, K.F. Lai, K.-S. Tan, On geometric main conjectures, appendix to [17].

[23] D. Burns, C. Popescu, J. Sands, D. Solomon (eds.), Stark's Conjecture: recent progress and new direction, Contemp. Math. vol. 358, Amer. Math. Soc., Providence, 2004.

[24] D. Burns, O. Venjakob, On descent theory and main conjectures in non-commutative Iwasawa theory, J. Inst. Math. Jussieu **10** (2011), 59–118.

[25] Ph. Cassou-Noguès, T. Chinburg, A. Fröhlich, M.J. Taylor, *L*-functions and Galois modules, (notes by D. Burns and N.P. Byott), In: '*L*-functions and Arithmetic', J. Coates, M.J. Taylor (eds.), London Math. Soc. Lecture Note Series vol. 153, 75–139, Cambridge University Press, Cambridge, 1991.

[26] T. Chinburg, On the Galois structure of algebraic integers and *S*-units, Invent. Math. **74** (1983), 321–349.

[27] T. Chinburg, Exact sequences and Galois module structure, Ann. Math. **121** (1985), 351–376.

[28] T. Chinburg, Galois module structure of de Rham cohomology, J. Th. Nombres Bordeaux **4** (1991), 1–18.

[29] J. Coates, T. Fukaya, K. Kato, R. Sujatha, O. Venjakob, The GL_2 main conjecture for elliptic curves without complex multiplication, Publ. IHÉS **101** (2005), 163–208.

[30] R. Crew, *L*-functions of *p*-adic characters and geometric Iwasawa theory, Invent. Math. **88** (1987), 395–403.

[31] P. Deligne, Séminaire de Géométrie Algébrique du Bois-Marie, SGA4$\frac{1}{2}$, Lecture Notes in Math. vol. 569, Springer, Berlin, Heidelberg, 1977.

[32] C. Deninger, On Artin–Verdier duality for function fields, Math. Z. **188** (1984), 91–100.

[33] A. Dress, Induction and structure theorems for orthogonal representations of finite groups, Ann. of Math. **102** (1975), 291–325.

[34] J.S. Ellenberg, Selmer groups and Mordell–Weil groups of elliptic curves over towers of function fields, Compos. Math. **142** (2006), 1215–1230.

[35] M. Emerton, M. Kisin, Unit *L*-functions and a conjecture of Katz, Ann. of Math. **153** (2001), 329–354.

[36] J.-M. Fontaine, W. Messing, *p*-adic periods and *p*-adic étale cohomology. Current trends in arithmetical algebraic geometry, Proc. Summer Res. Conf., Arcata, California 1985, Contemp. Math. vol. 67, Amer. Math. Soc., Providence, 1987, 179–207.

[37] A. Fröhlich, Galois Module Structure of Algebraic Integers, Ergeb. Math. vol. 1, Springer, New York, 1983.

[38] T. Fukaya, K. Kato, A formulation of conjectures on *p*-adic zeta functions in non-commutative Iwasawa theory, Proc. St. Petersburg Math. Soc. Vol. XII, 1–85, Amer. Math. Soc. Transl. Ser. 2, vol. 219, Amer. Math. Soc., Providence, 2006.

[39] T. Geisser, Arithmetic cohomology over finite fields and special values of ζ-functions, Duke Math. J. **133** (2006), 27–58.

[40] R. Gold, H. Kisilevsky, On geometric $\mathbb{Z}_p$-extensions of function fields, Manuscripta Math. **62** (1988), 145–161.

[41] R. Greenberg, On the structure of certain Galois groups, Invent. Math. **47** (1978), 85–99.

[42] C. Greither, Arithmetic annihilators and Stark-type conjectures, in [23], pp. 55–78.

[43] C. Greither, C. Popescu, Fitting ideals of ℓ-adic realizations of 1-motives and class groups of function fields, J. reine angew. Math. **675** (2013), 223–247.

[44] B.H. Gross, On the value of abelian *L*-functions at $s = 0$, J. Fac. Sci. Univ. Tokyo, Sect. IA, Math., **35** (1988), 177–197.

[45] A. Grothendieck, Formule de Lefschetz et rationalité des fonctions *L*, in Sem. Bourbaki vol. 1965–1966, Benjamin (1966), exposé 306.

[46] A. Grothendieck, Revêtements Étales et Groupe Fondamental, Lecture Notes in Math. vol. 224, Springer, Berlin, Heidelberg, 1971.

[47] D. Hayes, The refined $\mathfrak{p}$-adic abelian Stark conjecture in function fields, Invent. Math. **94** (1988), 137–184.

[48] H. Johnston, A. Nickel, Noncommutative Fitting invariants and improved annihilation results, J. London Math. Soc., **88** (2013), 137–160.

[49] M. Kakde, The main conjecture of Iwasawa theory for totally real fields, Invent. Math. **193** (2013), 539–626.

[50] K. Kato, Iwasawa theory of totally real fields for Galois extensions of Heisenberg type, preprint, 2006.

[51] K. Kato, F. Trihan, On the conjecture of Birch and Swinnerton-Dyer in characteristic $p > 0$, Invent. Math. **153** (2003), 537–592.

[52] F. Knudsen, Determinant functors on exact categories and their extensions to categories of bounded complexes, Michigan Math. J. **50** (2002), 407–444.

[53] K.-L. Kueh, K.F. Lai, K.-S. Tan, On Iwasawa theory over function fields, preprint, 2007. `arXiv:math/0701060v1[math.NT]`.

[54] K.-L. Kueh, K.F. Lai, K.-S. Tan, The Stickelberger elements for $\mathbb{Z}_p^d$-extensions of function fields, J. Number Theory **128** (2008), 2776–2783.

[55] K.F. Lai, I. Longhi, K.-S. Tan, F. Trihan, Iwasawa main conjecture of abelian varieties over function fields, preprint, `http://arxiv.org/abs/1205.5945` (2012).

[56] S. Lang, Cyclotomic Fields, I and II, Graduate Texts in Math., vol. 121, Springer, New York, 1990.

[57] S. Lichtenbaum, Values of Zeta and L-functions at zero, Journées Arithmétiques de Bordeaux, pp. 133–138, Astérisque, **24–25**, Soc. Math. France, Paris, 1975.

[58] S. Lichtenbaum, The Weil-étale topology on schemes over finite fields, Compos. Math. **141** (2005), 689–702.

[59] J.S. Milne, Étale Cohomology, Princeton University Press, Princeton, 1980.

[60] F. Muro, A. Tonks, The 1-type of a Waldhausen K-theory spectrum, Adv. Math. **216** (2007), 178–211.

[61] A. Nickel, Non-commutative Fitting invariants and annihilation of class groups, J. Algebra **323** (2010), 2756–2778.

[62] T. Ochiai, F. Trihan, On the Selmer groups of abelian varieties over functions fields of characteristic $p > 0$, Math. Proc. Cambridge Philos. Soc. **146** (2009), 23–43.

[63] R. Oliver, Whitehead Groups of Finite Groups, London Math. Soc. Lecture Note Series vol. 132, Cambridge University Press, Cambridge, 1988.

[64] R. Oliver, L. Taylor, Logarithmic descriptions of Whitehead groups and class groups for p-groups, Mem. Amer. Math. Soc. **392** (1988).

[65] A. Pacheco, Selmer groups of abelian varieties in extensions of function fields, Math. Z. **261** (2009), 787–804.

[66] A. Parker, Equivariant Tamagawa numbers and non-commutative Fitting invariants, PhD Thesis, King's College London, 2007.

[67] M. Rapoport, Th. Zink, Über die lokale Zetafunktion von Shimuravarietäten. Monodromiefiltration und verschwindende Zyklen in ungleicher Charakteristik, Invent. Math. **68** (1982), 21–101.

[68] J. Ritter, A. Weiss, On the 'main conjecture' of equivariant Iwasawa theory, J. Amer. Math. Soc. **24** (2011), 1015–1050.

[69] K. Rubin, A Stark Conjecture 'over $\mathbb{Z}$' for abelian L-functions with multiple zeros, Ann. Inst. Fourier **46** (1996), 33–62.

[70] G. Sechi, GL_2 Iwasawa Theory of Elliptic Curves over Global Function Fields, PhD thesis, University of Cambridge, 2006.

[71] R.G. Swan, Algebraic K-theory, Lecture Notes in Math. vol. 76, Springer, Berlin, Heidelberg, 1968.

[72] L. Taelman, Special L-values of Drinfeld modules, Ann. of Math. **175** (2012), 369–391.

[73] L. Taelman, A Herbrand–Ribet theorem for function fields, Invent. Math. **188** (2012), 253–275.

[74] K.-S. Tan, Selmer groups over $\mathbb{Z}_p^d$-extensions, `http://arxiv.org/abs/1205.3907` (2012), to appear in Math. Ann.

[75] J. Tate, The cohomology groups of tori in finite Galois extensions of number fields, Nagoya Math. J. **27** (1966), 709–719.

[76] J. Tate, Les Conjectures de Stark sur les Fonctions L d'Artin en $s = 0$ (notes par D. Bernardi et N. Schappacher), Progress in Math. vol. 47, Birkhäuser, Boston, 1984.

[77] J. Tate, Letter to Joongul Lee, 22 July 1997.

[78] J. Tate, Refining Gross' conjecture on the values of abelian L-functions, Contemp. Math. vol. 358, Amer. Math. Soc., Providence, 2004, 189–192.

[79] M.J. Taylor, Classgroups of Group Rings, London Math Soc. Lecture Note Series vol. 91, Cambridge University Press, Cambridge, 1984.

[80] F. Trihan, D. Vauclair, On the non commutative Iwasawa Main Conjecture for abelian varieties over function fields, manuscript in preparation.

[81] C. Ward, On geometric Zeta functions, epsilon constants and canonical classes, PhD thesis, King's College London, 2011.

[82] M. Witte, Noncommutative Iwasawa main conjecture for varieties over finite fields, PhD thesis, Universität Leipzig, 2008. Available at `http://www.dart-europe.eu/full.php?id=162794`.

[83] M. Witte, On a noncommutative Iwasawa main conjecture for varieties over finite fields, J. Eur. Math. Soc. **16** (2014), 289–325.

The Ongoing Binomial Revolution

David Goss

The Ongoing Binomial Revolution

David Goss

Introduction

The Binomial Theorem has played a crucial role in the development of mathematics, algebraic or analytic, pure or applied. It was very important in the development of calculus, in a variety of ways, and has certainly been as important in the development of number theory. It plays a dominant role in function field arithmetic. In fact, it almost appears as if function field arithmetic (*and* a large chunk of arithmetic in general) is but a commentary on this amazing result. In turn, function field arithmetic has recently returned the favor by shedding new light on the Binomial Theorem. It is our purpose here to recall the history of the Binomial Theorem, with an eye on applications in characteristic p, and finish by discussing these new results.

We obviously make no claims here to being encyclopedic. Indeed, to thoroughly cover the Binomial Theorem would take many volumes. Rather, we have chosen to walk a quick and fine line through the many relevant results.

This paper constitutes a serious reworking of my April 2010 lectures at the Centre de Recerca Matemàtica in Barcelona. It is my great pleasure to thank the organizers of the workshop and, in particular, Francesc Bars.

1 Early history

According to our current understanding, the Binomial Theorem can be traced to the 4th century B.C. and Euclid where one finds the formula for $(a + b)^2$. In the 3rd century B.C. the Indian mathematician Pingala presented what is now known as "Pascal's triangle" giving binomial coefficients in a triangle. Much later, in the 10th century A.D., the Indian mathematician Halayudha and the Persian mathematician al-Karaji derived similar results as did the 13th century Chinese mathematician Yang Hui. It is remarkable that al-Karaji appears to have used mathematical induction in his studies.

Indeed, binomial coefficients, appearing in Pascal's triangle, seem to have been widely known in antiquity. Besides the mathematicians mentioned above, Omar Khayyam (in the 11th century), Tartaglia, Cardano, Viète, Michael Stifel (in the 16th century), and William Oughtred, John Wallis, Henry Briggs, and Father Marin Mersenne (in the 17th century) knew of these numbers. In the 17th century, Blaise Pascal gave the binomial coefficients their now commonly used form: for a nonnegative integer n, one sets

$$\binom{n}{k} = \frac{n!}{k!(n-k)!} = \frac{n(n-1)(n-2)\cdots(n-k+1)}{k(k-1)(k-2)\cdots 1}. \tag{1.1}$$

With this definition we have the very famous, and equally ubiquitous, *Binomial Theorem*:

$$(x+y)^n = \sum_k \binom{n}{k} x^k y^{n-k}. \tag{1.2}$$

And of course, we also deduce the first miracle giving the integrality of the binomial coefficients $\binom{n}{k}$.

Replacing n by a variable s in Equation 1.1 gives the *binomial polynomials*

$$\binom{s}{k} = \frac{s(s-1)(s-2)\cdots(s-k+1)}{k(k-1)(k-2)\cdots 1} = \frac{s(s-1)(s-2)\cdots(s-k+1)}{k!}. \tag{1.3}$$

2   Newton, Euler, Abel, and Gauss

We now come to Sir Isaac Newton and his contribution to the Binomial Theorem. His contributions evidently were discovered in the year 1665 (while sojourning in Woolsthorpe, England to avoid an outbreak of the plague) and discussed in a letter to Oldenburg in 1676. Newton was highly influenced by work of John Wallis who was able to calculate the area under the curves $(1-x^2)^n$, for n a nonnegative integer. Newton then considered fractional exponents s instead of n. He realized that one could find the successive coefficients c_k of $(-x^2)^k$, in the expansion of $(1-x^2)^s$, by multiplying the previous coefficient by $(s-k+1)/k$ exactly as in the integral case. In particular, Newton formally computed the Maclaurin series for $(1-x^2)^{1/2}$, $(1-x^2)^{3/2}$ and $(1-x^2)^{1/3}$.

(One can read about this in the paper [Co1], where the author believes that Newton's contributions to the Binomial Theorem were relatively minor and that the credit for discussing fractional powers should go to James Gregory – who in 1670 wrote down the series for $b\left(1+\frac{d}{b}\right)^{a/c}$. This is a distinctly minority viewpoint.)

In any case, Newton's work on the Binomial Theorem played a role in his subsequent work on calculus. However, Newton did not consider issues of convergence. This was discussed by Euler, Abel, and Gauss. Gauss gave the first satisfactory proof of convergence of such series in 1812. Later Abel gave a treatment that would work for general complex numbers. The theorem on binomial series can now be stated.

Theorem 2.1. *Let* $s \in \mathbb{C}$. *Then the series* $\sum_{k=0}^{\infty} \binom{s}{k} x^k$ *converges to* $(1+x)^s$ *for all complex* x *with* $|x| < 1$.

Remark 2.2. Let $s = n$ be any integer, positive or negative. Then for all complex x and y with $|x/y| < 1$ one readily deduces from Theorem 2.1 a convergent expansion

$$(x + y)^n = \sum_{k=0}^{\infty} \binom{n}{k} x^k y^{n-k}. \tag{2.4}$$

It is worth noting that Gauss' work on the convergence of the binomial series marks the first time convergence involving *any* infinite series was satisfactorily treated!

Now let $f(x)$ be a polynomial with coefficients in an extension of $\mathbb{Q}$. The degree of $\binom{x}{k}$ as a polynomial in x is k. As such one can always expand $f(x)$ as a linear combination of the terms $\binom{x}{k}$. Such an expansion is called the *Newton series* and can be traced back to his *Principia Mathematica* (1687). The coefficients of such an expansion are given as follows.

Proposition 2.3. *If we set* $(\Delta f)(x) = f(x+1) - f(x)$, *then*

$$f(x) = \sum_{k} (\Delta^k f)(0) \binom{x}{k}. \tag{2.5}$$

3 The pth power mapping

Let p be a prime number and let $\mathbb{F}_p$ be the field with p elements. The following elementary theorem is then absolutely fundamental for number theory and arithmetic geometry. Indeed its importance cannot be overstated.

Theorem 3.1. *Let R be any $\mathbb{F}_p$-algebra. Then the mapping $x \mapsto x^p$ is a homomorphism from R to itself.*

As is universally known, the proof amounts to expanding by the Binomial Theorem and noting that, for $0 < i < p$, one has $\binom{p}{i} \equiv 0 \pmod{p}$ as the denominator of Equation 1.1 is prime to p.

According to Leonard Dickson's history (Chapter III of [Di1]), the first person to establish (a form of) Theorem 3.1 was Gottfried Leibniz on September 20, 1680. One can then rapidly deduce a proof of Fermat's Little Theorem, i.e., $a^p \equiv a \pmod{p}$ for all integers a and primes p. Around 1830 Galois used interates of the pth power mapping to construct general finite fields.

It was 216 years after Leibniz (1896) that the equally essential *Frobenius automorphism* (or *Frobenius substitution*) in the Galois theory of fields was born.

Much of modern number theory and algebraic geometry consists of computing invariants of the pth power mapping/Frobenius map.

Drinfeld modules are subrings of the algebra (under composition!) of "polynomials" in the pth power mapping; thus their very existence depends on the Binomial Theorem.

Remark 3.2. With regard to the proof of Theorem 3.1, it should also be noted that Kummer in 1852 established that the exact power of a prime p dividing $\binom{n}{k}$ is precisely the number of "carries" involved in adding $n - k$ and k when they are expressed in their canonical p-adic expansion.

4   The Theorem of Lucas

Basic for us, and general arithmetic in finite characteristic, is the famous Theorem of Lucas from 1878 [Lu1]. Let n and k be two nonnegative integers and p a prime. Write n and k p-adically as $n = \sum_i n_i p^i$, $0 \le n_i < p$ and $k = \sum_i k_i p^i$, $0 \le k_i < p$.

Theorem 4.1 (Lucas). *We have*

$$\binom{n}{k} = \prod_i \binom{n_i}{k_i} \quad (\mathrm{mod}\ p). \tag{4.6}$$

Proof. We have $(1+x)^n = (1+x)^{\sum_i n_i p^i} = \prod_i (1+x)^{n_i p^i}$. Modulo p, Theorem 3.1 implies that $(1+x)^n = \prod_i (1+x^{p^i})^{n_i}$. The result then follows by expressing both sides by the Binomial Theorem and the uniqueness of p-adic expansions. $\square$

5   The Theorem of Mahler

The binomial polynomials $\binom{s}{k}$ (given in Equation 1.3) obviously have coefficients in $\mathbb{Q}$ and thus can also be considered in the p-adic numbers $\mathbb{Q}_p$.

Proposition 5.1. *The functions* $\binom{s}{k}$, $k = 0, 1, \ldots$, *map* $\mathbb{Z}_p$ *to itself.*

Proof. Indeed, $\binom{s}{k}$ takes the nonnegative integers to themselves. As these are dense in $\mathbb{Z}_p$, and $\mathbb{Z}_p$ is closed, the result follows. $\square$

Let $y \in \mathbb{Z}_p$, and formally set $f_y(x) = (1+x)^y$. By the above proposition, $f_y(x) \in \mathbb{Z}_p[[x]]$. As such, we can consider $f_y(x)$ in *any* non-Archimedean field of *any* characteristic where it will converge on the open unit disc.

Let $\{a_k\}$ be a collection of p-adic numbers approaching 0 as $k \to \infty$ and put $g(s) = \sum_k a_k \binom{s}{k}$; it is easy to see that this series converges to a continuous function from $\mathbb{Z}_p$ to $\mathbb{Q}_p$. Moreover, given a continuous function $f \colon \mathbb{Z}_p \to \mathbb{Q}_p$, the Newton series (Equation 2.5) certainly makes sense formally.

Theorem 5.2 (Mahler). *The Newton series of a continuous function* $f \colon \mathbb{Z}_p \to \mathbb{Q}_p$ *uniformly converges to it.*

The proof can be found in [Ma1] (1958). The Mahler expansion of a continuous p-adic function is obviously unique.

Mahler's Theorem can readily be extended to continuous functions of $\mathbb{Z}_p$ into complete fields of characteristic p. One can also find analogs of it that work for functions on the maximal compact subrings of arbitrary local fields. In characteristic p, an especially important analog of the binomial polynomials was constructed by L. Carlitz as a byproduct of his construction of the *Carlitz module* (see, e.g., [Wa1]).

Carlitz' construction can be readily described. Let $e_k(x) = \prod(x - \alpha)$ where α runs over elements of $\mathbb{F}_q[t]$, $q = p^{m_0}$, of degree $< k$. As these elements form a finite-dimensional $\mathbb{F}_q$-vector space, it is easy to see that the functions $e_k(x)$ are $\mathbb{F}_q$-linear. Set $D_k = e_k(t^k) = \prod f$ where $f(t)$ runs through the *monic* polynomials of degree k. Carlitz then establishes that $e_k(g)/D_k$ is *integral* for $g \in \mathbb{F}_q[t]$.

Remark 5.3. The binomial coefficients $\binom{s}{k}$ appear in the power series expansion of $(1 + x)^s$. It is very important to note that the polynomials $e_k(x)/D_k$ appear in a completely similar fashion in terms of the expansion of the Carlitz module – an $\mathbb{F}_q[t]$-analog of $\mathbb{G}_m$; see, e.g., Corollary 3.5.3 of [Go1].

Now let k be any nonnegative integer written q-adically as $\sum_t k_t q^t$, $0 \le k_t < q$.

Definition 5.4. We write $G_k(x) = \displaystyle\prod_t \left(\frac{e_t(x)}{D_t} \right)^{k_t}$.

The set $\{G_k(x)\}$ is then an excellent characteristic p replacement for $\{\binom{s}{k}\}$ in terms of analogs of Mahler's Theorem, etc.; see [Wa1]. In 2000 K. Conrad [Con1] showed that Carlitz' use of digits in constructing analogs of $\binom{s}{k}$ can be applied quite generally.

In a very important refinement of Mahler's result, in 1964 Y. Amice [Am1] gave necessary and sufficient conditions on the Mahler coefficients guaranteeing that a function can be locally expanded in power series. In fact, Amice's results work for arbitrary local fields and are also essential for the function field theory. Indeed, as the function $(1 + x)^y$, $y \in \mathbb{Z}_p$, is clearly locally analytic, Amice's results show that its expansion coefficients tend to 0 very quickly, thus allowing for general analytic continuation of L-series and partial L-series [Go2].

In 2009, S. Jeong [Je1] established that the functions $u \mapsto u^y$, $y \in \mathbb{Z}_p$ precisely comprise the group of *locally-analytic* endomorphisms of the 1-units in a local field of finite characteristic.

6   Measure theory

Given a local field K with maximal compact R, one is able to describe a theory of integration for all continuous K-valued functions on R. A *measure on R with values in R* is a finitely-additive, R-valued function on the compact open subsets of R. Given a measure μ and a continuous K-valued function f on R, the Riemann sums for f (in terms of compact open subsets of R) are easily seen to converge to an element of K naturally denoted $\int_R f(z)\,d\mu(z)$.

Given two measures μ_1 and μ_2, we are able to form their convolution $\mu_1 * \mu_2$ in exactly the same fashion as in classical analysis. In this way, the space of measures forms a commutative K-algebra.

In the case of $\mathbb{Q}_p$ and $\mathbb{Z}_p$ one is able to use Mahler's Theorem (Theorem 5.2 above) to express integrals of general continuous functions in terms of the integrals of binomial coefficients.

Now $(1 + z)^{x+y} = (1 + z)^x (1 + z)^y$ giving an *addition formula* for the binomial coefficients. Using this in the convolution allows one to establish that the convolution algebra of measures (the *Iwasawa algebra*) is isomorphic to $\mathbb{Z}_p[[X]]$.

In finite characteristic, we obtain a *dual* characterization of measures that is still highly mysterious and *also* depends crucially on the Binomial Theorem. So let q be a power of a prime p as above. Let n be a nonnegative integer written q-adically as $\sum n_k q^k$. Thus, in characteristic p, we deduce that

$$(x + y)^n = \prod_k (x + y)^{n_k q^k} = \prod_k \left(x^{q^k} + y^{q^k} \right)^{n_k}. \tag{6.7}$$

Now recall the definition of the functions $G_n(x)$ (Definition 5.4 above) via digit expansions. As the functions $e_j(x)$ are also *additive* we immediately deduce from Equation 6.7 the next result.

Theorem 6.1. *We have*

$$G_n(x + y) = \sum_{j=0}^{n} \binom{n}{j} G_j(x) G_{n-j}(x). \tag{6.8}$$

In other words, the functions $\{G_n(x)\}$ *also* satisfy the Binomial Theorem!

Let $\mathfrak{D}_j$ be the hyperdifferential (= "divided derivative") operator given by $\mathfrak{D}_j z^i = \binom{i}{j} z^{i-j}$. Notice that $\mathfrak{D}_i \mathfrak{D}_j = \binom{i+j}{i} \mathfrak{D}_{i+j}$. Let $R\{\{\mathfrak{D}\}\}$ be the algebra of formal power series in the $\mathfrak{D}_i$ with the above multiplication rule, where R is any commutative ring. Note further that this definition makes sense for all R precisely since $\binom{i}{j}$ is always integral.

Let $A = \mathbb{F}_q[t]$ and let $f \in A$ be irreducible; set $R = A_f$, the completion of A at (f). Using the Binomial Theorem for the Carlitz polynomials we have the next result [Go3]:

Theorem 6.2. *The convolution algebra of R-valued measures on R is isomorphic to $R\{\{\mathfrak{D}\}\}$.*

Remark 6.3. The history of Theorem 6.2 is amusing. I had calculated the algebra of measures using the Binomial Theorem and then showed the calculation to Greg Anderson who, rather quickly(!), recognized it as the ring of hyperderivatives/divided power series.

Remark 6.4. One can ask why we represent the algebra of measures as operators as opposed to divided power series. Let μ be a measure on R (where R is as above) and let f be a continuous function; one can then obtain a new continuous function $\mu(f)$ by

$$\mu(f)(x) = \int_R f(x+y) \, d\mu(y). \tag{6.9}$$

The operation of passing from the expansion of f (in the Carlitz polynomials) to the expansion of $\mu(f)$ formally appears as if the differential operator attached to μ acted on the expansion. This explains our choice.

7 The group $S_{(p)}$ and binomial symmetries in finite characteristic

Let $q = p^{m_0}$, p prime, as above, and let $y \in \mathbb{Z}_p$. Write y q-adically as

$$y = \sum_{k=0}^{\infty} y_k q^k, \tag{7.10}$$

where $0 \le y_k < q$ for all k. If y is a nonnegative integer (so that the sum in Equation 7.10 is obviously finite), then we set $\ell_q(y) = \sum_k y_k$.

Let ρ be a permutation of the set $\{0, 1, 2, \ldots\}$.

Definition 7.1. We define $\rho_*(y)$, $y \in \mathbb{Z}_p$, by

$$\rho_*(y) = \sum_{i=0}^{\infty} y_k q^{\rho(i)}. \tag{7.11}$$

Clearly $y \mapsto \rho_*(y)$ is a bijection of $\mathbb{Z}_p$. Let $S_{(q)}$ be the group of bijections of $\mathbb{Z}_p$ obtained this way. Note that if q_0 and q_1 are powers of p, and $q_0 \mid q_1$, then $S_{(q_1)}$ is naturally realized as a subgroup of $S_{(q_0)}$.

Proposition 7.2. *Let $\rho_*(y)$ be defined as above.*

1. *The mapping $y \mapsto \rho_*(y)$ is a homeomorphism of $\mathbb{Z}_p$.*
2. *("Semi-additivity") Let x, y, z be three p-adic integers with $z = x + y$ and where there is no carry over of q-adic digits. Then $\rho_*(z) = \rho_*(x) + \rho_*(y)$.*
3. *The mapping $\rho_*(y)$ stabilizes both the nonnegative and nonpositive integers.*
4. *Let n be a nonnegative integer. Then $\ell_q(n) = \ell_q(\rho_*(n))$.*
5. *Let n be an integer. Then $n \equiv \rho_*(n) \pmod{q-1}$.*

For the proof, see [Go4].

Proposition 7.3. *Let* $\sigma \in S_{(p)}$, $y \in \mathbb{Z}_p$, *and* k *a nonnegative integer. Then we have*

$$\binom{y}{k} \equiv \binom{\sigma y}{\sigma k} \pmod{p}. \tag{7.12}$$

Proof. This follows immediately from the Theorem of Lucas (Theorem 4.1).   □

Corollary 7.4. *Modulo* p*, we have* $\binom{\sigma y}{k} = \binom{y}{\sigma^{-1} k}$.

Corollary 7.5. *We have* $p \mid \binom{y}{k}$ *if and only if* $p \mid \binom{\sigma y}{\sigma k}$.

Proposition 7.6. *Let* i *and* j *be two nonnegative integers. Let* $\sigma \in S_{(p)}$*. Then*

$$\binom{i + j}{i} \equiv \binom{\sigma i + \sigma j}{\sigma i} \pmod{p}. \tag{7.13}$$

Proof. The theorems of Lucas and Kummer show that if there is any carry over of p-adic digits in the addition of i and j, then $\binom{i+j}{i}$ is 0 modulo p. However, there is carry over of the p-adic digits in the sum of i and j if and only if there is carry over in the sum of σi and σj; in this case both sums are 0 modulo p. If there is no carry over, then the result follows from Part 2 of Proposition 7.2 and Proposition 7.3.   □

Let R be as in the previous section.

Corollary 7.7. *The mapping* $\mathfrak{D}_i \mapsto \mathfrak{D}_{\sigma i}$ *is an automorphism of* $R\{\{\mathfrak{D}\}\}$.

It is quite remarkable that the group $S_{(q)}$ very much *appears* to be a symmetry group of characteristic p L-series. Indeed, in examples, this group preserves the orders of trivial zeroes as well as the denominators of special zeta values (the "Bernoulli–Carlitz" elements). Moreover, given a nonnegative integer i, one has the "special polynomials" of characteristic p L-series arising at $-i$. It is absolutely remarkable, and highly nontrivial to show, that the degrees of these special polynomials are *invariant* under the action of $S_{(q)}$ on i. Finally, the action of $S_{(q)}$ even appears to extend to the *zeroes* themselves of these characteristic p functions. See [Go4] for all this.

8   The future

We have seen how the Binomial Theorem has impacted the development of both algebra and analysis. In turn these developments have provided the foundations for characteristic p arithmetic. Furthermore, as in Section 7, characteristic p arithmetic has contributed results relating to the Binomial Theorem of both an algebraic (automorphisms of $\mathbb{Z}_p$ and binomial coefficients) and analytic (automorphisms of algebras of divided derivatives) nature. Future research should lead to a deeper understanding of these recent offshoots of the Binomial Theorem as well as add many, as yet undiscovered, new ones.

Bibliography

[Am1] Y. AMICE: Interpolation p-adique, *Bull. Soc. Math. France* **92** (1964), 117–180.

[Con1] K. CONRAD: The digit principle, *J. Number Theory* **84** (2000), 230–257.

[Co1] J. L. COOLIDGE: The story of the Binomial Theorem, *Amer. Math. Monthly* **56(3)** (Mar. 1949), 147–157.

[Di1] L. DICKSON: *History of the Theory of Numbers*, Vol. I, Chelsea Publishing Co., New York (1971).

[Go1] D. GOSS: *Basic Structures of Function Field Arithmetic*, Springer-Verlag, Berlin (1996).

[Go2] D. GOSS: Applications of non-Archimedean integration to the L-series of τ-sheaves, *J. Number Theory* **110** (2005), 83–113.

[Go3] D. GOSS: Fourier series, measures and divided power series in the arithmetic of function fields, *K-theory* **1** (1989), 533–555.

[Go4] D. GOSS: ζ-phenomenology, *Noncommutative Geometry, Arithmetic, and Related Topics: Proceedings of the Twenty-First Meeting of the Japan-U.S. Mathematics Institute*, The Johns Hopkins University Press, Baltimore, MD (to appear).

[Gr1] A. GRANVILLE: Arithmetic properties of binomial coefficients, I, Binomial coefficients modulo prime powers, *Organic mathematics (Burnaby, BC, 1995)*, 253–276, CMS Conf. Proc., 20, Amer. Math. Soc., Providence, RI, 1997.

[Je1] S. JEONG: On a question of Goss, *J. Number Theory* **129(8)** (2009), 1912–1918.

[Lu1] E. LUCAS: Théorie des fonctions numériques simplement périodiques, *Amer. J. Math.* **1(2)** (1878), 184–196.

[Ma1] K. MAHLER: An interpolation series for continuous functions of a p-adic variable, *J. reine angew. Math.* **199** (1958), 23–34.

[Wa1] C. G. WAGNER: Interpolation series for continuous functions on π-adic completions of $GF(q, x)$, *Acta Arith.* **17** (1971), 389–406.

Arithmetic of Gamma, Zeta and Multizeta Values for Function Fields

Dinesh S. Thakur

Arithmetic of Gamma, Zeta and Multizeta Values for Function Fields

Dinesh S. Thakur

Introduction

In the advanced course given at Centre de Recerca Matemàtica, consisting of twelve hour lectures from 22 February to 5 March 2010, we described results and discussed some open problems regarding the gamma and zeta functions in the function field context. The first four lectures of these notes, dealing with gamma, roughly correspond to the first four lectures of one and half hour each, and the last three lectures, dealing with zeta, cover the last three two-hour lectures. Typically, in each part, we first discuss elementary techniques, then easier motivating examples with Drinfeld modules in detail, and then outline general results with higher-dimensional t-motives. Lecture 4 is independent of Lecture 3, whereas the last part (last three lectures) is mostly independent of the first part, except that the last two lectures depend on Lecture 3. At the end, we include a guide to the relevant literature.

We will assume that the reader has basic familiarity with the language of function fields, cyclotomic fields and Drinfeld modules and t-motives, though we will give quick reviews at the appropriate points. We will usually just sketch the main points of the proofs, leaving the details to references.

We will use the following setting and notation, although sometimes it will be specialized.

$\mathbb{F}_q$: a finite field of characteristic p having q elements
X: a smooth, complete, geometrically irreducible curve over $\mathbb{F}_q$
K: the function field of X
∞: a closed point of X, i.e., a place of K
d_∞: the degree of the point ∞
A: the ring of elements of K with no pole outside ∞

Supported in part by NSA grants H98230-08-1-0049 and H98230-10-1-0200.

K_∞: the completion of K at ∞

C_∞: the completion of an algebraic ('separable', equivalently) closure of K_∞

$\overline{K}$, $\overline{K_\infty}$: the algebraic closures of K, K_∞ in C_∞

$\mathbb{F}_\infty$: the residue field at ∞

A_v: the completion of A at a place $v \neq \infty$

g: the genus of X

h: the class number of K

h_A: the class number of A $(= h d_\infty)$

We will consider ∞ to be the distinguished place at infinity and call any other place v a *finite* place. It can be given by a (non-zero) prime ideal $\wp$ of A. As we have not fixed ∞ above, we have already defined K_v, $K_\wp$, $\mathbb{F}_\wp$, C_v, d_v etc.

Since $\mathbb{F}_\infty^*$ has $q^{d_\infty} - 1$ elements, in defining zeta and gamma when $d_\infty > 1$, sign conditions giving analogs of positivity have to be handled more carefully by choosing sign representatives in $\mathbb{F}_\infty^*/\mathbb{F}_q^*$ rather than just saying monic. Having said this, *we will assume $d_\infty = 1$ throughout* for simplicity, leaving the technicalities of the more general case to the references.

Basic analogs are

$$K \longleftrightarrow \mathbb{Q}, \quad A \longleftrightarrow \mathbb{Z}, \quad K_\infty \longleftrightarrow \mathbb{R}, \quad C_\infty \longleftrightarrow \mathbb{C}.$$

Instead of $\mathbb{Q}$ we can also have an imaginary quadratic field with its unique infinite place and the corresponding data.

The Dedekind domain A sits discretely in K_∞ with compact quotient, in analogy with $\mathbb{Z}$ inside $\mathbb{R}$ or the ring of integers of an imaginary quadratic field inside $\mathbb{C}$. We will see that analogies are even stronger when X is the projective line over $\mathbb{F}_q$ and ∞ is of degree 1, so that $K = \mathbb{F}_q(t)$, $A = \mathbb{F}_q[t]$, $K_\infty = \mathbb{F}_q((1/t))$. Comparing sizes of A^* and $\mathbb{Z}^*$, which are $q-1$ and 2 respectively, in our situation, we call multiples of $q-1$ 'even' and other integers in $\mathbb{Z}$ 'odd'.

It should be kept in mind that, unlike in number field theory, there is neither a canonical base like the prime field $\mathbb{Q}$, nor a canonical (archimedean) place at infinity. Unlike $\mathbb{Q}$, the field K has many automorphisms. Another crucial difference – a big plus point, as we will see – is that we can combine (even isomorphic copies of our) function fields by using independent variables.

In many respects, in classical function field theory one works over $\overline{\mathbb{F}_q}$, which is the maximal 'cyclotomic' extension of $\mathbb{F}_q$, and uses Frobenius to descend. Classical study of function fields, their zeta functions, geometric class field theory goes smoothly (once we understand genus well with Riemann–Roch) for all function fields, any q, and treats all places similarly, while we do distinguish a place at infinity. Some aspects, such as zeta zero distributions, show much more complexity in our case, when q is not a prime.

We denote by H the Hilbert class field for A, i.e., the maximal abelian, unramified extension of K in which ∞ splits completely. There are h sign-normalized, non-isomorphic, rank one Drinfeld A-modules; they are Galois conjugates and are defined over H, which is a degree h extension of K. We use e and ℓ for the exponential and the logarithm for a Drinfeld module ρ. Recall that $e(az) = \rho_a(e(z))$, $\ell(\rho_a(z)) = a\ell(z)$. We write Λ_a for the corresponding a-torsion. We denote by Λ the lattice corresponding to ρ, namely the kernel of the exponential of ρ. Recall that $e(z) = z \prod'(1 - z/\lambda)$, where the λ runs over the non-zero elements of Λ. If ρ corresponds to the principal ideal class (e.g., for the Carlitz module for $A = \mathbb{F}_q[t]$ or for class number one A's), then we write $\Lambda = \widetilde{\pi}A$ and think of $\widetilde{\pi}$ as analog of $2\pi i$. The Carlitz module C is given by $C_t(z) = tz + z^q$.

Warning on the notation. We use the same notation for the same concepts, for example, for various gamma in function fields, or the complex gamma. It will be always made clear, sometimes at the start of a lecture or a section, which concept we are talking about.

Lecture 1

Gamma: Definitions, Properties and Functional Equations

Once we fix a place at infinity, we have two kinds of families of cyclotomic extensions. The first family, the one mentioned above, is the family of constant field extensions. These are cyclotomic abelian (everywhere unramified) extensions obtained by solving $x^n - 1 = 0$, for (non-zero) integers n. (The classical Iwasawa theory exploits these analogies.) The second cyclotomic family is obtained by adjoining a-torsion of appropriate rank one Drinfeld A-modules (rather than n-torsion of the multiplicative group as above) for (non-zero) integers $a \in A$.

We will see two kinds of gamma functions closely connected to these two theories. The second one is often called geometric as (or when) there is no constant field extension involved. The first one is then called arithmetic.

1.1 Arithmetic gamma for $\mathbb{F}_q[t]$: Definitions and analogies

The easiest way to introduce the Carlitz factorial Π associated to $A = \mathbb{F}_q[t]$ is to define it, for $n \in \mathbb{Z}_{\geq 0}$, by

$$\Pi(n) := n! := \prod_{\wp \text{ monic prime}} \wp^{n_\wp} \in \mathbb{F}_q[t], \quad n_\wp := \sum_{e \geq 1} \left\lfloor \frac{n}{\text{Norm}\,\wp^e} \right\rfloor$$

in analogy with the well-known prime factorization of the usual factorial. We define $\Gamma(n) = \Pi(n-1)$ as usual.

The following formula, which is the key to everything that follows, also gives a much faster way to compute it:

$$\Pi(n) = \prod_i D_i^{n_i} \quad \text{for } n = \sum n_i q^i \text{ with } 0 \leq n_i < q,$$

where

$$D_i = \left(t^{q^i} - t\right)\left(t^{q^i} - t^q\right) \cdots \left(t^{q^i} - t^{q^{i-1}}\right).$$

Here is a quick sketch of how one can see the equivalence. The case $n = q^i$ immediately follows from the following fact:

Claim: D_i is the product of all monic polynomials of degree i.

The τ-version (i.e., q-linearized version) of the more familiar Vandermonde determinant $|x_j^{i-1}| = \prod_{i>j}(x_i - x_j)$ is the Moore determinant (a very useful tool)

$$M(x_i) := \left|\tau^{i-1}(x_j)\right| = \left|x_j^{q^{i-1}}\right| = \prod_i \prod_{f_j \in \mathbb{F}_q} \left(x_i + f_{i-1}x_{i-1} + \cdots + f_1 x_1\right).$$

(The proof of the last equality is similar to that of the Vandermonde identity.) Now $M(1, t, \ldots, t^d)$ is a Moore determinant as well as Vandermonde, so that the two evaluations give us the claim (after taking the ratio of terms for d and $d - 1$).

The general case follows from $\sum \lfloor n/\mathrm{Norm}(\wp)^e \rfloor = \sum n_i \lfloor q^i/\mathrm{Norm}(\wp)^e \rfloor$, where $n = \sum n_i q^i$ is the base q expansion of n, i.e., $0 \le n_i < q$.

Example 1.1. For $q = 3$, for $i = 0, 1, 2$ we have $\Pi(i) = 1$, $\Pi(3 + i) = t^3 - t$, $\Pi(6 + i) = (t^3 - t)^2$ and $\Pi(9 + i) = (t^9 - t)(t^9 - t^3)$.

Note that $D_i = (q^i)!$ fits in with the analogy:

$$e^z = \sum z^n/n!, \quad e(z) = \sum z^{q^n}/D_n,$$

where we compare the usual exponential e^z with the Carlitz exponential $e(z)$. This can be seen by substituting $e(z) = \sum e_i z^{q^i}$ in $e(tz) = te(z) + e(z)^q$ coming from the functional equation of the Carlitz module action and solving for $e_i = 1/D_i$ with initial value $e_0 = 1$ to get $D_i = (t^{q^i} - t)D_{i-1}^q$.

The general factorial is obtained then by these basic building blocks by digit expansion, a phenomenon which we will see again and again in various contexts. It can be motivated in this case by the desire to have integral binomial coefficients. For example, $q^{n+1}!/q^n! = D_{n+1}/D_n = (t^{q^{n+1}} - t)D_n^{q-1} = (t^{q^{n+1}} - t)q^n!^{q-1}$, which suggests that the factorial of $q^{n+1} - q^n = (q - 1)q^n$ should be D_n^{q-1}, as it is for the Carlitz factorial.

Here are some naive analogies. The defining polynomial $[n] = t^{q^n} - t$ of $\mathbb{F}_{q^n}$, which can also be described as the product of monic irreducible polynomials of degree dividing n, does sometimes play some role analogous to the usual n, or rather q^n.

We have $q^n! = D_n = ([n] - [0])([n] - [1]) \cdots ([n] - [n - 1])$ in this context looking like a factorial of $[n]$. One has twisted recursions

$$\Pi(q^{n+1}) = [n + 1]\Pi(q^n)^q, \quad [n + 1] = [n]^q + [1]$$

in place of the usual $(n + 1)! = (n + 1)n!$ and $(n + 1) = n + 1$ respectively.

In this vein, note that $[n][n-1]\cdots[1] =: L_n$ and

$$([k+1]-[k])([k+2]-[k])\cdots([n]-[k]) = L_{n-k}^{q^k}$$

also play fundamental roles, as we will see.

These vague analogies are made much more precise by a definition, due to Manjul Bhargava [Bha97, Bha00], of a factorial in a very general context.

Let X be an arbitrary nonempty subset of a Dedekind ring R (i.e., noetherian, locally principal and with all non-zero primes maximal). Special cases would be Dedekind domains $\mathcal{O}_S$ coming from global fields or their quotients. Bhargava associates to a natural number k an ideal $k!_X := \prod \wp^{v_k(X,\wp)}$ of R, with the exponents v_k of the primes $\wp$ of R defined as follows. Let a_0 be any element of X. Choose a_k to be an element of X which minimizes the exponent of the highest power of a prime $\wp$ dividing $(a_k - a_0)(a_k - a_1)\cdots(a_k - a_{k-1})$ and $v_k(X,\wp)$ be this exponent. It can be proved that it is well defined, independently of the choices involved. The sequence a_i is called a $\wp$-*ordering*. If a_i is $\wp$-ordering for all $\wp$, then the ideal is the principal ideal generated by $k! = (a_k - a_0)\cdots(a_k - a_{k-1})$.

Examples 1.2.

(1) The sequence $0, 1, 2, \ldots$ in $X = R = \mathbb{Z}$ gives a simultaneous p-ordering for all p and leads to the usual factorial, once we choose the positive generator of the corresponding factorial ideal. For X consisting of q-powers for an integer $q > 1$ in $R = \mathbb{Z}$, we have $k!_X = (q^k - 1)\cdots(q^k - q^{k-1})$. For the set of $(q^j - 1)/(q-1)$'s, we get the q-factorial.

(2) For $X = R = \mathbb{F}_q[t]$, we can consider the following simultaneous $\wp$-ordering: Let $0 = a_0, a_1, \ldots, a_{q-1}$ be the elements in $\mathbb{F}_q$ and put $a_n = \sum a_{n_i} t^i$ where $n = \sum n_i q^i$ is the base q expansion of n. Hence the monic generator of the factorial ideal is $n! = (a_n - a_0)\cdots(a_n - a_{n-1}) = \prod[i]^{n_i + n_{i+1}q + \cdots + n_h q^{h-i}}$, which is the Carlitz factorial of n.

If q is a prime, then as a nice mnemonic we can think of associating to the base q expansion $n = n(q) = \sum n_i q^i$ a polynomial $a_n = n(t) = \sum n_i t^i$ and with this ordering the Carlitz factorial can be described by the usual formula $n! = (n-0)(n-1)\cdots(n-(n-1))$. (But keep in mind that addition of n's is like integers, with carry-overs and not like polynomials!) The same works for general q, except that we have to identify n_i between 0 and $q-1$ with elements of $\mathbb{F}_q$ by force then.

(3) Let $R = \mathbb{F}_q[t]$. We saw that, if $X = R$, then $D_i = (q^i)!_X$. We also have $D_i = i!_X$ for $X = \{t^{q^j} : j \geq 0\}$ or for $X = \{[j] : j \geq 0\}$, thus justifying the naive analogies mentioned above.

Bhargava shows that the generalized factorial, although its values are ideals which may not be principal even for $R = A$, retains the most important divisibility properties of the usual factorial, such as the integrality of binomial coefficients.

On the other hand, the Carlitz factorial (but not its generalizations below for general A) even satisfies the following analog of the well-known theorem of Lucas.

Theorem 1.3. *Let $A = \mathbb{F}_q[t]$, let $\binom{m}{n}$ denote the binomial coefficient for the Carlitz factorial, and let $\wp$ be a prime of A of degree d. Then we have*

$$\binom{m}{n} \equiv \prod \binom{m_i(d)}{n_i(d)} \mod \wp,$$

where $m = \sum m_i(d)q^{di}$ and $n = \sum n_i(d)q^{di}$ are the base q^d expansions of m and n respectively, so that $0 \le m_i(d), n_i(d) < q^d$.

In particular, if $m > n$, the left side is zero modulo $\wp$ if and only if there is a carry over of q^d-digits in the sum $n + (m - n)$.

Proof. First observe that if there is no carry over of base q digits, then all the binomial coefficients above are equal to one, because of the digit expansion definition of Carlitz factorial. Now suppose there is a carry over at (base q) exponents i, $i+1,\ldots,j-1$, but not at $i-1$ or j. Let $\sum m_k q^k$, $\sum n_k q^k$ and $\sum \ell_k q^k$ be the base q expansions of $m, n, m - n$ respectively. Then $n_k + \ell_k$ is $m_i + q$, $m_k + q - 1$ or $m_j - 1$ according to whether k is i, $i + 1 \le k \le j - 1$ or $k = j$. Thus the contribution of this block of digits to the binomial coefficient expression using the digit expansion is

$$\frac{D_j}{D_{j-1}^{q-1} \cdots D_{i+1}^{q-1} D_i^q} = [j] \cdots [i].$$

On the other hand, the congruence class of $[k]$ modulo $\wp$ depends on the congruence class of k modulo d, and both are zero if d divides k. $\square$

1.2   Arithmetic gamma for s$\mathbb{F}_q[t]$: Interpolations

Goss made interpolations of the factorial at all places of $\mathbb{F}_q[t]$ as follows.

Since $D_i = t^{iq^i} - t^{(i-1)q^i + q^{i-1}} + $ lower degree terms, the unit part

$$\overline{D_i} := D_i/t^{\deg D_i} = 1 - 1/t^{(q-1)q^{i-1}} + \cdots$$

tends to 1 in $\mathbb{F}_q((1/t))$ as i tends to ∞. So the unit part of $\Pi(n)$ interpolates to a continuous function $\overline{\Pi}(n)$ called ∞-*adic factorial:*

$$\overline{\Pi} \colon \mathbb{Z}_p \longrightarrow \mathbb{F}_q((1/t)), \quad \sum n_i q^i \longmapsto \prod \overline{D_i}^{n_i}.$$

Let v (sometimes we use symbol $\wp$) be a prime of A of degree d. Since D_i is the product of all monic elements of degree i, we have a Morita-style v-adic factorial $\Pi_v \colon \mathbb{Z}_p \to \mathbb{F}_q[t]_v$ for finite primes v of $\mathbb{F}_q[t]$ given by

$$\Pi_v(n) = \prod (-D_{i,v})^{n_i}$$

where $D_{i,v}$ is the product of all monic elements of degree i, which are relatively prime to v, and n_i are the digits in the q-adic expansion of n. This makes sense since $-D_{i,v} \to 1$, v-adically, as $i \to \infty$. This is because, if $m = \lfloor i/d \rfloor - \ell$ for sufficiently large fixed ℓ, then $D_{i,v}$ is a q-power of the product of all elements of $(A/v^m A)^*$ and hence is $-1 \bmod v^m$, for large i, by an analog of the usual group-theoretic proof of Wilson's theorem $(p-1)! \equiv -1 \bmod p$.

Here is a direct proof in the case where v is of degree one. Using the automorphism sending t to $t + \alpha$, $\alpha \in \mathbb{F}_q$, we can assume without loss of generality that $v = t$. Now, for a general monic prime v of degree d, we have $D_{i,v} = D_i / v^w D_{i-d}$, where w is such that $D_{i,v}$ is a unit at v. So, in our case,

$$-D_{n,t} = \left(\prod_{i=0}^{n-1} \left(1 - t^{q^n - q^i}\right) \right) \Big/ \left(\prod_{i=0}^{n-2} \left(1 - t^{q^{n-1} - q^i}\right) \right) \longrightarrow 1 \quad \text{as } n \to \infty.$$

In fact, it is easy to evaluate the t-adic gamma value:

$$\Pi_t \left(\frac{1}{1-q} \right) = \lim \prod_{i=0}^{N} (-D_{i,t}) = - \lim \prod_{j=0}^{N-1} \left(1 - t^{q^N - q^j} \right) = -1$$

because the product telescopes after the first term $-D_{0,t} = -1$.

1.3 Arithmetic gamma for general A: Definitions and interpolations

The different analogies that we have discussed for $\mathbb{F}_q[t]$ diverge for general A, giving different possible generalizations and we have to choose the ones with best properties. It turns out that the factorial coming from the prime factorization analogy, which is the same as the Bhargava factorial, though excellent for divisibility and combinatorial properties, is local in nature and in general there is no simultaneous p-ordering. No good interpolation for this ideal-valued factorial is known. To get good global properties connecting with Drinfeld modules and cyclotomic theory, we proceed as follows. (We will deal with the exponential analogy in Lecture 4.)

The arithmetic of gamma is closely connected with cyclotomic theory, thus with rank one Drinfeld A-modules, thus with A-lattices, or projective rank one A-modules, and thus with ideals.

Let $\mathcal{A}$ be an ideal of A, and D_i be the product of all monic elements a of $\mathcal{A}$ of degree i. (Note that even for $h_A = 1$ cases, now D_i need not divide D_{i+1}, unlike the $\mathbb{F}_q[t]$ case.) So $D_i \in \mathcal{A} \subset A$. Also let d_i be the number of these elements. We choose a uniformizer $u = u_\infty$ at ∞. The one-unit part $\overline{D}_i$ with respect to u satisfies $\overline{D}_i \to 1$ as $i \to \infty$. We then define $\overline{\Pi}$ and $\overline{\Gamma}$ similarly.

By the Riemann–Roch theorem, $d_i = q^{i+c}$, which tends to zero, q-adically as i tends to infinity. Thus (following a suggestion by Gekeler), we can recover the degree of gamma as follows.

The map $\mathbb{N} \to \mathbb{Z}$ given by $z \to \deg \Pi(z)$ interpolates to a continuous function $\deg \Pi \colon \mathbb{Z}_p \to \mathbb{Z}_p$ given by $\sum z_i q^i \to \sum i z_i d_i$. Hence we p-adically complete $K_\infty^\times$, i.e., we define $\widehat{K}_\infty^\times := \varprojlim K_\infty^\times / K_\infty^{\times p^n}$. Since finite fields are perfect, signs in $K_\infty^\times$ project to 1 in $\widehat{K}_\infty^\times$. Then we define the ∞-*adic interpolation* $\Pi = \Pi_\infty \colon \mathbb{Z}_p \to \widehat{K}_\infty^\times$ with

$$\Pi(z) = \overline{\Pi}(z) u^{-\deg \Pi(z)}.$$

We use the symbol Π again, as we have recovered the degree part.

Let v be a finite place of A relatively prime to $\mathcal{A}$, and of degree d. We form $\tilde{D}_i = D_{i,v}$ as usual by removing the factors divisible by v.

Definition 1.4. Let $\tilde{D}_i$ be the product of monic elements a of degree i and $v(a) = 0$.

Again, a generalized Wilson theorem type argument, which we omit, shows that $-\tilde{D}_i \to 1$, so we put

Definition 1.5.

$$\Pi_v \left(\sum z_i q^i \right) := \prod (-\tilde{D}_i)^{z_i}$$

so that $\Pi_v \colon \mathbb{Z}_p \to K_v$.

1.4 Functional equations for arithmetic gamma

We will now see how the structure of the functional equations for the factorial functions, for all places and all A's follows just from manipulation of p-adic digits of the arguments. So the proofs of functional equations reduce to this plus calculation of one single value: the value of gamma at 0, which we take up later.

After the more familiar reflection and multiplication formula, we will prove a general functional equation directly. We will see later how the cyclotomy and the Galois groups play a role in this structure.

Recall that the classical gamma function Γ satisfies

(1) a reflection formula: $\Gamma(z)\Gamma(1 - z) = \pi/\sin \pi z$, and
(2) a multiplication formula:

$$\Gamma(z)\Gamma\left(z + \frac{1}{n}\right) \cdots \Gamma\left(z + \frac{n-1}{n}\right)/\Gamma(nz) = (2\pi)^{(n-1)/2} n^{1/2 - nz}.$$

We will prove analogs of these and also their p-adic counterparts in the function field case, by first proving relations in the abstract setting below.

Consider a function f defined on $\mathbb{Z}_p$ via base q expansions by

$$f\left(\sum n_j q^j \right) := \prod f_j^{n_j}$$

for some f_j's. One can think of f_j's as independent variables with the evident manipulation rules. Put $g(z) = f(z - 1)$. The various factorial functions ('f') and gamma functions ('g') introduced above, and below in Lecture 4, are all of this form. We want to get formal relations satisfied by f. In particular, we would like to know when $\prod f(x_i)^{n_i} = 1$ formally, i.e., independently of f_i's.

First we have a reflection formula:

Theorem 1.6. $g(z)g(1-z) = g(0)$ *or equivalently (after a suitable change of variables)* $f(z)f(-1-z) = f(-1)$.

Proof. Let the digit expansion of z be $z = \sum z_j q^j$. Since $-1 = \sum (q-1)q^j$ and $0 \le q - 1 - z_j < q$, $-1 - z = \sum (q - 1 - z_j)q^j$ is a digit expansion. Hence the relation with f's follows. $\square$

Next we have a multiplication formula:

Theorem 1.7. *For $z \in \mathbb{Z}_p$ and $(n,q) = 1$,*

$$g(z)g\left(z + \frac{1}{n}\right) \cdots g\left(z + \frac{n-1}{n}\right) \Big/ g(nz) = g(0)^{(n-1)/2}.$$

Here, if n is even, so that q is odd, then we mean by $g(0)^{1/2}$ the element $\prod_{j=0}^{\infty} f_j^{(q-1)/2}$ whose square is $g(0)$.

Proof. If $(n,q) = 1$, then $-1/n$ has a purely recurring q base expansion of r recurring digits where r is minimal such that n divides $q^r - 1$. The recurring digits for $-a/n$'s are related in such a way that the sum of the ith digits of all of them is constant, independent of i. (This can be seen by considering orbits under multiplication by q, but we omit the details.) This constant is easily seen to be equal to $(q-1)(n-1)/2$, as $-1/n + \cdots + -(n-1)/n = -(n-1)/2 = ((n-1)/2)\sum (q-1)q^j$. $\square$

We now give a more general functional equation, explained in a uniform framework in the next lecture.

Let N be a positive integer prime to p. For $x \in \mathbb{Q}$, define $\langle x \rangle$ by $x \equiv \langle x \rangle$ modulo $\mathbb{Z}$, $0 \le \langle x \rangle < 1$. If $\underline{a} = \sum m_i[a_i]$ $(m_i \in \mathbb{Z}, a_i \in \frac{1}{N}\mathbb{Z} - \{0\})$ is an element of the free abelian group with basis $\frac{1}{N}\mathbb{Z} - \{0\}$, put $n(\underline{a}) := \sum m_i \langle a_i \rangle$. Also, for $u \in (\mathbb{Z}/N\mathbb{Z})^\times$, let $\underline{a}^{(u)} := \sum m_i[ua_i]$.

Theorem 1.8. *If $n(\underline{a}^{(q^j)})$ is an integer independent of j, then*

$$\prod f(-\langle a_i \rangle)^{m_i} = f(-1)^{n(\underline{a})}.$$

We skip the proof involving digit manipulations (see [T04, §4.6] or [T91a]). We will discuss the interesting value $f(-1) = g(0)$ in the next lecture, by relating it to periods.

Remarks 1.9.

(1) By an integer translation, any proper fraction in $\mathbb{Z}_p$ can be brought strictly between -1 and 0 and therefore it is possible to write it as $\sum_{i=0}^{n-1} a_i q^i/(1-q^n)$,

with $0 \leq a_i < q$. The proof of the above theorem reduces ultimately, after these reduction steps and some combinatorics, to

$$\left(\frac{\sum a_i q^i}{1 - q^n} \right)! = \prod \left(\frac{q^i}{1 - q^n} \right)!^{a_i}.$$

This is in turn a simple consequence of $(z_1 + z_2)! = z_1! z_2!$, if there is no carry over base p of digits of z_1 and z_2. There are uncountably many pairs satisfying these conditions, which would have implied a limit point and thus a functional identity in the real or complex case, but does not imply it in our case because of the differences in function theory.

(2) We will see (in the discussion after Theorem 6.6) that the 'basis elements' $(q^i/(1 - q^n))!$ with $0 \leq i < n$ are algebraically independent for a fixed n.

1.5   Geometric gamma: Definitions and interpolations

The gamma function we studied so far has domain in characteristic zero, even though the values are in characteristic p. With all its nice analogies, it has one feature strikingly different than the classical gamma function: It has no poles. The usual gamma function has no zeros and has simple poles exactly at 0 and negative integers, which we interpret as negative of the positive integers and replace 'positive' by 'monic'. (Monicity is an analog of positivity, but positivity is closed under both addition and multiplication, while monicity only under multiplication. Also, for $p = 2$, positive is the same as negative and for $q = 2$ all integers are negative!) Having thus decided upon the location of the poles, note that in our non-archimedean case the divisor determines the function up to a multiplicative constant. (This follows easily from the Weierstrass preparation theorem by associating a distinguished polynomial to a power series, or from the Newton polygon method.) The simplest constant that we choose below also seems to be the best for the analogies we describe later.

   We denote by A_+ (A_{d+} respectively) the set of monic elements (monic of degree d respectively) in A. Similarly, we define $A_{<d}$ etc. Hence we define the geometric gamma function as a meromorphic function on C_∞ by

Definition 1.10.

$$\Gamma(x) := \frac{1}{x} \prod_{a \in A_+} \left(1 + \frac{x}{a} \right)^{-1} \in C_\infty \cup \{\infty\}, \quad x \in C_\infty.$$

   From this point of view of divisors, the factorial Π should be defined as $\Pi(x) := x\Gamma(x)$.

Remarks 1.11.

(1) Classically, we have $x\Gamma(x) = \Pi(x) = \Gamma(x + 1)$, whereas in our situation the first equality is natural for the gamma and factorial defined here and the

second equality is natural for the ones considered previously. Consequently, the geometric gamma and factorial now differ by more than just a harmless change of variable. Also, in characteristic p, addition of p brings us back, so giving the value at $x + 1$ in terms of that at x will not cover all integers by recursion anyway.

(2) Unlike the arithmetic gamma case, where we started with the values at positive integers and interpolated them, in the geometric gamma case the values at integers do not even exist for $q = 2$. For general q, in the $A = \mathbb{F}_q[t]$ case, the reciprocals of the values at integers are integral. In general, even this property is lost! At least, the values at integers (when they exist) are rational, as the gamma product is then finite and the terms for a given degree contribute 1 to the product when the degree is large enough.

This rationality of values at integers gives hope for interpolation à la Morita.

Definition 1.12. For $a \in A_v$, let $\bar{a} := a$ or 1 according to whether $v(a) = 0$ or $v(a) > 0$ respectively, and when $x \in A$, put

$$\Pi_v(x) := \prod_{j=0}^{\infty} \left(\prod_{n \in A_{j+}} \frac{\overline{n}}{\overline{x+n}} \right).$$

Note that the terms are 1 for large j. Hence $\Pi_v(a) \in K$ for $a \in A$.

Lemma 1.13. Π_v *interpolates to* $\Pi_v \colon A_v \to A_v^*$ *and is given by the same formula as in the definition, even if* $x \in A_v$. *Similarly,* $\Gamma_v(x) := \Pi_v(x)/\overline{x}$ *interpolates to a function on* A_v.

Proof. It is easy to see that if $x \equiv y \bmod v^l$, then $\Pi_v(x) \equiv \Pi_v(y) \bmod v^l$. $\qquad\square$

1.6 Functional equations for geometric gamma

Reflection formula

For $q = 2$, all non-zero elements are monic, so

$$\Gamma(x) = \frac{1}{e_A(x)} = \frac{\widetilde{\pi}}{e(\widetilde{\pi}x)},$$

where e_A is the exponential corresponding to the lattice A and e is the exponential corresponding to the sign-normalized Drinfeld module with the period lattice $\widetilde{\pi}A$. Hence, for $x \in K - A$, $\Gamma(x)$ has algebraic (even 'cyclotomic') ratio with $\widetilde{\pi}$, and so, as we will see, $\Gamma(x)$ is transcendental.

From the point of view of their divisors, $e(\widetilde{\pi}x)$ being analogous to $\sin(\pi x)$ (i.e., both have simple zeros at integers and no poles), this observation suggests a relation between Γ and sine. We reformulate the reflection relation as follows to make the analogy more visible.

The classical reflection formula can be stated as

$$\prod_{\theta \in \mathbb{Z}^\times} \Pi(\theta x) = \frac{\pi x}{\sin(\pi x)},$$

and here, for general q, we clearly have

Theorem 1.14.

$$\prod_{\theta \in A^\times} \Pi(\theta x) = \frac{\widetilde{\pi} x}{e(\widetilde{\pi} x)}.$$

Multiplication formula for geometric Π

Theorem 1.15. *Let $g \in A$ be monic of degree d and let α run through a full system of representatives modulo g. Then*

$$\prod_\alpha \Pi\left(\frac{x + \alpha}{g}\right) = \Pi(x)\widetilde{\pi}^{(q^d - 1)/(q-1)}((-1)^d g)^{q^d/(1-q)} R(x)$$

where

$$R(x) = \frac{\prod_{\beta \in A_{\leq m+}} \beta + x}{\prod_\alpha \prod_{a \in A_{\leq m+d+}} ga + \alpha + x},$$

with m being any integer larger than $\max(\deg \alpha, 2g_K) + d$.

Here $R(x)$ takes care of the irregularity in Riemann–Roch at low degrees. For example, $R(x) = \prod_{\alpha \text{ monic}}(x + \alpha)$, when $A = \mathbb{F}_q[t]$ and $\{\alpha\}$ is the set of all polynomials of degree not more than d.

Multiplication and reflection formula for Π_v

Theorem 1.16. (1) *Let α, g be as in the previous theorem and with $(g, v) = 1$. Then*

$$\prod_\alpha \Pi_v\left(\frac{x + \alpha}{g}\right) \Big/ \Pi_v(x) \in K(x)^\times.$$

(2) *For $a \in A_v$,*

$$\prod_{\theta \in \mathbb{F}_q^\times} \Pi_v(\theta a) \Big/ \overline{a} \in \mathbb{F}_q^*$$

and it can be prescribed by congruence conditions. For example, it is equal to $-\text{sgn}(a_v)^{-1}$ or 1 respectively, according to whether a_v, the $\mod v$ representative of a of degree less than $\deg(v)$ (if it exists, as it always does when $A = \mathbb{F}_q[t]$) is zero or not.

We have omitted proofs, which follow by manipulating cancellations in the product expansions.

Lecture 2

Special Γ-values, Relations with Drinfeld Modules and Uniform Framework

2.1 Arithmetic gamma: $\mathbb{F}_q[t]$ case

First we deal with the case $A = \mathbb{F}_q[t]$ and relate the special values of $\overline{\Gamma}$ to the period $\widetilde{\pi}$ of the Carlitz module. Later, we will describe how to derive more general results in a different fashion.

For $0 \neq f \in \mathbb{F}_q((1/t))$, $f/t^{\deg f}$ will be denoted by $\overline{f}$.

The most well-known gamma value at a fraction is $\Gamma(1/2) = \sqrt{\pi}$. In our case, when $p \neq 2$, so that $1/2 \in \mathbb{Z}_p$, we have $\Gamma(1/2) = \Pi(-1/2) = \Pi(-1)^{1/2}$, where the last equality follows directly by the digit expansion consideration. We will now prove an analog of this fundamental evaluation, which will also complete our functional equation in this case of the arithmetic gamma for $\mathbb{F}_q[t]$.

Carlitz [Car35] (see also [T04, § 2.5]) proved that

$$\widetilde{\pi} = (-1)^{1/(q-1)} \lim [1]^{q^k/(q-1)}/[1] \cdots [k]$$

so $\widetilde{\pi}^{q-1} \in \mathbb{F}_q((1/t))$ and $\overline{\widetilde{\pi}^{q-1}}$ makes sense. By $\overline{\widetilde{\pi}}$ we will denote its unique $(q-1)$th root which is a one unit in $\mathbb{F}_q((1/t))$.

Theorem 2.1. *Let $A = \mathbb{F}_q[t]$. For $0 \leq a \leq q - 1$, we have*

$$\overline{\Gamma}\left(1 - \frac{a}{q-1}\right) = (\overline{\widetilde{\pi}})^{a/(q-1)}.$$

In particular, we have $\overline{\Gamma}(0) = \overline{\widetilde{\pi}}$, and if $q \neq 2^n$, then

$$\overline{\Gamma}(1/2) = \sqrt{\overline{\widetilde{\pi}}}.$$

Proof. Since $-1 = \sum (q - 1)q^i$, we have

$$\overline{\Gamma}(0) = \overline{\Pi}(-1) = \lim \overline{(D_0 \cdots D_n)^{q-1}}.$$

Now

$$(D_0 \cdots D_n)^{q-1} = D_{n+1}/[1] \cdots [n+1].$$

Hence

$$\overline{\Gamma}(0)^{q-1}/\overline{\widetilde{\pi}^{q-1}} = \lim \overline{D_{n+1}^{q-1}/[1]^{q^{n+1}}} = 1,$$

since we have already seen that $\overline{D_i} \to 1$, and since $\overline{[1]}^{q^n} \to 1$, because any one unit raised to the q^nth power tends to 1 as $n \to \infty$. Hence $\overline{\Gamma}(0) = \overline{\widetilde{\pi}}$.

(We will not prove here the Carlitz formula above, but will see in 3.3 another formula which also leads to the same calculation. Also, our proof below, for general A, gives another approach. The reason for giving this incomplete proof here is that the same idea generalizes in the next Chowla–Selberg analog.)

Since $a/(1-q) = \sum aq^i$ for $0 \le a \le q-1$, we get the theorem.   □

Corollary 2.2. *For $A = \mathbb{F}_q[t]$, we have $\Gamma(0)^{q-1} = -\widetilde{\pi}^{q-1}$.*

Proof. Both sides have degree $q/(q-1)$ and $\widetilde{\pi}^{q-1} \in \mathbb{F}_q((1/t))$ has sign -1 as we see from any of the formulas for it, whereas any $\Gamma(z)$ has sign one by construction.   □

To investigate the nature of gamma values at all fractions (with denominator not divisible by p), it is sufficient to look at all $\overline{\Pi}(q^j/(1-q^k))$ for $0 \le j < k$, since a general value is (up to a harmless translation of the argument by an integer resulting in rational modification in the value) a monomial in these basic ones. They can be related to the periods $\widetilde{\pi}_k$ of the Carlitz module for $\mathbb{F}_{q^k}[t]$, a rank k A-module with complex multiplication by this cyclotomic ring. For example,

Theorem 2.3. *If $A = \mathbb{F}_q[t]$, then*

$$\overline{\widetilde{\pi}_k} = \frac{\overline{\Pi}(q^{k-1}/(1-q^k))^q}{\overline{\Pi}(1/(1-q^k))}.$$

Proof. We have

$$\frac{\overline{\Pi}(1/(1-q^k))}{\overline{\Pi}(q^{k-1}/(1-q^k))^q} = \lim \frac{\overline{D_{kn} D_{k(n-1)} \cdots D_0}}{(D_{kn-1})^q \cdots (D_{k-1})^q}$$

$$= \lim \overline{[kn][k(n-1)] \cdots [k]}$$

$$= (\overline{\widetilde{\pi}_k})^{-1}.   □$$

This is the Chowla–Selberg formula for constant field extensions, as will be explained in 2.5. Similarly, it can be shown, for example, that

$$\overline{\Pi}(1/(1-q^2))^{q^2-1} = \overline{\widetilde{\pi}}^q \overline{\widetilde{\pi}}_2^{-(q-1)}$$

$$\overline{\Pi}(q/(1-q^2))^{q^2-1} = \overline{\widetilde{\pi}} \, \overline{\widetilde{\pi}}_2^{q-1}.$$

2.2 Arithmetic gamma: General A case

Next, we relate, for general A, $\Gamma(0)$ with a period $\widetilde{\pi}$, defined up to multiplication from $\mathbb{F}_q^*$, of a sign-normalized rank one Drinfeld A-module ρ with corresponding rank one lattice $\Lambda = \widetilde{\pi}\mathcal{A}$, with $\mathcal{A}$ an ideal of A, and exponential $e_\rho = e_\Lambda$.

Let x be an element of A of degree > 0, say of degree d and with $\mathrm{sgn}(x) = 1$. The coefficient of the linear term of ρ_x is x, and ρ is sign-normalized. Hence x is the product of the non-zero roots of the polynomial ρ_x:

$$x = \prod_{a \in \mathcal{A}/\mathcal{A}x} {}'\widetilde{\pi} e_\mathcal{A}(a/x) = \widetilde{\pi}^{q^d - 1} \prod {}'e_\mathcal{A}(a/x).$$

So

$$\widetilde{\pi}^{1-q^d} = \frac{1}{x} \prod_{a \in \mathcal{A}/\mathcal{A}x} {}'e_\mathcal{A}(a/x).$$

By using the product expansion of the exponential, careful grouping and manipulation of signs, degrees, limits, etc., one can prove the following theorem (we omit the details).

Theorem 2.4. *We have* $\Gamma(0) = \mu\widetilde{\pi}$, *where* μ *is* $(q-1)$th *root of* -1.

We note that all rank 1 normalized Drinfeld A-modules are isogenous, so periods for different choices of $\mathcal{A}$ are algebraic multiples of each other.

2.3 Special values of arithmetic Γ_v

We first prove strong results for the $\mathbb{F}_q[t]$ case, but only weak results for general A. We will show, in Lecture 6, how comparable strong results follow after developing more machinery.

The Gross–Koblitz formula, based on crucial earlier work by Honda, Dwork and Katz, expresses Gauss sums lying above a rational prime p in terms of values of Morita's p-adic gamma function at appropriate fractions.

Honda conjectured and Katz proved a formula for Gauss sums made up from pth roots of unity in terms of p-adic limits involving factorials, combining two different calculations of Frobenius eigenvalues on p-adic cohomology (Crystalline or Washnitzer–Monsky) of Fermat and Artin–Schreier curves. Gross and Koblitz interpreted this as a special value of Morita's then recently developed p-adic interpolation of the classical factorial.

For a prime $\wp \in A$ of degree d, and $0 \le j < d$, an analog g_j of Gauss sums was defined in the author's thesis (see Lecture 4 for details) as a 'character sum' with multiplicative character coming from the theory of cyclotomic extensions of constant field extension type and an analog of additive character coming from the Carlitz–Drinfeld cyclotomic theory. It was shown that for a prime $\wp$ of degree d

the corresponding Gauss sum has the property that $g_j \in K(\Lambda_\wp)(\zeta_{q^d-1})$ and the corresponding Jacobi sum is such that $g_j^{q^d-1} \in K(\zeta_{q^d-1})$ (so that, in particular, the cyclotomic extension $K(\Lambda_\wp)$ is a Kummer extension of $K(\zeta_{q^d-1})$ given by a root of the Gauss sum), and also an analog of Stickelberger factorization and congruence was proved for these Gauss sums.

Theorem 2.5 ('Analog of the Gross–Koblitz formula'). *Let $A = \mathbb{F}_q[t]$ and $\wp$ be a monic prime of A of degree d. Then for $0 \le j < d$ we have*

$$g_j = -\lambda^{q^j} / \Pi_\wp\left(\frac{q^j}{1-q^d}\right),$$

where λ is a (q^d-1)th root of $-\wp$ (fixed by a congruence condition that we omit). In particular, these values are algebraic (in fact, cyclotomic).

Proof. We have $\tilde{D}_a = D_a/D_{a-d}\wp^l$, where l is such that $\tilde{D}_a$ is a unit at $\wp$. Hence, using the base q expansion $q^j/(1-q^d) = \sum q^{j+id}$, we get

$$\Pi_\wp\left(\frac{q^j}{1-q^d}\right) = \lim(-1)^{m+1}\tilde{D}_j \cdots \tilde{D}_{j+md} = \lim(-1)^{m+1}D_{j+md}/\wp^{w_m}$$

where $w_m = \mathrm{ord}_\wp D_{j+md}$. Moreover, the recursion formula for D_i gives

$$D_{j+md} = [j+md][j-1+md]^q \cdots [j+1+(m-1)d]^{q^{d-1}}D_{j+(m-1)d}^{q^d}.$$

Without loss of generality, we can assume that $\wp \ne t$. Thus t is a unit in $K_\wp$ and we can write it in the form $t = au$, namely as a product of its 'Teichmüller representative' $a = \lim t^{q^{md}}$ and its one-unit part u. As $a^{q^{md}} = a$ and $u^{q^n} \to 1$ as $n \to \infty$, we have $[l+md] = ((au)^{q^{md+l}} - t) \to (a^{q^l} - t)$ as $m \to \infty$, which is just $-\wp_{1-l}$, the negative of one of the monic primes $\wp_j$'s above $\wp$. Using this in the limit above and counting powers of $\wp$, using the description of $[i]$ given above, we see that

$$\Pi_\wp\left(\frac{q^j}{1-q^d}\right)^{1-q^d} = (-\wp_{1-j})(-\wp_{2-j})^q \cdots (-\wp_{-j})^{q^{d-1}}/\wp^{q^j}.$$

Comparing with the Stickelberger factorization (note the naive analogy with $\sum a\sigma_a$ where $a \in (\mathbb{Z}/n\mathbb{Z})^*$, when the cyclotomic Galois group $(\mathbb{Z}/n\mathbb{Z})^*$ is replaced by $q^{\mathbb{Z}/n\mathbb{Z}}$), we see that the factorizations are the same and we fix the root of unity by comparing the congruences. We omit the details. □

This proof is quite direct and does not need a lot of machinery, unlike the proof in the classical case.

For general A, we evaluate below only a few simple values.

Theorem 2.6. *We have $\Gamma_v(0) = (-1)^{\deg v - 1}$ if v is prime to $\mathcal{A}$. For $0 \le a \le q-1$, $\Gamma_v(1-a/(q-1))$ are roots of unity and $\Gamma_v(b/(q-1))$ is algebraic for $b \in \mathbb{Z}$.*

Proof. The first statement of the theorem implies the first part of the second statement just by using the definitions in terms of relevant digit expansions. This implies the last claim immediately from the definitions, by considering the effect of integral translations of the arguments.

Let $d = \deg v$. The first statement will follow if we show that

$$\left(\prod m\right)^{q-1} \equiv (-1)^{d-1} \quad \mathrm{mod}\ v^{l_i}$$

where m runs through monic polynomials prime to v and of degree not more than t_i and with $l_i, t_i \to \infty$ as $i \to \infty$. Given l_i, choose t_i so that $\{am : a \in \mathbb{F}_q^\times\}$ spans the reduced residue class system $\mathrm{mod}\ v^{l_i}$ (for example, in the $\mathbb{F}_q[t]$ case, $t_i = dl_i - 1$ works). Then it is easy to see that $\{am\}$ covers each reduced residue class equal number (which is a power of q) of times. Hence, again by the usual Wilson theorem argument, we have

$$-1 \equiv \left(\prod a\right)^{\#\{m\}} \left(\prod m\right)^{q-1} \quad \mathrm{mod}\ v^{l_i}.$$

But $\prod a = -1$, so we are done if $p = 2$. If p is not two, then we have to show that $\#\{m\} \equiv d \bmod 2$. But for some c we have

$$\#\{m\} = \left(q^{t_i-c} - q^{t_i-c-d}\right)/(q-1) \equiv (q^d - 1)/(q-1) \equiv d \bmod 2. \qquad \square$$

2.4 Geometric gamma values

Functional equations (see Section 1.6) express many monomials at fractions as period times cyclotomic numbers. In particular, for $q = 2$, all geometric gamma values at a proper fraction are algebraic multiples of $\tilde{\pi}$ and all v-adic gamma values are algebraic.

Now we look at special value results in the simplest case, that of $A = \mathbb{F}_q[t]$, q any prime power, and v a prime of degree one, to give a flavor of what can be done. It can be shown by Moore determinant calculations as before that

Lemma 2.7. *Let $D_{r,\eta,t}$ denote a product of monic polynomials of degree r which are congruent to $\eta \in \mathbb{F}_q^*$ modulo t. Then*

$$D_{r+1,\eta,t} = D_r t^{q^r} \left(1 - \eta/(-t)\right)^{(q^{r+1}-1)/(q-1)}.$$

Now notice that we are dealing with denominator t, and the tth torsion of Carlitz module is $\lambda_t = (-t)^{1/q-1}$. So the formula in the previous lemma can be rewritten as

$$\prod_{n \in A_{d+}} (1 + \eta/nt) = 1 - \eta\lambda_t/\lambda_t^{q^{d+1}}.$$

Consider the Carlitz module over $B = \mathbb{F}_q[\lambda_t]$, the integral closure of $A = \mathbb{F}_q[t]$ in the tth cyclotomic field. This is a Drinfeld module of rank $q - 1$ over A and of rank one over B. Hence this can be considered as a tth Fermat motive, having complex multiplications by the tth cyclotomic field. Denote its period by π_B.

Theorem 2.8. *With this notation, we have* $\pi_B = \lambda_t^{q/(q-1)} \Pi(1/t)$.

Proof. The displayed formula after the lemma shows that

$$\Pi(1/t) = \prod_{d=1}^{\infty} (1 - \lambda_t/\lambda_t^{q^d})^{-1}.$$

On the other hand, using the Carlitz period formula (see 2.1) in the case of base B, we see that

$$\pi_B/\Pi(1/t) = \lim(\lambda_t^q - \lambda_t)^{q^n/(q-1)}/(\lambda_t^q)^{(q^n-1)/(q-1)} = \lambda_t^{q/(q-1)},$$

as $\lim(1 - \lambda_t^{1-q})^{q^n/(q-1)} = 1$.　　　　　　　　　　　　　□

We remark that $\Pi(-1/t)$ turns out to be a 'quasi-period'.

Now let us look at the v-adic values. First we set up some preliminary notation. The Galois group of $K(\lambda_t)$ over K can be identified with $\mathbb{F}_q^*$, with $\eta \in \mathbb{F}_q^*$ acting as $\sigma_\eta(\lambda_t) = \eta\lambda_t$. If v is a prime of degree d, then the Galois group of $K\mathbb{F}_v$ over K can be identified with $\mathbb{Z}/d\mathbb{Z}$, with $j \in \mathbb{Z}/d\mathbb{Z}$ acting as $\tau_j(\zeta_{q^d-1}) = \zeta_{q^d-1}^{q^j}$. Let $\overline{\lambda}$ be the Teichmüller representative of λ_t, so that $\lambda_t^{q^{di}} \to \overline{\lambda} \in \mathbb{F}_v \subset K_v$ as $i \to \infty$. Write $\wp = \lambda_t - \overline{\lambda}$. Then the $\wp^\eta$'s are the primes of $K(\lambda_t)\mathbb{F}_v$ above v.

Theorem 2.9. *Let* $A = \mathbb{F}_q[t]$. *Let* v *be a monic prime of degree* d *of* A *which is congruent to* 1 *modulo* t, *and suppose that* $\eta \in \mathbb{F}_q^*$. *Then*

$$\Pi_v(\eta/t) = (-\overline{\lambda})^{(q^d-1)/(q-1)}(\wp^{\sigma_\eta})^{-\sum \tau_j} \text{ for } \eta \neq 1, \text{ and } \Pi_v(1/t) = \frac{v}{t}(\wp)^{-\sum \tau_j}.$$

In particular, $\Pi_v(a/t)$ *is algebraic for every* $a \in A$.

Proof. Write $w_j := \prod_{n \in A_{j+}} \overline{n}/\overline{n + \eta/t}$ and $x_j := \prod_{n \in A_{j+}} n/(n + \eta/t)$. Since v is monic congruent to 1 modulo t, we have $w_j = x_j/x_{j-d}v^{r_j}$ for $j \geq d$, where r_j is such that the right-hand side is a unit at v.

For $0 \leq j < d$, we have $w_j = x_j$ unless $j = d - 1$ and $\eta = 1$, in which case they differ in a term $n = (v - 1)/t$, giving $w_{d-1}t/v = x_{d-1}$. Hence, for $\eta \neq 1$, the product telescopes and we get

$$\Pi_v(\eta/t) = \prod w_j = \lim \prod_{i=0}^{d-1} \left(1 - \eta\lambda_t/\lambda_t^{q^{nd+i}}\right)^{-1} = \prod_{i=0}^{d-1} \left(1 - \eta\lambda_t/\overline{\lambda}^{q^i}\right).$$

We calculate similarly the case $\eta = 1$, where the answer is a unit at v.　　□

2.5 Uniform framework

Now we compare and explain analogies by giving a unified treatment for the gamma functions in the three cases: classical, arithmetic and geometric.

Classical	Arithmetic	Geometric
$\Gamma(\mathbb{Z}_+) \subset \mathbb{Z}_+$	$\Gamma(\mathbb{Z}_+) \subset A_{>0}$	$\Gamma(A - A_{\leq 0}) \subset K^*$
$\Gamma: \mathbb{C} - \mathbb{Z}_{\leq 0} \longrightarrow \mathbb{C}^*$	$\Gamma: \mathbb{Z}_p \longrightarrow \hat{K}_\infty^*$	$\Gamma: C_\infty - A_{\leq 0} \longrightarrow C_\infty^*$
$\Gamma_p: \mathbb{Z}_p \longrightarrow \mathbb{Z}_p^*$	$\Gamma_v: \mathbb{Z}_p \longrightarrow K_v^*$	$\Gamma_v: A_v \longrightarrow A_v^*$

We look at the questions of algebraicity, transcendence, and relations to the periods of the special values of gamma functions at fractional arguments. Note how special value combinations occurring in the reflection and multiplication formulas, for z a fraction, introduce cyclotomic ('$\sin \pi a/b$') and Kummer ('$n^{1/2 - nz}$') extensions, and keep in mind the vague connection $\Gamma(1/a) \leftrightarrow e(\widetilde{\pi}/a) \leftrightarrow a$-torsion of ρ. A unified treatment requires unified notation and identification of similar objects.

	Classical	Arithmetic	Geometric
I: integers in domain	$\mathbb{Z}$	$\mathbb{Z}$	A
F: fractions in domain	$\mathbb{Q}$	$\{a/b \in \mathbb{Q}: (p,b) = 1\}$ $= \{a/(q^m - 1)\}$	K
$\overline{\mathcal{A}}$: algebraic nos. in range	$\overline{\mathbb{Q}}$	$\overline{K}$	$\overline{K}$
underlying objects	$\mathbb{G}_m$, CM elliptic curves	$\mathbb{G}_m$, CM ρ's	CM ρ's
exponential e	classical e	classical for domain, e_ρ for range	e_ρ
period θ	$2\pi i$	$2\pi i$ for domain, $\widetilde{\pi}$ for range	$\widetilde{\pi}$
B: base field	$\mathbb{Q}$	K	K
$\mathcal{O}$: base ring	$\mathbb{Z}$	A	A
extension $B(e(\theta f))$	usual cyclotomic	constant field	Drinfeld cyclotomic
Galois group G identified with	$\mathrm{Gal}(\mathbb{Q}(\mu_N)/\mathbb{Q})$ $= (\mathbb{Z}/N\mathbb{Z})^*$	$\mathrm{Gal}(K(\mu_N)/K)$ $= q^{(\mathbb{Z}/r\mathbb{Z})}$	$\mathrm{Gal}(K(\Lambda_N)/K)$ $= (A/NA)^*$

In this table, $f = a/N$, with $a, N \in I$, and r was defined by $\mathbb{F}_{q^r} = \mathbb{F}_q(\mu_N)$ and in the geometric case we restricted to $h = d_\infty = 1$ or even to $A = \mathbb{F}_q[t]$ first for simplicity. See below (and [T04, 4.12.5]) for comments on a more general situation.

We restrict our attention to gamma values at proper fractions. In all the three cases, if $f \in F - I$, $i \in I$, then $\Gamma(f + i)/\Gamma(f) \in B \subset \overline{A}$. So for our question of algebraicity it is enough to look at $f \in (F - I)/I$. To extract this information, we define for $f \in F - I$ a rational number $\langle f \rangle \in \mathbb{Q}$ such that $\langle f \rangle$ depends only on $f \bmod I$.

Definition 2.10.

(1) For classical and arithmetic gamma: For the unique $n \in \mathbb{Z}$ with $0 < f - n < 1$, put $\langle f \rangle := f - n$.

(2) For geometric gamma: Let $A = \mathbb{F}_q[t]$; for the unique $a \in A$ with $f - a = -a_1/a_2$, where $a_1, a_2 \in A$, $\deg a_1 < \deg a_2$, and a_2 monic, put $\langle f \rangle := 1$ or 0, according to whether a_1 is monic or not.

Definition 2.11. Consider a finite formal sum $\underline{f} = \oplus m_i[f_i]$, $m_i \in \mathbb{Z}$, $f_i \in F - I$. This is nothing but an integral linear combination of symbols $[f_i]$, i.e., a divisor. Let N be a common denominator for the f_i's. Put $\Gamma(\underline{f}) := \prod \Gamma(f_i)^{m_i}$ and make similar definitions for Π, Γ_v, Π_v etc. Also put $m(\underline{f}) := \sum m_i \langle -f_i \rangle \in \mathbb{Q}$.

For $\sigma \in G$, let $\underline{f}^{(\sigma)} := \oplus m_i[f_i^{(\sigma)}]$, where $f_i^{(\sigma)}$ is just multiplication of f_i by σ as an element of the identification of the Galois group given in the table.

For example, in the case of arithmetic gamma, if σ corresponds to q^j, then $f_i^{(\sigma)} = q^j f_i$.

For a finite place v, let $\mathrm{Frob}_v \in G$ be the Frobenius, so that it makes sense to talk of Frob_v-orbits of f_i or of $\underline{f}$.

Consider the following hypotheses/recipe/conjecture:

(H1)   If $m(\underline{f}^{(\sigma)})$ is independent of $\sigma \in G$, then $\Gamma(\underline{f})/\theta^{m(\underline{f})}$ and $\Gamma_v(\underline{f})$ belong to $\overline{A}$.

(H2)   If $\underline{f}$ is a linear combination of Frob_v-orbits, then $\Gamma_v(\underline{f}) \in \overline{A}$ (or rather algebraic in an appropriate v-adic context).

Example 2.12. With $\underline{f} = \langle f \rangle + \langle 1 - f \rangle$, $m(\underline{f}) = 1$ and $\underline{f} = \oplus_{j=0}^{n-1} \langle z + j/n \rangle - \langle nz \rangle$, $m(\underline{f}) = (n-1)/2$, the hypothesis of (H1) is satisfied in the first two cases. These correspond to the reflection and multiplication formulas. In case II, the theorems above show that $\Gamma(-\underline{f})/\tilde{\pi}^{n(\underline{f})} \in \overline{A}$. This implies that $\Gamma(\underline{f})/\tilde{\pi}^{\sum m_i - n(\underline{f})} \in \overline{A}$, but in case of reflection and multiplication, $\sum m_i = 2n(\underline{f})$. Hence $\Gamma(\underline{f})/\tilde{\pi}^{\sum m_i \langle -f_i \rangle}$ works in both cases under the independence hypothesis.

What is known about (H1) and (H2), and are they best possible?

(H1) is true in the three cases, but still we have written it as a hypothesis, because we will consider other situations later which fit in this framework: For gamma of the fourth lecture, it is true with $\theta = \Gamma(0)$, but we do not understand yet the nature, or connection with the period, of this value.

In the classical case, $\underline{f}$ satisfying the condition in (H1) is a linear combination, with rational coefficients, of the examples above (i.e., the conclusion follows from the known reflection and multiplication formulas) by a result of Koblitz and Ogus. With a proof analogous to that of Koblitz and Ogus, the same can be proved for the geometric case. (We do need rational coefficients, so that we get some non-trivial χ_a's for a's that are not integral linear combinations of our functional equations. See [Sin97a] for an analysis of rational versus integral linear combinations of functional equations and Anderson's 'ϵ-generalization' of the Kronecker–Weber theorem [A02] in the classical case, inspired by this analysis.) But, since the independence condition in the arithmetic case is much weaker than the one in the classical case, the general functional equations above which prove (H1) directly in this case are more general than those generated by the multiplication and reflection formulas.

(H2) follows from the Gross–Koblitz theorem in the classical case, and its analog above for the arithmetic case with $A = \mathbb{F}_q[t]$. We will see a full proof for the geometric case when $A = \mathbb{F}_q[t]$. There is some evidence in other cases.

In the complex multiplication situation, we also have a Chowla–Selberg type formula (up to multiplication by an element of $\overline{A}$) for the period, in terms of gamma values at appropriate fractions. This formula is predicted via a simple calculation involving the 'brackets' introduced above. It works as follows.

Let E be a Drinfeld A-module (or an elliptic curve for the case I) over B^{sep} with complex multiplication by the integral closure of $\mathcal{O}$ in the appropriate abelian extension L of B as in the table, e.g., a constant field extension for the arithmetic case and a Drinfeld cyclotomic extension for the geometric case, etc. (In fact, we can get much more flexibility using the higher-dimensional A-motives of Anderson, e.g., solitons give rise to higher-dimensional A-motives with Drinfeld cyclotomic CM whose periods are values of $\Gamma(z)$, for $z \in K$, in the geometric case with $A = \mathbb{F}_q[t]$. But here we will be content with this simple case.)

For $f \in F - I$, let $h(f)\colon \mathrm{Gal}(B^{\mathrm{sep}}/B) \to \mathbb{Q}$ be defined by $h(f)(\sigma) := \langle -f' \rangle$, where $e(\theta f)^\sigma = e(\theta f')$. (Note that, in the arithmetic case, the exponential used here is the classical exponential according to the table.)

Let $\chi_{L/B}$ be the characteristic function of $\mathrm{Gal}(L/B)$. In other words, the function $\chi_{L/B}\colon \mathrm{Gal}(B^{\mathrm{sep}}/B) \to \mathbb{Z}$ is such that $\chi_{L/B}(\sigma)$ is 1 if σ is the identity on L and it is 0 otherwise.

We then have the hypothesis/recipe/conjecture:

(H3)   If $\chi_{L/B} = \sum m_f h(f)$, $m_f \in \mathbb{Z}$, then $\mathrm{Period}_E / \prod \Gamma(f)^{m_f} \in \overline{A}$.

This is known in the classical case. In function fields, there is some evidence.

Example 2.13. $\mathcal{O} = A = \mathbb{F}_q[t]$, $K = B = \mathbb{F}_q(t)$, $L = \mathbb{F}_{q^k}(t)$. Then the rank k Drinfeld A-module $u \mapsto tu + u^{q^k}$ is the rank one Carlitz module over $\mathbb{F}_{q^k}[t]$. Hence we want to express its period $\widetilde{\pi}_k$ in terms of the gamma values at fractions with the gamma function built up from the base. If τ is the qth power Frobenius, then $\chi_{L/B}(\tau^n)$ is 1 if k divides n and it is 0 otherwise. The same is true for

$$\left[qh\left(\frac{q^{k-1}}{1-q^k}\right) - h\left(\frac{1}{1-q^k}\right)\right](\tau^n) = q\left\langle\frac{q^{k+n-1}}{q^k-1}\right\rangle - \left\langle\frac{q^n}{q^k-1}\right\rangle.$$

Hence (H3) predicts that $\Gamma(q^{k-1}/(1-q^k))^q/\Gamma(1/(1-q^k))$ is an algebraic multiple of $\widetilde{\pi}_k$, as we have seen before.

Remarks 2.14.

(1) In case I, the Γ function is closely connected with the ζ and L functions. It is in some sense a factor at infinity for the ζ function and hence it appears in functional equations. Such connections are missing in the other cases. We only have $\Pi'(x)/\Pi(x) = \zeta(x,1)$ in the third case.

(2) The Hurwitz formula $\zeta(x,s) = \langle -x \rangle - 1/2 + s \log \Gamma(x) + o(s^2)$ around $s = 0$ connects partial zeta, 'brackets' and logarithm of gamma. For case III, the partial zeta value for $x = a_1/a_2$ (with $\deg a_1 < \deg a_2$ and $a_2 \in A_+$) is obtained by summing $q^{-s\deg a}$ over a monic a congruent to a_1 modulo a_2. Hence, at $s = 0$, it equals $1 - 1/(q-1)$ or $-1/(q-1)$ according to whether a_1 is monic or not. This is the motivation for the definition, due to Anderson, of the brackets as above in the geometric case.

(3) See [T04, Ch. 4] for the connection of this set-up with function field Gauss and Jacobi sums as mentioned before. Chris Hall made the following interesting remark about the connection of the condition in (H1) with the classical Gauss sums and their occurrence in the work related to Ulmer's parallel lecture series on ranks of elliptic curves over function fields. The classical Gauss and Jacobi sums arise, e.g., in Weil's work, as (reciprocal) zeros of zeta functions of Fermat varieties $X/\mathbb{F}_q$. When such a sum arises as a (reciprocal) eigenvalue of Frobenius acting on the even-index cohomology of X and when the nth power of the sum is a power of q, then the Tate conjectures predict that there are corresponding algebraic cycles on $X/\mathbb{F}_{q^n}$. While finding such algebraic cycles is usually very difficult, the seemingly simpler question of determining whether or not some power of a Gauss or Jacobi sum is a power of q also appears to be quite difficult. (In the literature – cf. papers of R. Evans and N. Aoki – such sums are called 'pure'). One can use Stickelberger's theorem to reformulate the condition for a particular sum to be pure in terms of a fractional sum (cf. Section 2.3 of [Ulmer, Math. Res. Lett. 14(3) (2007), 453–467]), and the resulting condition resembles hypothesis (**) of Theorem 1.8 or (H1).

We will see in Lecture 4 that (H1) is best possible in the arithmetic case by an automata method. In both the function field cases, by a motivic method which was developed a little later, we can even prove (see Lectures 3 and 6) stronger algebraic independence results. In the classical case, much less is known (see Section 4.1).

Lecture 3

Solitons, t-motives and Complete Gamma Relations for $\mathbb{F}_q[t]$

In this lecture, we restrict to $A = \mathbb{F}_q[t]$. Our examples in the last lecture can be considered as simple cases of Fermat motives: Drinfeld modules with complex multiplication by rings of integers of 'cyclotomic fields'. In these cases, we saw connections between periods and gamma values and Frobenius eigenvalues at finite primes, Gauss sums connecting to values of interpolated gammas at that prime. Now we will outline a complete generalization using t-motives, a higher-dimensional generalization of Drinfeld modules.

So far, we handled gamma values by connecting them to Drinfeld modules with complex multiplications by (the integral closure of A in) cyclotomic fields, such as constant field extensions $K(\mu_n)$ of K or Carlitz–Drinfeld cyclotomic extensions $K(\Lambda_\wp)$. Drinfeld modules can handle only one point at infinity. For the geometric gamma function, this restricted us to $\Gamma(1/\wp)$ (and its v-adic counterparts), with a degree one prime $\wp$. To handle $\Gamma(1/f)$ for any $f(t) \in A$, we need objects with multiplication from $K(\Lambda_f)$. Thus we need 'higher-dimensional t-motives' which can handle many infinite places. We will focus on the geometric gamma in this lecture and return to the general case in Lecture 6.

3.1 Anderson's solitons: General overview

If one looks at the proof of Theorems 2.8 and 2.9, the crucial point was to produce special functions (called 'solitons' by Anderson – the reason will be explained below) on $X_f \times X_f$, where X_f is the fth cyclotomic cover of the projective line, namely the curve corresponding to $K(\Lambda_f)$, with the property that they specialize on the graph of the dth power Frobenius to the dth degree term in the product defining the gamma value.

Example 3.1. For $f(t) = t$, as we saw before by Moore determinants, this corresponds to a 'compact' formula for the left-hand side, which has exponentially growing size:

$$\left(\prod_{\substack{a \in A_{i+} \\ a \equiv 1 \bmod t}} a \right) \Big/ \left(t^{q^{i-1}} D_{i-1} \right) = 1 - \frac{\zeta_t}{\zeta_t^{q^i}} = 1 - (-t)^{-(q^i-1)/(q-1)},$$

with corresponding 'soliton' $\phi = (t/T)^{1/q-1} = \zeta_t/\zeta_T$, if one uses two independent variables t and T for the two copies of the line. The additivity of the solitons (as functions of a/f) was established using Moore determinants, and this formula was used (essentially with the method of partial fractions) to get formulas for (what is now understood as) the solitons when f was a product of distinct linear factors.

In general, we get 'compact' formulas for the product of the monic polynomials in $\mathbb{F}_q[t]$ of degree i in a given congruence class (essentially the same as asking for a formula for the term in the definition of the geometric gamma function at a fraction). Here is another example:

$$\left(\prod_{\substack{a \in A_{i+} \\ a \equiv 1 \bmod t^2}} a \right) \Big/ \left((t^2)^{q^{i-2}} D_{i-2} \right) = 1 - \frac{\zeta_{t^2}(-t)^{q^{i-1}/(q-1)} - \zeta_{t^2}^{q^{i-1}}(-t)^{1/(q-1)}}{(-t)^{2q^{i-1}/(q-1)}}.$$

In this case, $f(t) = t^2$ and

$$\phi = (\zeta_{t^2} C_T(\zeta_{T^2}) - \zeta_{T^2} C_t(\zeta_{t^2}))/C_T(\zeta_{T^2})^2.$$

Remark 3.2. For $A = \mathbb{F}_q[t]$, we have good analogs of binomial coefficient polynomials $\left\{ \begin{smallmatrix} x \\ n \end{smallmatrix} \right\}$ obtained by digit expansions, just as in the factorial case, from the basic ones for $n = q^d$, defined by analogy as $e(x\ell(z)) = \sum \left\{ \begin{smallmatrix} x \\ q^d \end{smallmatrix} \right\} z^{q^d}$. We have the exponential $e(z) = \sum z^{q^i}/d_i$ and the logarithm $\ell(z) = \sum z^{q^i}/\ell_i$, in general, so that $D_i = d_i$ and $L_i = (-1)^i \ell_i$, in our case of a Carlitz module for $A = \mathbb{F}_q[t]$.

It can be easily shown that

$$\left\{ \begin{matrix} x \\ q^k \end{matrix} \right\} = \sum x^{q^i}/(d_i \ell_{k-i}^{q^i}) = e_k(x)/d_k =: \left(\begin{matrix} x \\ q^k \end{matrix} \right),$$

with $e_k(x) = \prod(x-a)$, where the product runs over all $a \in A$ of degree less than k. The same definitions easily generalize to all A's, but the curly and round bracket binomials differ in general and we will return to this issue. For several analogies, we refer to [T04, Section 4.14].

It is easy to verify that the degree d term in the geometric gamma product definition for $\Gamma(a/f)$ is nothing but $1 + \left(\begin{smallmatrix} a/f \\ q^d \end{smallmatrix} \right)$. So solitons are 'interpolations' of these binomial coefficients for $x = a/f$ as d varies.

Some functions of this kind were constructed by the author using Moore determinants. Anderson had the great insight that such determinant techniques are analogs of techniques found in the so-called tau-function theory or soliton theory. By applying deformation techniques from soliton theory that occur in Krichever's theory for solving, using theta functions, certain integrable systems of partial differential equations such as KdV to the arithmetic case of Drinfeld's dictionary, Anderson obtained in [A92, A94, ABP04], for any f, the required functions, which he called solitons. Anderson used tau- functions, theta functions, and explicit constructions (using exponential, torsion and adjoints) respectively in the three papers mentioned. We refer to these papers and [T01] for motivation and analogies.

We will only describe the last construction. See [T04, 8.5] or [T99] for an alternative approach of the author giving Frobenius semilinear difference equations analogous to partial differential equations.

By interpolating partial gamma products by such algebraic functions on the cyclotomic curve times itself, Anderson constructed t-motives and showed that their periods are essentially the gamma values, and that they also give ideal class annihilators. This vastly generalizes very simple examples we looked at. The applications to special values and transcendence go way beyond the classical counterparts this time, because the t-motives occurring can have arbitrary fractions as weights, in contrast to the classical motives.

We saw that the gamma product for $\Gamma(a/f)$, restricted to degree d, is essentially the binomial coefficient $\{a/f, q^d\}$, which is the coefficient of t^{q^d} in $e(\ell(t)a/f)$, which can be thought of as a deformation of torsion of $e(\ell(0)a/f)$, since $\tilde{\pi}$ is a value of $\ell(0)$. In function fields, we have the luxury of introducing more copies of variables (tensor products, products of curves, generic base change) with no direct analog for number fields. (In other words, we do not know how to push the polynomials-versus-numbers analogy to multi-variable polynomials.)

Before describing constructions of solitons, we sketch how they are used to construct t-motives which have as periods and quasi-periods the corresponding gamma values.

3.2 *t*-modules, *t*-motives and dual *t*-motives

Let F be an A-field, $\iota\colon A \to F$ being the structure map (not needed in the number field case, where we have the canonical base $\mathbb{Z}$). We assume F is perfect. This is needed for technical reasons, but for our transcendence applications, we can even assume it to be algebraically closed. The kernel of ι is called the *characteristic*. We will usually stay in 'generic characteristic' situation of zero kernel.

Recall that a Drinfeld A-module ρ over F is roughly a non-trivial embedding of (a commutative ring) A in the (non-commutative ring of) $\mathbb{F}_q$-linear endomorphisms of the additive group over F. It yields, for a given $a \in A$, a polynomial $\rho_a \in F\{\tau\}$ (where τ is the qth power map) with constant term $\iota(a)$. Now ρ_t, for a single non-constant $t \in A$, determines ρ_a for any $a \in A$ by commutation relations.

So, without loss of generality, we can just look at t-modules, which one can think of as the special case $A = \mathbb{F}_q[t]$, or the general case can be thought of as having extra multiplications. In other words, a t-module G is an algebraic group isomorphic to the additive group together with a non-trivial endomorphism denoted by t, such that at the Lie algebra level t acts by a scalar $\theta = \iota(t) \in F$.

(*Warning:* Since we have stated everything with t-variable at the start, we will do so for the end results also, but in proofs we will distinguish t and θ, just as we distinguish the role of the coefficient field and the multiplication field, even when they are the same. Our answers will thus often be in terms of θ, which we then replace by t! As we progress, we will see the advantage of having two isomorphic copies of $\mathbb{F}_q(x)$ together, something which cannot be done in the number field case.)

We can generalize to d dimensions by replacing the additive group by its dth power, and Anderson realized that rather than requiring t to act by a scalar matrix θ at the Lie algebra level, one should relax the condition to t having all eigenvalues θ or equivalently $t - \theta$ should be a nilpotent matrix. We will see the need for the relaxation clearly when we calculate $C^{\otimes n}$ in Lecture 6.

Anderson called $M(G) := \mathrm{Hom}(G, \mathbb{G}_a)$ the corresponding t-motive. By composing with t-action on the left and τ-action on the right, this is a module for (the non-commutative ring) $F[t, \tau]$. This dual notion is called a motive, because it is a nice concrete linear/τ-semilinear object from which cohomology realizations of Drinfeld modules and t-modules can be obtained by simple linear algebra operations.

We will not go into details here (see [G94] or [T04, 7.5]), but just note that by efforts of Drinfeld, Deligne, Anderson, Gekeler, and Yu, cohomologies of Betti, v-adic, de Rham, christalline types, comparison isomorphisms etc. were developed, and that the recent works of Pink, Böckle, Hartl, Vincent Lafforgue, Genestier etc. have developed Hodge-theoretic and christalline aspects much further into a mature theory.

The dimension d and rank r of M are, by definition, the ranks of M (freely generated) over $F[\tau]$ and $F[t]$ respectively. Choosing a basis, a t-motive can thus be described by a size r matrix M_τ with entries in $F[t]$. Unlike the one-dimensional case of Drinfeld modules, in general the corresponding exponential function need not be surjective. The surjectivity or uniformizability criterion can be expressed as existence of solutions of a matrix equation (see below) in the rigid analytic realm (uniformizability is equivalent to the rank of a lattice being r). The period matrix is then obtained by residue operation (at $t = \theta$) from this solution.

Anderson later used the 'adjoint' duality [T04, 2.10] developed by Ore, Poonen, and Elkies between τ objects and $\sigma := \tau^{-1}$-objects, and considered 'dual t-motives', which are $F[t, \sigma]$-modules. To a t-module G one associates $M^*(G) := \mathrm{Hom}(\mathbb{G}_a, G)$. When F is perfect, e.g., algebraically closed (as is sufficient for transcendence purposes), and when we do not bother about finer issues of field of definition etc., then the qth root operation is fine too.

The technical advantage is a quick and clean description, as in [ABP04], as $M^*/(\sigma-1)M^*$ for points of G and $M^*/\sigma M^*$ for its Lie algebra in terms of M^*, and the possibility of replacing residue operation by the simpler evaluation operation.

In [ABP04], this was put to great use and transcendence theory papers now use this language and call the dual t-motives just t-motives or Anderson t-motives. Just be warned that this basic term means slightly different (but related) concepts in different papers. In transcendence literature, the rigid analytically trivial (equivalently uniformizable) dual t-motive is often called t-motive.

We just remark that in the first part of these notes, dealing with gamma values, we use 'pure' complex multiplication (rank one over higher base) t-motives (coming from solitons), while dealing with zeta and multizeta values in the second part, we use mixed motives of higher rank and the language of Tannakian theory in terms of motivic Galois group, which we could have used to simplify once developed, in the first part as well.

3.3 Period recipe and examples

Let Φ be a size r square matrix with entries in $F[t]$ giving the σ action $\sigma m = m^{(-1)}$ on the (dual!) t-motive M. Consider the Frobenius-difference equation $\Psi^{(-1)} = \Phi\Psi$. If it has a solution Ψ with entries in the fraction field of power series convergent in the closed unit disc (fraction field of the Tate algebra), then Anderson proved that M is uniformizable with period matrix the inverse of Ψ evaluated at $t = \theta$. (Here the nth twist $^{(n)}$ represents entry-wise q^nth power on elements of F and identity on t.)

For example, for the Carlitz module we have $\Phi = t - \theta$. Then we have

$$\Omega := \Psi = (-\theta)^{-q/(q-1)} \prod_{i=1}^{\infty} \left(1 - t/\theta^{q^i}\right)$$

and $\tilde{\pi} = 1/\Psi(\theta)$. We will denote this particular Ψ by Ω, as is the common notation.

Remark 3.3. We remark here that, although C_∞ being of infinite degree over K_∞ is quite different than $[\mathbb{C} : \mathbb{R}] = 2$, the formulas for $\tilde{\pi}$ show that for many purposes the role of C_∞ is played by the degree $q - 1$ extension

$$K_\infty(\tilde{\pi}) = K_\infty(\zeta_t) = K_\infty((-t)^{1/(q-1)}) = K_\infty((-[1])^{1/(q-1)})$$
$$\longleftrightarrow \mathbb{C} = \mathbb{R}(i) = \mathbb{R}(2\pi i).$$

The algebraic elements of $K_\infty(\tilde{\pi})$ are separable over K.

Note that, ignoring simple manipulations needed for the convergence in the right place, we always have a formal solution $\Psi = \prod_{i=1}^{\infty} \Phi^{(i)}$, so that in the rank one case the (reciprocal of the) period is given as a product of Φ evaluated at graphs of powers of Frobenius, as we mentioned above.

So given a soliton Φ, one gets a motive with period the corresponding gamma value at the proper fraction, up to a simple factor.

3.4   Explicit construction

We follow the notations of [ABP04]. Thus, T (respectively t) below corresponds to our t (respectively θ).

Let $\mathbf{e}(z) = e(\tilde{\pi}z)$. Now put $\Omega^{(-1)}(T) = \sum a_i T^i$, $a_i \in K_\infty(\check{t})$. Let Res: $K_\infty \to \mathbb{F}_q$ be the unique $\mathbb{F}_q$-linear functional with kernel $\mathbb{F}_q[t] + 1/t^2 \mathbb{F}_q[[t]]$ and such that $\mathrm{Res}(1/t) = 1$, i.e., the usual residue function for parameter t. For $x \in K_\infty$, put

$$\mathbf{e}_*(x) := \sum_{i=0}^{\infty} \mathrm{Res}(t^i x) a_i.$$

We have

$$\Omega^{(-1)}(T) = \sum_{i=0}^{\infty} \mathbf{e}_* \left(\frac{1}{t^{i+1}} \right) T^i, \qquad \frac{1}{\Omega^{(-1)}(T)} = \sum_{i=0}^{\infty} \mathbf{e} \left(\frac{1}{t^{i+1}} \right) T^i,$$

so $\mathbf{e}_*(1/t) = 1/\check{t}$. From the functional equation above, we get a recursion on the a_i's which implies that

$$t\mathbf{e}_*(x)^q + \mathbf{e}_*(x) = \mathbf{e}_*(tx)^q.$$

Comparison with the adjoint $C_t^* = t + \tau^{-1}$ of the Carlitz module shows that $\mathbf{e}_*(a/f)$ are qth roots of f-torsion points of this adjoint.

Definition 3.4. For $f \in A_+$, we say that two families $\{a_i\}$, $\{b_j\}$ $(i,j = 1$ to $\deg f)$ of elements in A are f-dual if $\mathrm{Res}(a_i b_j / f) = \delta_{ij}$.

Theorem 3.5. *Fix $f \in A_+$ and f-dual families $\{a_i\}$, $\{b_j\}$ as in the definition above. Fix $a \in A$ with $\deg a < \deg f$. Then*

$$\sum_{i=1}^{\deg f} \mathbf{e}_*(a_i/f)^{q^{N+1}} \mathbf{e}(b_i a/f) = -\binom{a/f}{q^N}$$

for all $N \geq 0$. Also, if $a \in A_+$, then

$$\sum_{i=1}^{\deg f} \mathbf{e}_*(a_i/f)\mathbf{e}(b_i a/f)^{q^{\deg f - \deg a - 1}} = 1.$$

Thus we define a soliton by

Definition 3.6. For $x = a_0 / f \in f^{-1}A - A$, put

$$g_x := 1 - \sum_{i=1}^{\deg f} \mathbf{e}_*(a_i/f)(C_{a_0 b_i}(z)|_{t=\theta}).$$

If y is the 'fractional part' of ax, then

$$g_x^{(N+1)}(\xi_a) = 1 + \binom{y}{q^N} = \prod_{n \in A_N+} (1 + y/n),$$

where $\xi_a := (t, \mathbf{e}(a/f))$. Thus on the graph of the $(N+1)$-st Frobenius power it gives the Nth term of the gamma product.

For $x \in K_\infty = \mathbb{F}_q((1/t))$ and integers $N \geq 0$, let us define $\langle x \rangle_N$ to be 1 if the fractional part power series of x starts with $(1/t)^{n+1}$, and 0 otherwise. Then for the bracket defined in a uniform framework we have $\langle x \rangle = \sum_{N=0}^{\infty} \langle x \rangle_N$.

Theorem 3.7. *For $x \in f^{-1}A - A$, we have equality of divisors of the fth cyclotomic cover X_f of the projective line*

$$(g_x) = -\frac{1}{q-1}\infty_{X_f} + \sum_{a \in A_{<\deg f}, (a,f)=1} \sum_{N=0}^{\infty} \langle ax \rangle_N \xi_a^{(N)}.$$

The divisor D on the right (note that the sum is finite) being the Stickelberger divisor of the cyclotomic theory that one gets through the partial zeta function corresponding to a/f, we obtain the soliton specialization as the Stickelberger element, generalizing Coleman's constructions [A92] corresponding to few denominators f.

Specializing the solitons at appropriate geometric points, Anderson proved [A92] a two-dimensional version of Stickelberger's theorem. For a connection with the arithmetic of zeta values and theta functions, see [A94, T92b]. The reason for the name soliton is that the way it arises in the theory of Drinfeld modules or shtukas, when dealing with the projective line with some points identified (in the support of f), it is analogous to the way in which soliton solutions occur in Krichever's theory of algebro-geometric solutions of differential equations, as explained in [Mum78, pp. 130, 145] and [A92, A94, T01].

3.5 Analog of Gross–Koblitz for geometric gamma: The $\mathbb{F}_q[t]$ case

Using Anderson's analog of Stickelberger, the proof of the Gross–Koblitz analog generalizes from the simplest case considered in 2.3 to the general $\mathbb{F}_q[t]$ case by a proof exactly as in previous cases.

Theorem 3.8. *Let $A = \mathbb{F}_q[t]$. Let v be a monic prime of degree d such that f (of positive degree) divides $v - 1$. Let a be an element of degree less than that of f. Then $\Gamma_v(a/f)$ is an (explicit) rational multiple of the a/f-Stickelberger element applied to v. In particular, it is algebraic. (So If $\underline{a}$ consists of Frob$_v$-orbits up to translation by elements in A, then $\Gamma_v(\underline{a})$ is algebraic.)*

3.6   Fermat *t*-motives

With soliton function $g = g_x$, consider $M_f := $ as a $\overline{K}[t, \tau]$-module by the left multiplication action of $\overline{K}[t]$ and by the τ-action given by g. The underlying space here is of rank one over a 'cyclotomic base' (see [ABP04] for details).

Let $f \in A_+$, $I := \{a \in A : (a, f) = 1, \deg(a) < \deg(f)\}$, and $I_+ = I \cup A_+$. Sinha, in his University of Minnesota thesis, proved the following result.

Theorem 3.9. *With the notation as above, M_f is a uniformizable abelian t-motive over $\overline{K}$ of dimension $\phi(f)/(q-1)$ and rank $\phi(f) := |(A/fA)^*|$. In fact, the corresponding t-module is an HBD module with multiplications by $A[\zeta_f]_+$ and is of CM type with complex multiplication by $A[\zeta_f]$. Its period lattice is free of rank one over $A[\zeta_f]$, with the ath coordinate (for $a \in I_+$) of any non-zero period being (with an appropriate explicitly given $C_\infty\{\tau\}$-basis of M to give coordinates) the (explicit non-zero algebraic multiple of) value $\Gamma(a/f)$.*

Brownawell and Papanikolas [BP02] proved:

Theorem 3.10. *The coordinates of periods and quasi-periods of M_f (in the same coordinates as above) are exactly the (explicit non-zero algebraic multiples of) values $\Gamma(a/f)$, with $a \in I$.*

We refer to [Sin97b, BP02, ABP04] for a very nice and clean treatment of these issues considered in 3.4 and 3.6–3.8.

3.7   ABP criterion: Period relations are motivic

Here we use $F = K$, C_∞ with variable θ, and t an independent variable. Let $C_\infty\{t\}$ be the ring of power series over C_∞ convergent in the closed unit disc.

Theorem 3.11 ([ABP04]). *Consider $\Phi = \Phi(t) \in \mathrm{Mat}_{r \times r}(\overline{K}[t])$ such that $\det \Phi$ is a polynomial in t vanishing (if at all) only at $t = \theta$ and $\psi = \psi(t) \in \mathrm{Mat}_{r \times 1}(C_\infty\{t\})$ satisfying $\psi^{(-1)} = \Phi\psi$.*

If $\rho\psi(\theta) = 0$ for $\rho \in \mathrm{Mat}_{1 \times r}(\overline{K})$, then there is $P = P(t) \in \mathrm{Mat}_{1 \times r}(\overline{K}[t])$ such that $P(\theta) = \rho$ and $P\psi = 0$.

Thus $\overline{K}$-linear relations between the periods are explained by $\overline{K}[T]$-level linear relations (which in our set-up are the motivic relations and thus 'algebraic relations between periods are motivic', in analogy with Grothendieck's conjecture for the motives that he defined). In terms of special functions of our interest, this makes precise the vague hope that 'there are no accidental relations and the relations between special values come from known functional equations', and proves it.

Motives are thus simple, concrete linear algebra objects and have tensor products via which algebraic relations between periods, i.e., linear relations between powers and monomials in them reduce to linear relations between periods (of some other motives). In this sense, the new [ABP] criterion below is similar to the Wüstholz type sub-*t*-module theorem proved by Jing Yu, as remarked in [ABP, 3.1.4]. The great novelty is, of course, the direct simple proof as well as its perfect adaptation to the motivic set-up here. Because our motives are concrete linear algebra objects, it is 'easy' (compared to the cycle-theoretic difficulties one encounters in the classical theory) to show that their appropriate category (up to isogeny) is a neutral Tannakian category over $\mathbb{F}_q(t)$, with fiber functor to the category of vector spaces over $\mathbb{F}_q(t)$ given by lattice Betti realization, because the formalism of such categories is based on linear algebra motivation anyway. (Soon afterwards, Beukers [Be06] proved a similar criterion for dependence of values of *E*-functions, but it does not have such strong applications to relations between periods of classical motives, because of differences in period connections in this case.) Each motive M also generates such a (sub-) category and is thus equivalent to a category of finite-dimensional representations over $\mathbb{F}_q(t)$ of an affine group scheme Γ_M over $\mathbb{F}_q(t)$, called the motivic Galois group of M, which can be described as the group of tensor automorphisms of the fiber functor.

After developing this machinery [P08], the following very useful theorem from [P08] (which is analog to the Grothendieck period conjecture for our abelian *t*-modules) follows [P08, pp. 166–167] easily from the ABP criterion and goes one step further in the quantitative direction:

Theorem 3.12. *If M is a uniformizable t-motive over $\overline{K}$, then the transcendence degree of the field extension of $\overline{K}$ generated by its periods is the dimension of the motivic Galois group of M (i.e., the group corresponding to the Tannakian category generated by M).*

In [P08], we have further a description of the motivic Galois group as 'difference equations Galois group' for the 'Frobenius semilinear difference equation' $\Psi^{(-1)} = \Phi\Psi$. This allows the calculation of dimensions and proofs of theorems described in Lecture 6.

Here is the proof from [ABP04] of the theorem for the simplest $r = 1$ case. Without loss of generality we can assume that $\rho \neq 0$, so we have to conclude that ψ vanishes identically from $\psi(\theta) = 0$. For $v \geq 0$, we have

$$\left(\psi(\theta^{q^{-v}})\right)^{q^{-1}} = \psi^{(-1)}\left(\theta^{q^{-v-1}}\right) = \Phi\left(\theta^{q^{-v-1}}\right)\psi\left(\theta^{q^{-v-1}}\right).$$

Since $\Phi(\theta^{q^{-v-1}}) \neq 0$, ψ has thus infinitely many zeros $\theta^{q^{-v}}$ in the disc $|t| \leq |\theta|$ and hence it vanishes identically.

The general case makes a similar beautiful use of the functional equation of the hypothesis by manipulating suitable auxiliary functions to vanish identically (so as to recover P). This is done by applying standard transcendence theory tools such as Siegel's lemma (to solve a system of linear equations thus arising) and the Schwarz–Jensen and Liouville inequalities (to estimate bounds needed). We refer the reader to the clean treatment in [ABP].

3.8 Complete determination of geometric gamma relations for $\mathbb{F}_q[t]$

Brownawell and Papanikolas [BP02] developed a complex multiplication theory and a theory of quasi-periods for general t-motives. By analyzing the connection between CM types of soliton t-motives, they showed that the only $\overline{K}$-linear relations among the gamma values at fractions are those explained by the bracket criterion in our uniform framework, as they mirror relations coming from CM type relations.

Anderson, Brownawell and Papanikolas [ABP04], using tensor powers to realize all monomials as periods (no quasi-periods are then needed) and the ABP criterion, proved the following result:

Theorem 3.13. *Let Γ stand for the geometric gamma function for $A = \mathbb{F}_q[T]$. By a Γ-monomial we mean an element of the subgroup of C_∞^* generated by $\widetilde{\pi}$ and Γ-values at proper fractions in K. Then a set of Γ-monomials is $\overline{K}$-linearly dependent exactly when some pair of Γ-monomials is, and pairwise $\overline{K}$-linear dependence is entirely decided by the bracket criterion* (H1).

In particular, for any $f \in A_+$ of positive degree, the extension of $\overline{K}$ generated by $\widetilde{\pi}$ and $\Gamma(x)$ with x ranging through proper fractions with denominator f (not necessarily reduced), is of transcendence degree $1 + (q-2)|(A/f)^*|/(q-1)$ over $\overline{K}$.

We will talk about important applications of Papanikolas' theorem to understanding relations between gamma (arithmetic or geometric) and zeta values in Lecture 6.

Lecture 4

Automata Method, General A, the v-adic Situation and Another Gamma Mystery

This lecture is independent of the previous one.

4.1 Automata and transcendence for arithmetic gamma

In this section, we restrict to $A = \mathbb{F}_q[t]$ and look at the arithmetic gamma. We will show the following.

Theorem 4.1 (Arithmetic gamma). *In the case $A = \mathbb{F}_q[t]$, we have $m(\underline{f}^{(\sigma)}) = 0$ for all $\sigma = q^j$ if and only if $\Gamma(\underline{f})$ is algebraic.*

The 'only if' part is (H1) in this case and was already explained, so the new result in this theorem is that (H1) is best possible for arithmetic gamma.

This was proved by the 'automata method'. Before this, the methods based on the transcendence of periods and the Chowla–Selberg formula in Lecture 2 gave transcendence results (due to Thiery [Thi] and Jing Yu [Y92] independently) parallel to those known in the number field case. In the number field case, transcendence is known only for gamma values at fractions with denominators dividing 4 or 6, while in our case it was known only for denominators dividing $q^2 - 1$, the analogy being that 4 and 6 are numbers of roots of unity in imaginary quadratic fields, whereas $q^2 - 1$ is the number of roots of unity in the quadratic extension field $\mathbb{F}_{q^2}(t)$, and the reason being the Chowla–Selberg formula and complex multiplication from these fields, as we saw. The automata method settled completely [T96, All96] the question of which monomials in gamma values at fractions are algebraic and which are transcendental. (In Lecture 6 we will mention recent results [CPTY] proving a much stronger result than the theorem above). It is based on the following theorem by Christol and others [Chr79, CKMR80].

Theorem 4.2. *The following statements are equivalent:*

(i) $\sum f_n x^n$ *is algebraic over $\mathbb{F}_q(x)$;*

(ii) $f_n \in \mathbb{F}_q$ is produced by a q-automaton;

(iii) there are only finitely many subsequences of the form $f_{q^k n + r}$ with $0 \le r < q^k$.

For our proof, we will only need (and prove) that (i) implies (iii). But (ii) is the reason for the name of the method and the reason for various other techniques from this viewpoint and results, for which we refer to the surveys [All87, AS03, T98, T04]. Thus we also say a few words about the concept of automata and how (ii) quickly implies (iii), although it is logically not necessary. The concept of automata will not be used in these notes after the next paragraph, except for the references to the method.

Here, an m-automaton (we shall usually use $m = q$, a prime power in the applications) consists of a finite set S of states, a table of how the base m digits operate on S, and a map Out from S to $\mathbb{F}_q$ (or some alphabet in general). For a given input n, fed in digit by digit from the left, each digit changing the state by the rule provided by the table, the output is Out$(n\alpha)$ where α is some chosen initial state. So instead of our ideal Turing machine which has infinite tape and no restriction on input size (and is still a good approximation to computers because of the enormous memories available these days) the finite automaton has a restricted memory, so an integer has to be fed in digit by digit, with the machine retaining no memory of previous digits fed except through its changed states.

Proof. (ii) implies (iii): There are only finitely many possible maps $\beta \colon S \to S$ and any $f_{q^k n + r}$ is of the form Out$(\beta(n\alpha))$.

(i) implies (iii): For $0 \le r < q$, define C_r (twisted Cartier operators) by $C_r(\sum f_n x^n) = \sum f_{qn+r} x^n$. Considering the vector space over $\mathbb{F}_q$ generated by the roots of the polynomial satisfied by f, we can assume that $\sum_{i=0}^{k} a_i f^{q^i} = 0$ with $a_0 \ne 0$. Using $g = \sum_{r=0}^{q-1} x^r (C_r(g))^q$ and $C_r(g^q h) = g C_r(h)$, we see that

$$\left\{ h \in \mathbb{F}_q((x)) : h = \sum_{i=0}^{k} h_i (f/a_0)^{q^i}, \ h_i \in \mathbb{F}_q[x], \ \deg h_i \le \max\left(\deg a_0, \deg a_i a_0^{q^{i-2}} \right) \right\}$$

is a finite set containing f and stable under C_r's. □

The particular case of Theorem 4.1 is the transcendence at any proper fraction, proved by Allouche generalizing the author's result for any denominator, but with restrictions on the numerator. Mendès-France and Yao [MFY97] generalized to gamma values at p-adic integers from the values at fractions, and simplified further. We present below an account based on their method.

Lemma 4.3. *For positive integers a, b, c, the number $q^c - 1$ divides $q^a(q^b - 2) + 1$ if and only if c divides the greatest common divisor (a, b) of a and b.*

The proof, which is short, straightforward and elementary, is omitted.

Theorem 4.4. *If the sequence $n_j \in \mathbb{F}_q$ is not ultimately zero, then the element $\sum_{j=1}^{\infty} n_j/(t^{q^j} - t) \in K_\infty$ is transcendental over K.*

Proof. We have $\sum t\, n_j / (t^{q^j} - t) = \sum c(m) t^{-m}$, where

$$c(m) = \sum_{(q^j-1)\mid m} n_j.$$

Consider the subsequences $c_t(m) := c(q^t m + 1)$. As there are infinitely many non-zero n_j's, it is enough to show, by Christol's theorem, that $c_a \neq c_b$ for any $a > b$ such that n_a and n_b are non-zero.

Let h be the least positive integer s dividing a, but not dividing b and with $n_s \neq 0$. (Note that $s = a$ satisfies the three conditions, so h exists.) By the lemma,

$$c_a(q^h - 2) - c_b(q^h - 2) = \sum_{(q^l-1)\mid(q^a(q^h-2)+1)} n_l - \sum_{(q^l-1)\mid(q^b(q^h-2)+1)} n_l$$

$$= \sum_{l\mid\gcd(a,h)} n_l - \sum_{l\mid\gcd(b,h)} n_l$$

$$= n_h \neq 0.$$

Hence $c_a \neq c_b$ and the theorem follows. □

Theorem 4.5. *Let* $A = \mathbb{F}_q[t]$. *If* $n \in \mathbb{Z}_p$ *is not a non-negative integer, then* $n! = \Gamma(n+1)$ *is transcendental over* K. *In particular, the values of the gamma function at proper fractions and at non-positive integers are transcendental.*

Proof. If a power series f is algebraic, so is its derivative f', and hence also the logarithmic derivative f'/f. In other words, transcendence of the logarithmic derivative implies transcendence. This is a nice tool to turn products into sums, sometimes simplifying the job further because now exponents matter modulo p only. (This nice trick, due to Allouche, got rid of the size restrictions on the numerator that the author had before.)

Write the base q expansion $n = \sum n_j q^j$ as usual, so that $n! = \prod D_j^{n_j}$, and hence

$$\frac{n!'}{n!} = \sum n_j \frac{D'_j}{D_j} = -\sum \frac{n_j}{t^{q^j} - t}.$$

Now if all sufficiently large digits n_j are divisible by p^k, then by modifying the first few digits (which does not affect transcendence) we can arrange that all are divisible by p^k and then take p^kth roots (which does not affect transcendence). So, without loss of generality, we can assume that the sequence n_j is not ultimately zero (modulo p). Hence the previous theorem applies. □

Now we describe the proof of Theorem 4.1.

Proof. By taking a common divisor and using Fermat's little theorem, any monomial of gamma values at proper fractions can be expressed as a rational function times a monomial in $(q^j/(1-q^d))!$'s, where d is fixed and $0 \leq j < d$. It was shown in the proof of Theorem 1.4 [T04, p. 107] that if the hypothesis of (H1) is not met then this monomial is non-trivial. Hence, taking again p-powers out as necessary, as in the proof of the previous theorem, we can assume that the exponents are not all divisible by p. But the exponents matter only modulo p when we take the logarithmic derivative. So this logarithmic derivative is a logarithmic derivative of some gamma value also, and hence it is transcendental by the previous theorem.   □

Remark 4.6. We cannot expect to have an analogous result for the classical gamma function, because its domain and range are archimedean, and continuity is quite a strong condition in the classical case. In other words, a non-constant continuous real-valued function on an interval cannot fail to take on algebraic values.

Morita's p-adic gamma function has domain and range $\mathbb{Z}_p$, which being non-archimedean is closer to our situation. Let us now look at interpolation of $\Pi(n)$ at a finite prime v of $A = \mathbb{F}_q[t]$.

We have proved, as a corollary to the analog of the Gross–Koblitz theorem, that if d is the degree of v then $\Pi_v(q^j/(1 - q^d))$, $0 \leq j < d$, is algebraic. Straight manipulation with digits then shows that $\Pi_v(n)$ is algebraic if the digits n_j are ultimately periodic of period d. The converse, in case n is a fraction, is a question raised earlier, namely whether (H2) is best possible. In analogy with the theorem above, Yao has conjectured the converse for $n \in \mathbb{Z}_p$.

Things become quite simple [T98] when v is of degree one, so that without loss of generality we can assume that $v = t$. Yao used a result in [MFY97] to simplify and generalize again.

Theorem 4.7. *Let $A = \mathbb{F}_q[t]$. If v is a prime of degree 1, then $\Pi_v(n)$ is transcendental if and only if the digits n_j of n are not ultimately constant.*

Proof. Using the automorphism $t \to t + \theta$ for $\theta \in \mathbb{F}_q$ of A, we can assume without loss of generality that $v = t$. Then $\Pi_v(n) = \prod(-D_{j,v})^{n_j}$, for $n = \sum n_j q^j$. Since $q = 0$ in characteristic p, when we take logarithmic derivative it simplifies to give

$$t\frac{\Pi_v(n)'}{\Pi_v(n)} = \sum n_j \left(\frac{t^{q^j}}{1 - t^{q^j - 1}} - \frac{t^{q^{j-1}}}{1 - t^{q^{j-1} - 1}} \right) = \sum \frac{n_j - n_{j+1}}{(1/t)^{q^j} - (1/t)}.$$

This power series in $\mathbb{F}_q((t))$ is transcendental over $\mathbb{F}_q(t) = \mathbb{F}_q(1/t)$ by the previous theorem, by just replacing t by $1/t$, because the hypothesis implies that $n_j - n_{j+1}$ is not ultimately zero. Then the theorem follows as before.   □

Remark 4.8. What should be the implications for the Morita p-adic gamma function? The close connection to cyclotomy leads us to think that the situation for values at proper fractions should be parallel. But then this implies that the algebraic values in the image not taken at fractions should be taken at irrational p-adic integers. Thus we do not expect a Mendès-France–Yao type result for Morita's p-adic gamma function, but it may be possible to have such a result for Π_v's, for any v. This breakdown of analogies seems to be due to an important difference: in the function field situation, the range is a 'huge' finite characteristic field of Laurent series over a finite field, and the resulting big difference in the function theory prevents analogies being as strong for non-fractions.

We will see in Lecture 6 that by powerful techniques of [ABP04, P08] there is a much stronger and complete independence result [CPTY] for values at fractions for arithmetic gamma, whereas the non-fraction result (Theorem 4.5) or the v-adic result above (Theorem 4.7) are still not provable by other methods.

4.2 General A

What happens for general A? In short, it seems that several analogies that coincide for $A = \mathbb{F}_q[t]$ now diverge, but still generalize to different concepts with some theorems generalizing well. So we can think that this divergence helps us focus on core relations in the concepts by throwing out accidents.

In 4.3–4.9 we discuss a new gamma and return in 4.10 to gammas of Lecture 1, for general A. We start with a summary in 4.3, followed by some details in 4.4–4.9.

4.3 Another gamma function coming from the exponential analogy: Summary

Consider the exponential $e_\rho(z) = \sum z^{q^i}/d_i$ for a sign-normalized rank one Drinfeld A-module ρ. Now $d_i \in K_\infty$ lie in the Hilbert class field H for A. It turns out that still we can define a factorial by digit expansion and interpolate its unit parts to get gamma functions at ∞ and v, by generalizing our earlier approach for the arithmetic gamma in one particular way. The fundamental two-variable function $t - \theta$ specializing to $[n]$'s on the graphs of Frobenius power, in the $\mathbb{F}_q[t]$ case, now generalizes to the shtuka function corresponding to ρ by Drinfeld–Krichever correspondence. Our proof of the Gross–Koblitz formula generalizes very nicely, connecting the values of v-adic interpolation at appropriate fractions to Gauss sums that the author had defined, even though these Gauss sums in general no longer have Stickelberger factorization, but have quite strange factorizations, which now get explained by the geometry of theta divisors! This fundamental correspondence with a shtuka and these v-adic results suggest that there should be a very interesting special value theory at the infinite place too.

To me, the most interesting mystery in the gamma function theory is that we do not even understand the nature of the basic value $\Gamma(0)$ for this gamma at the infinite place. Is it related to the period? The question is tied to the switching symmetry (or quantifying the lack of it) of two variables of the shtuka function.

4.4   Some details: Drinfeld correspondence in the simplest case

We describe Drinfeld's geometric approach for the simplest case when the rank and the dimension is one, namely for Drinfeld modules of rank one.

Let F be an algebraically closed field containing $\mathbb{F}_q$ and of infinite transcendence degree. Let $\overline{X}$ denote the fiber product of X with $\mathrm{Spec}(F)$ over $\mathbb{F}_q$. We identify closed points of $\overline{X}$ with F-valued points of X in the obvious way. For $\xi \in X(F)$, let $\xi^{(i)}$ denote the point obtained by raising the coordinates of ξ to the q^ith power. We extend the notation to the divisors on $\overline{X}$ in the obvious fashion. For a meromorphic function f on $\overline{X}$, let $f|_\xi$ denote the (possibly infinite) value of f at ξ and let $f^{(i)}$ denote the pull-back of f under the map $\mathrm{id}_X \times \mathrm{Spec}(\tau^i)\colon \overline{X} \to \overline{X}$, where $\tau := x \to x^q \colon F \to F$. (Note that our earlier mention of two copies of the curve is taken care of $\overline{X}$, as F can even contain several copies of (extensions of) function fields of X.) For a meromorphic differential $\omega = f\,dg$, let $\omega^{(i)} := f^{(i)}d(g^{(i)})$.

Fix a local parameter $t^{-1} \in \mathcal{O}_{\overline{X},\infty}$ at ∞. For a non-zero $x \in \mathcal{O}_{\overline{X},\infty}$, define $\deg(x) \in \mathbb{Z}$ and $\mathrm{sgn}(x) \in F$ to be the exponent in the highest power of t and the coefficient of the highest power, respectively, in the expansion of x as a Laurent series in t^{-1} with coefficients in F.

Drinfeld proved, by a nice argument analyzing cohomology jumps provided by Frobenius twists and multiplications by f's of sections and using Riemann–Roch the following result.

Theorem 4.9. *Let $\xi \in X(F)$ be given, together with a divisor V of $\overline{X}$ of degree g and a meromorphic function f on $\overline{X}$ such that*

$$V^{(1)} - V + (\xi) - (\infty) = (f).$$

If $\xi \neq \infty$, then
$$H^i\big(\overline{X}, \mathcal{O}_{\overline{X}}(V - (\infty))\big) = 0 \qquad (i = 0, 1).$$

In particular, ξ does not belong to the support of V.

We review the language of Drinfeld modules again with this notation. Let $\iota\colon A \to F$ be an embedding of A in F. By a Drinfeld A-module ρ relative to ι we will mean such a ρ normalized with respect to sgn, of rank one and generic characteristic, but we will drop these words. Let H be the Hilbert class field. Write its exponential as $e(z) = \sum e_i z^{q^i}$.

Fix a transcendental point $\xi \in X(F)$. Then evaluation at ξ induces an embedding of A into F. In our earlier notation, $\xi = \iota$. By solving the corresponding equation on the Jacobian of X, we see that $V^{(1)} - V + (\xi) - (\infty)$ is principal for some divisor V. A *Drinfeld divisor* relative to ξ is defined to be an effective divisor V of degree g such that $V^{(1)} - V + (\xi) - (\infty)$ is principal. From the theorem and Riemann–Roch, it follows that a Drinfeld divisor is the unique effective divisor in its divisor class. (In particular, there are h such divisors.) Hence there exists a unique function $f = f(V)$ with $\mathrm{sgn}(f) = 1$ and such that $(f) = V^{(1)} - V + (\xi) - (\infty)$. By abuse of terminology, we call f a shtuka. (In fact, in our context, a shtuka is a line bundle $\mathcal{L}$ on $\overline{X}$ with $\mathcal{L}^{(1)}$ being isomorphic to $\mathcal{L}(-\xi + \overline{\infty})$, and in our case, with $\mathcal{L} = \mathcal{O}_{\overline{X}}(V)$, f realizes this isomorphism.)

Drinfeld bijection

The set of Drinfeld divisors V (relative to ξ) is in natural bijection with the set of Drinfeld A-modules ρ (relative to ξ) as follows. (See [Mum78] for details of the proof.) Let $f = f(V)$ be as above. Then

$$1, f^{(0)}, f^{(0)}f^{(1)}, f^{(0)}f^{(1)}f^{(2)}, \ldots$$

is an F-basis of the space of sections of $\mathcal{O}_{\overline{X}}(V)$ over $\overline{X} - (\infty \times_{\mathbb{F}_q} F)$. Define $\rho_{a,j} \in F$ by the rule

$$a := \sum_j \rho_{a,j} f^{(0)} \cdots f^{(j-1)}.$$

Then the ρ corresponding to V is given by $\rho_a := \sum \rho_{a,j} \tau^j$.

Theorem 4.10. *We have*

$$e(z) = \sum_{n=0}^{\infty} \frac{z^{q^n}}{(f^{(0)} \cdots f^{(n-1)})|_{\xi^{(n)}}}.$$

Proof. To see that the right-hand side satisfies the correct functional equations for $e(z)$, divide both sides of the equation above defining $\rho_{a,j}$ by $f^{(0)} \cdots f^{(n-1)}$ and evaluate at $\xi^{(n)}$. □

Given ρ, we can recover f and V from the exponential.

Theorem 4.11. *We have*

$$l(z) = \sum_{n=0}^{\infty} \left(\mathrm{Res}_{\xi} \frac{\omega^{(n+1)}}{f^{(0)} \cdots f^{(n)}} \right) z^{q^n}.$$

Example 4.12. For the Carlitz module for $\mathbb{F}_q[t]$, the genus is zero and thus the Drinfeld divisor is empty, so that $f = t - t|_{\xi}$ and $\omega = dt$. The reader should verify the formula for the coefficients of the exponential and the logarithm.

We give one more example in 4.7. For more examples, see [T04, 8.2].

4.5   Some details: Definition of new gamma

We can write $e_\rho(z) = \sum z^{q^i}/d_i$.

Definition 4.13. For $n \in \mathbb{N}$, we define the factorial $\Pi(n)$ of n as follows: Write $n = \sum n_i q^i$, $0 \le n_i < q$ and put $\Pi(n) := \prod d_i^{n_i}$.

Let $\wp$ be a prime of A of degree d and let θ be an $\overline{\mathbb{F}_q}$-point of $\overline{X}$ above $\wp$. If $w \in K_\wp$ is a local parameter at θ, we put

$$\tilde{d}_i := \tilde{d}_{i,w} := \frac{d_i}{d_{i-d}w^{l_i}}$$

where l_i is chosen so that $\tilde{d}_i$ is a unit at θ.

We can show that as i tends to infinity, $\overline{d}_i$ (the one-unit part with respect to the chosen uniformizer at infinity) tends to one ∞-adically and the degree of d_i tends to zero p-adically. So we can interpolate the one-unit part and also put the degree back to get the gamma interpolation at ∞ exactly as in the first lecture. We just note here that when $d_\infty > 1$ there are several sign issues to be taken care of, as the coefficients are now in a field bigger than the Hilbert class field, etc.

It can be shown that if u is any local parameter at θ, then there is a non-zero $c \in \mathbb{F}_{q^d}$ such that, with $w = cu$, $\tilde{d}_i$ tends to one θ-adically as i tends to infinity.

If w is a local parameter at θ such that $\tilde{d}_i$ is a one unit for large i, then this implies that $\tilde{d}_i$ tends to one.

Definition 4.14. Define the $\wp$-*adic factorial* $\Pi_\wp(z) := \Pi_w(z)$ for $z \in \mathbb{Z}_p$ as follows. Write $z = \sum z_i q^i$, $0 \le z_i < q$ and put $\Pi_\wp(z) := \prod \tilde{d}_i^{z_i}$.

When $h_A = 1$, if we choose w to be a monic prime $\wp$ of A of degree d, then $-\tilde{d}_i \to 1$ as i tends to infinity.

4.6   Some details: Gauss sums

First we recall the definition of Gauss and Jacobi sums. Let $\wp$ be a prime of A of degree d. Choose an A-module isomorphism $\psi: A/\wp \to \Lambda_\wp$ (an analog of an additive character) and let χ_j, for $j \bmod d$, be $\mathbb{F}_q$-homomorphisms $A/\wp \to F$, indexed so that $\chi_j^q = \chi_{j+1}$ (special multiplicative characters which are q^j-powers of the 'Teichmüller character'). We can identify $\chi_j(z)$ with $a|_{\theta^{(j)}}$ for some geometric point θ above $\wp$ if $a \in A$ is such that $a \bmod \wp$ is z. Then we define the *basic Gauss sums*

$$g_j := g(\chi_j) := - \sum_{z \in (A/\wp)^*} \chi_j(z^{-1})\psi(z).$$

The g_j are non-zero. We define the *Jacobi sums* J_j by $J_j := g_{j-1}^q/g_j$. Then J_j is independent of the choice of ψ. We put $J := J_0$.

For the Carlitz module, it was shown that $J_j = -(t - \chi_j(t))$ and the Stickelberger factorization of g_j (mentioned in the proof of Theorem 10) was easily obtained from this. Note for example that $g_j^{q^d-1} = J_j J_{j-1}^q \cdots J_{j-d+1}^{q^{d-1}}$.

We now describe Anderson's explicit Gauss–Jacobi sums [ATp, A92] in the geometric case, and the corresponding Stickelberger theorem already mentioned in 3.4. (The Stickelberger theorem in this context was first proved by Hayes [H85], but with a different construction.) We use the notation of 4.6. Let $f \in A_+$ be of positive degree, $x \in K$ be of the form $x = a/f$ with $a \in A$, $(a, f) = 1$, and let $\wp$ be a prime of $A[\mathbf{e}(1/f)]$ not dividing f. Let χ_j be the q^jth power of the Teichmüller character at $\wp$ as above. Let a_i, b_i be f-dual families as in 4.6. We consider Gauss and Jacobi sums

$$g_j(\wp, x) := 1 - \sum_{i=1}^{\deg f} \chi_j\big(\mathbf{e}_*(a_i/f)\big)\mathbf{e}(b_i x)) \in \overline{\mathbb{F}}_q A[\zeta_f],$$

$$g(\wp, x) := \prod_j g_j(\wp, x) \in A[\zeta_f].$$

For a fractional ideal I of $A[\mathbf{e}(1/f)]$ supported away from f, we define $g(I, x)$ multiplicatively from the above sums for primes. Let $G_f = \mathrm{Gal}(K(\mathbf{e}(1/f))/K)$ and $\sigma_{f,a} \in G_f$ be the element corresponding to a under the usual identification. Then we have a Stickelberger type result [ATp]:

$$(g(I, x)) = I^{\sum \sigma_{f,ab}-1},$$

where the sum is over $b \in A_+$ prime to f and of degree less than $\deg f$.

4.7 Some details: The shtuka connection and an analog of Gross–Koblitz

Now we show that the Jacobi sums made up from $\wp$ torsion of a Drinfeld module can be interpreted as specializations at geometric points above $\wp$ of a meromorphic function, obtained from the shtuka corresponding to the Drinfeld module on the curve cross its Hilbert cover. Hence the strange factorizations of the Gauss sums get related to the divisor of this function and, since the divisor is encoding cohomology jumps, to the theta divisor.

Theorem 4.15. *Let V and f be the Drinfeld divisor and the shtuka respectively, corresponding to a sign-normalized Drinfeld module ρ (of rank one and generic characteristic) via the Drinfeld bijection. Then with the Jacobi sums J_j defined using $\wp$-torsion of ρ and normalized as above, we have*

$$f|_{\theta(j)} = J_j.$$

We put $\overline{x} := x|_\xi$ and $\overline{y} := y|_\xi$, while giving the divisor of J, by abuse of notation, instead of the quantities corresponding to θ, those corresponding to ξ.

(i) $A = \mathbb{F}_2[x,y]/y^2 + y = x^3 + x + 1$: We have

$$f = \frac{\overline{x}(x+\overline{x}) + y + \overline{y}}{x + \overline{x} + 1}.$$

If $\xi + 1$ is the point where x is $\overline{x} + 1$ and y is $\overline{x} + \overline{y} + 1$, then $V = (\xi + 1)$.

(Note that this point corresponds to the automorphism σ above.) By the recipe above it follows that

$$(J) = 2(\xi+1)^{(-1)} - (\xi+1) + (\xi) - 2(\infty)$$

where $\xi + 1$ is the point where x is $\overline{x} + 1$ and y is $\overline{x} + \overline{y}$.

We showed how to write down f 'parametrically' in terms of the coefficients of sign-normalized ρ in the case $g = d_\infty = 1$.

Let F be an algebraic closure of $K_\wp$, ξ be the tautological point, i.e., the F-valued point of X corresponding to $K \hookrightarrow K_\wp$, and let θ be the Teichmüller representative in the residue disc of ξ. Note that even though we used the ∞-adic completion earlier, the Taylor coefficients of $e(z)$, being in H, can be thought of as elements of F.

Theorem 4.16. *Let $0 \leq j < d$. If μ is the valuation of g_j at ξ, then we have $g_j = \zeta w^\mu / \Pi_\wp(q^j/(1-q^d))$, where ζ is a $(q^d - 1)$th root of unity.*

4.8　Some details: Value for ∞-adic gamma

Let us use the short-form $f_{ij} := f^{(i)}|_{\xi^{(j)}}$. In the Carlitz case of genus 0, we have switching-symmetry $f_{ij} = -f_{ji}$. In general, up to simple algebraic factors arising from residue calculations, $\widetilde{\pi}^{-1}$ is $\prod_{i=1}^{\infty} f_{i0}$ and $\Pi_\infty(-1)^{-1}$ is $\prod_{j=1}^{\infty} f_{0j}$, as can be seen from formulas giving d_i and l_i in terms of specializations of f. So the switching-symmetry mentioned above connects the two in the Carlitz case.

4.9　Some details: Non-vanishing of exponential coefficients

We proved (for $d_\infty = 1$) that the coefficients of z^{q^i} in the expansion of $e(z)$ are non-zero, so that we can consider d_i as the reciprocal coefficient. In general, this non-vanishing is equivalent to the nice geometric condition that $\xi^{(i)}$ does not belong to the support of the Drinfeld divisor, and it is known for many situations [T04, 8.3.1] but not in full generality, when $d_\infty > 1$.

4.10 General A: Arithmetic and geometric gamma

The gamma functions we defined earlier for general A are expected to have good properties, and we have established some of them, such as period connections for simple values. But we saw very strong complete results only for the case when $A = \mathbb{F}_q[t]$. In general, the hypotheses (Hi) have not been scrutinized much in general. For simple examples of higher genus solitons, in the arithmetic as well as in the geometric case, see [T04, § 8.7]. But to get general results, one needs to expand both the technology of A-modules and solitons for general A. The crucial step of providing solitons for general A, at least conjecturally with a lot of evidence, has been taken by Greg Anderson in another spectacular paper [A06] 'A two variable refinement of the Stark conjecture in the function field case', building on his earlier work [A94, A96]. It uses the adelic framework of Tate's thesis.

I hope that somebody (hopefully a reader of these notes!) takes up this issue and settles the general case soon.

Lecture 5

Zeta Values: Definitions, First Properties and Relations with Cyclotomy

The Riemann zeta function, the Dedekind zeta function in the number field case, and Artin's analog in the function field case, can all be defined as

$$\zeta(s) = \sum \mathrm{Norm}(I)^{-s} \in \mathbb{C},$$

where the sum is over non-zero ideals and $s \in \mathbb{C}$, with $\mathrm{Re}(s) > 1$. Artin's zetas satisfy many analogies. However, they are simple rational functions of the variable $T = q^{-s}$. For example, $\zeta(s) = 1/(1 - qT)$, for $A = \mathbb{F}_q[t]$, as one can see by counting and summing the resulting geometric series. So the special values theory in this case is completely trivial.

The norm depends only on the degree and we thus ignore all the information except the degree. We can take the actual ideal or polynomial or a relative norm to the base instead. Carlitz used another function field analogy and considered special zeta values $\zeta(s) := \sum n^{-s}$, where now n runs through monic polynomials in $\mathbb{F}_q[t]$, i.e., monic generators of non-zero ideals. (With this basic idea, we can then consider various zeta and L-values; for example, those attached to finite characteristic-valued representations by using product of characteristic polynomials of 'Frobenius at $\wp$ multiplied by $\wp^{-s}$'. We refer to [G96] and references therein for various such definitions.)

Now polynomials can be raised to integral powers, and, in particular, if s is a natural number, then the sum converges (as the terms tend to zero) to a Laurent series in K_∞.

Goss showed that, if s is a non-positive integer then, by just grouping the terms for the same degree together, the sum reduces to a finite sum giving $\zeta(s) \in A$.

Example 5.1. For $A = \mathbb{F}_3[t]$, we have $\zeta(0) = 1 + 3 + 3^2 + \cdots = 1 + 0 + 0 + \cdots = 1$ and $\zeta(-3^n) = \zeta(-1)^{3^n} = 1$ because

$$\zeta(-1) = 1 + (t + (t+1) + (t-1)) + \cdots = 1 + 0 + \cdots = 1.$$

On the other hand, for $A = \mathbb{F}_2[t]$ we have $\zeta(-2^n) = 0$ because $\zeta(-1) = 1 + (t + (t+1)) + 0 + \cdots = 1 + 1 = 0$. We leave it to the reader to verify $\zeta(-5) = 1 + t - t^3$ for $A = \mathbb{F}_3[t]$ using either the bound in the lemma below or the formula below. Note that these examples fit with the naive analogy with the Riemann zeta, which vanishes at negative integers exactly when they are even, as 1 is 'even' exactly when $q = 2$.

In fact, the weaker version consisting of the first three lines of the proof of the following theorem, which is essentially due to Lee, implies that grouped terms vanish for large degree. (See [T90, T95] for the relevant history and [She98, T09a] for more recent results.)

For a non-negative integer $k = \sum k_i q^i$ with $0 \leq k_i < q$, we let $\ell(k) := \sum k_i$, i.e., $\ell(k)$ is the sum of base q digits of k. Here is a very useful general vanishing theorem.

Theorem 5.2. *Let W be an $\mathbb{F}_q$-vector space of dimension d inside a field (or ring) $\mathcal{F}$ over $\mathbb{F}_q$. Let $f \in \mathcal{F}$. If $d > \ell(k)/(q-1)$, then $\sum_{w \in W}(f + w)^k = 0$.*

Proof. Let $w_1, \ldots, w_d$ be an $\mathbb{F}_q$-basis of W. Then $(f+w)^k = (f+\theta_1 w_1 + \cdots + \theta_d w_d)^k$, $\theta_i \in \mathbb{F}_q$. When one multiplies out the k brackets, terms involve at most k of θ_i's, hence if $d > k$ then the sum in the theorem is zero, since we are summing over some θ_i, a term not involving it, and $q = 0$ in characteristic p. The next observation is that, in characteristic p, $(a + b)^k = \prod (a^{q^i} + b^{q^i})^{k_i}$, hence the sum is zero if $d > \ell(k)$, by the argument above. Finally, note that $\sum_{\theta \in \mathbb{F}_q} \theta^j = 0$ unless $q - 1$ divides j. Expanding the sum above by the multinomial theorem, we are summing multiples of products $\theta_1^{j_1} \cdots \theta_d^{j_d}$ and hence the sum is zero; since the sum of the exponents is $\ell(k) < (q-1)d$, not all the exponents can be multiples of $q - 1$. □

For the applications, note that $A_{<i} = \{a \in A : \deg(a) < i\}$ form $\mathbb{F}_q$-vector spaces whose dimensions are given by the Riemann–Roch theorem. Similarly, $A_{i+} = \{a \in A : \deg(a) = i, a \text{ monic}\}$ are made up of affine spaces as in the theorem.

Remark 5.3. Following Goss, we can interpolate the zeta values above to $\zeta(s)$ with $s = (x, y)$ in the much bigger space $C_\infty^* \times \mathbb{Z}_p$ by defining $a^s := x^{\deg a} \langle a \rangle^y$, with $\langle a \rangle$ denoting the one-unit part of a, which can be raised to a p-adic yth power, just as in the gamma case. These p-adic power homomorphisms are the only locally analytic endomorphisms of one-unit groups [J09]. At least for an integral y, the usual integral power a^y is recovered as $a^{(t^y, y)}$. In general, $\zeta(s)$ is a power series in the x-variable, which keeps track of the degrees. For more on these analytic issues, see [G96] or Böckle's lecture notes in this volume.

Focusing on just the special values at integers, we use the theorem above and can ignore the convergence questions. Thus we work in the following simpler set-up.

Let L be a finite separable extension of K and let $\mathcal{O}_L = \mathcal{O}$ denote the integral closure of A in L. Now we define the relevant zeta functions.

Definition 5.4. For $s \in \mathbb{Z}$, define the 'absolute zeta function' as

$$\zeta(s, X) := \zeta_A(s, X) := \sum_{i=0}^{\infty} X^i \sum_{a \in A_{i+}} \frac{1}{a^s} \in K[[X]]$$

$$\zeta(s) := \zeta_A(s) := \zeta(s, 1) \in K_\infty.$$

By the theorem above, $\zeta(s) \in A$ for every integer $s \leq 0$. The results on the special values and the connection with Drinfeld modules later on justify the use of elements rather than ideals even when the class number is more than one. Also, we can look at a variant where a runs through elements of some ideal of A or we can sum over all ideals by letting s be a multiple of the class number and letting a^s be the generator of I^s with, say, a monic generator. In the latter case, we have an Euler product for such s.

If L contains H (note that this is no restriction if $hd_\infty = 1$, e.g., for $A = \mathbb{F}_q[t]$), then it is known that the norm of an ideal $\mathcal{I}$ of $\mathcal{O}$ is principal. Let $\operatorname{Norm} \mathcal{I}$ denote the monic generator. Let us now define relative zeta functions in this situation.

Definition 5.5. For $s \in \mathbb{Z}$, define the 'relative zeta function' as

$$\zeta_{\mathcal{O}}(s, X) := \zeta_{\mathcal{O}/A}(s, X) := \sum_{i=0}^{\infty} X^i \sum_{\deg(\operatorname{Norm} \mathcal{I})=i} \frac{1}{\operatorname{Norm} \mathcal{I}^s} \in K[[X]]$$

$$\zeta_{\mathcal{O}}(s) := \zeta_{\mathcal{O}/A}(s) := \zeta_{\mathcal{O}}(s, 1) \in K_\infty.$$

Finally, we define the vector-valued zeta function, which generalizes both definitions above and works without assuming that L contains H. We leave the simple task of relating these definitions to the reader.

Definition 5.6. Let $\mathcal{C}_n$ $(1 \leq n \leq hd_\infty)$ be the ideal classes of A and choose an ideal I_n in $\mathcal{C}_n^{-1}$. For $s \in \mathbb{Z}$, define the vector $Z_{\mathcal{O}}(s, X)$ by defining its nth component $Z_{\mathcal{O}}(s, X)_n$ via

$$Z_{\mathcal{O}}(s, X)_n := Z_{\mathcal{O}/A}(s, X)_n := \sum_{i=0}^{\infty} X^i \sum \frac{1}{(I_n \operatorname{Norm} \mathcal{I})^s} \in K[[X]]$$

where $I_n \operatorname{Norm} \mathcal{I}$ stands for the monic generator of this ideal and the second sum is over $\mathcal{I}$ whose norm is in $\mathcal{C}_n$ and is of degree i:

$$Z_{\mathcal{O}}(s)_n := Z_{\mathcal{O}/A}(s)_n := Z_{\mathcal{O}}(s, 1)_n \in K_\infty.$$

Also note that Z depends very simply on the choice of I_n's, with components of Z, for different choices of I_n's, being non-zero rational multiples of each other. In particular, the questions we are interested in, such as when it (i.e., all the components) vanishes, when it is rational or algebraic, etc. are independent of such choice.

We use X as a deformation parameter for otherwise discretely defined zeta values and hence we define the order of vanishing ord_s of ζ, $\zeta_{\mathcal{O}}$ or $Z_{\mathcal{O}}$ at s to be the corresponding order of vanishing of $\zeta(s, X)$, $\zeta_{\mathcal{O}}(s, X)$ or $Z_{\mathcal{O}}(s, X)$ at $X = 1$. This procedure is justified by the results described below. The order of vanishing is the same for s and ps, as the characteristic is p.

Using these basic ideas, we can immediately define L-functions in various settings and can study their values and analytic properties [G96], but we focus here on the simplest case.

5.1   Values at positive integers

In contrast to the classical case where we have a pole at $s = 1$, here $\zeta(1)$ makes sense and can be considered as an analog of Euler's constant γ.

Euler's famous evaluation, $\zeta(m) = -B_m(2\pi i)^m/2(m!)$ for even m, has the following analog:

Theorem 5.7 (Carlitz [Car35]). *Let $A = \mathbb{F}_q[t]$. Then for 'even' m (i.e., a multiple of $q - 1$), we have $\zeta(m) = -B_m \widetilde{\pi}^m/(q-1)\Pi(m)$.*

Proof. First note that $q - 1 = -1$ in the formula. Multiplying the logarithmic derivative of the product formula for $e(z)$ by z, we get

$$\frac{z}{e(z)} = 1 - \sum_{\lambda \in \Lambda - \{0\}} \frac{z/\lambda}{1 - z/\lambda} = 1 - \sum_{n=1}^{\infty} \sum_{\lambda} \left(\frac{z}{\lambda}\right)^n = 1 + \sum_{n \text{ 'even'}} \frac{\zeta(n)}{\widetilde{\pi}^n} z^n$$

since $\sum_{c \in \mathbb{F}_q^\times} c^n = -1$ or 0 depending on whether n is 'even' or not. But $z/e(z) = \sum B_n z^n/\Pi(n)$. $\qquad\square$

For general A and corresponding sign-normalized ρ, noting that now the coefficients of the exponential are in H, we get, similarly

Theorem 5.8. *If s is a positive 'even' integer (i.e., a multiple of $q - 1$), then $\zeta_A(s)/\widetilde{\pi}^s \in H$.*

Example 5.9. By comparing the coefficients of z^{q-1} in the equation above, we get $\zeta(q-1)/\widetilde{\pi}^{q-1} = -1/d_1 \in H$.

Now we turn to the relative zeta functions.

Theorem 5.10. *Let L be an abelian totally real (i.e., completely split at ∞) extension of degree d of K containing H and let s be a positive 'even' integer (i.e., a multiple of $q - 1$). Then $R_s := \zeta_{\mathcal{O}}(s)/\widetilde{\pi}^{ds}$ is algebraic and, in fact, $R_s^2 \in H$.*

The idea of the proof is to factor such zeta values as products of L-values that are linear combinations of partial zeta values, which are handled as in the absolute case above, and use the Dedekind determinant formula to get better control of the fields involved. There is a more refined version when $d_\infty > 1$. We refer to [T04, 5.2.8].

Remarks 5.11.

(1) It is not possible to replace H with K in the theorem [T95], not even when $d_\infty = 1$. Classically, for an arbitrary totally real number field F (not necessarily abelian over $\mathbb{Q}$), it is known that $(\zeta_F(s)/(2\pi i)^{r_1 s})^2 \in \mathbb{Q}$. This uses Eisenstein series.

(2) For the history and references on this general case as well as the abelian case using L-functions, as we have done here, see the references in [T95], and see [G92] for ideas about carrying over the proof in the general case. On the other hand, such an algebraicity result for the ratio of the relative zeta value with an appropriate power of the period $2\pi i$ is not expected for number fields which are not totally real. In our case, we can have such a result even if L is not totally real, as was noted in [G87]: Let L be a Galois extension of degree p^k of K; then, since all the characters of the Galois group are trivial, the L-series factorization shows that $\zeta_{\mathcal{O}}(s) = \zeta(s)^{p^k}$ and the result follows then from the theorem above on the absolute case. A more elementary way to see this, when the degree is p and $\mathcal{O}$ is of class number one, is to note that for $\alpha \in \mathcal{O} - A$ there are p conjugates with the same norms which then add up to zero, whereas for $\alpha \in A$ the norm is α^p.

5.2 Values at non-positive integers

We have seen how just grouping together the terms of the same degree gives $\zeta(-k) = \sum_{i=0}^{\infty}(\sum_{n \in A_{i+}} n^k) \in A$, for $k > 0$. Hence we have stronger integrality rather than rationality in the number field case, reflecting the absence of a pole at $s = 1$ in our case.

There is also the following vanishing result, giving the 'trivial zeros':

Theorem 5.12. *For a negative integer s, we have $\zeta((q-1)s) = 0$.*

Proof. If $k = -(q-1)s$, then

$$\zeta(-k) = \sum_{i=0}^{\infty} \sum_{a \in A_{i+}} a^k = -\sum_{i=0}^{\infty} \sum_{a \in A_i} a^k = 0,$$

where the second equality holds since $\sum_{\theta \in \mathbb{F}_q} \theta^{q-1} = -1$ and the third equality is seen by using that the sum is finite and applying the theorem above with $W = A_{<m}$ and $f = 0$, for some large m. $\square$

For $A = \mathbb{F}_q[t]$, another proof giving a formula as well as a non-vanishing result parallel to the case of the Riemann zeta function can be given: For $k \in \mathbb{Z}_+$, $\zeta(-k) = 0$ if and only if $k \equiv 0 \bmod(q-1)$. Also, $\zeta(0) = 1$.

The proof [G79, T90] follows by writing a monic polynomial n of degree i as $th + b$ with h of degree $i - 1$ and $b \in \mathbb{F}_q$ and using the binomial theorem to get the induction formula

$$\zeta(0) = 1, \quad \zeta(-k) = 1 - \sum_{f=0,\,(q-1)|(k-f)}^{k-1} \binom{k}{f} t^f \zeta(-f),$$

which shows that $\zeta(-k) \in A$ since $\zeta(0) = 1$. If $(q - 1)$ divides k, then induction shows that $\zeta(-k) = 1 - 1 + 0 = 0$. If $(q-1)$ does not divide k, then since there is no term in the summation corresponding to $f = 0$, we have that $\zeta(-k) = 1 - tp(t) \neq 0$, where $p(t) \in A$.

We do not know for general A if the values at odd integers are non-zero. But we have the following result.

Theorem 5.13. *Let s be a negative odd integer. Then $\zeta(s)$ is non-zero if A has a degree one rational prime $\wp$ (e.g., if $g = 0$, which reproves the $\mathbb{F}_q[t]$ case, or if q is large compared to the genus of K) or if $h = 1$. (Together they take care of the genus 1 case.) In fact, in the first case, $\zeta(s)$ is congruent to $1 \bmod \wp$.*

Proof. The first part follows by looking at the zeta sum modulo $\wp$ and noticing that $\sum_{\theta \in \mathbb{F}_q} \theta^{-s} = 0$, hence the only contribution is 1 from the $i = 0$ term. Apart from $g = 0$, there are only 4 other A's with $hd_\infty = 1$ (these have no degree 1 primes), which we check by a calculation omitted here.   □

Remark 5.14. This can be easily improved, but the full non-vanishing is not known. Note that for the case of number fields, the Dedekind zeta functions have Euler product thus showing non-vanishing in the right half-plane, and the functional equation (missing here!) allows us to conclude the non-vanishing at negative odd integers by analyzing the relevant gamma factors.

Let us now turn to relative zeta functions.

Theorem 5.15 (Goss [G92]). *For a negative integer s, we have $\zeta_\mathcal{O}(s) \in A$. In fact, $Z_\mathcal{O}(s)_n \in A$ for every n. For a negative integer s, we have $\zeta_\mathcal{O}((q - 1)s) = 0$. In fact, $Z_\mathcal{O}((q - 1)s)_n = 0$ for every n.*

The idea is to decompose the sum carefully in $\mathbb{F}_q$-vector spaces and use the theorem above to show that, for large k, the kth term of the zeta sum is zero. The second part follows as before.

Remark 5.16. The values of the Dedekind zeta function at negative integers are all zero if the number field is not totally real. By the remarks above, a similar result does not hold in our case. In fact, we do not even need the degree to be a power of the characteristic, as will be seen from examples below. The ramification possibilities for the infinite places are much more varied in the function field case.

For $A = \mathbb{F}_q[t]$, Goss defined a modification $\beta(k) \in A$ of $\zeta(-k)$, for $k \in \mathbb{Z}_{\geq 0}$, as follows:

$$\beta(k) := \zeta(-k) \text{ if } k \text{ is odd}, \quad \beta(k) := \sum_{i=0}^{\infty}(-i) \sum_{n \in A_i+} n^k \text{ if } k \text{ is even}.$$

In other words, the deformation $\zeta(-k, X)$ of $Z(-k, 1) = \zeta(-k)$ has a simple zero at $X = 1$ if k is even and hence one considers $\frac{d\zeta}{dx}(-k, X)\big|_{X=1} = \beta(k)$ instead. For $k \in \mathbb{Z}_+$, we have $\beta(k) \neq 0$ and in fact

$$\beta(0) = 0, \quad \beta(k) = 1 - \sum_{f=0,\,(q-1)|(k-f)}^{k-1} \binom{k}{f} t^f \beta(f).$$

For general A, Goss similarly defines $\beta(k)$ by removing the 'trivial zero' at k, of order d_k say (see the next lecture for the discussion of this order), by $\beta(k) := (1 - X)^{-d_k}\zeta_A(-k, X)|_{X=1}$. One gets Kummer congruences and $\wp$-adic interpolations for $\beta(k)$'s, as for zeta values.

Since $\zeta(-k)$ turns out to be a finite sum of n^k's, by Fermat's little theorem we see that the $\zeta(-k)$'s satisfy Kummer congruences enabling us to define a $\wp$-adic interpolation $\zeta_\wp$ from $s \in \mathbb{Z}_p$, by removing the Euler factor at $\wp$. For interpolations at $\wp$ and ∞ on much bigger places, analytic properties, etc., we refer to [G96].

Instead of the Kummer congruences, the B_m's satisfy analogs of the von Staudt congruences and the Sylvester–Lipschitz theorem. We now have two distinct analogs of B_k/k, namely $-\zeta(-k+1)$ for $k - 1$ 'odd' on one hand and $\Pi(k-1)\zeta(k)/\tilde{\pi}^k$ for k 'even' on the other. But the shift by one does not transform 'odd' to 'even' unless $q = 3$, and we do not know any reasonable functional equation linking the two.

5.3 First relations with cyclotomic theory: At positive integers

Next we give the connection of these Bernoulli numbers with the class groups of cyclotomic fields, giving some analogs of Herbrand–Ribet theorems. We restrict to $A = \mathbb{F}_q[t]$.

Let $\wp$ be a monic prime of A of degree d. Recall the analogies

$$\Lambda_\wp = e(\tilde{\pi}/\wp) \longleftrightarrow \zeta_p = e^{2\pi i/p}, \quad K(\Lambda_\wp) \longleftrightarrow \mathbb{Q}(\zeta_p).$$

One also has a 'maximal totally real' subfield

$$K(\Lambda_\wp)^+ = K\left(\prod_{\theta \in \mathbb{F}_q^*} e\left(\frac{\theta\tilde{\pi}}{\wp}\right)\right) \longleftrightarrow \mathbb{Q}(\zeta_p)^+ = \mathbb{Q}\left(\sum_{\theta \in \mathbb{Z}^*} e^{\theta 2\pi i/p}\right).$$

The classical cyclotomic Galois group $(\mathbb{Z}/p\mathbb{Z})^*$ now gets replaced by $(A/\wp A)^*$ and the Frobenius action is similar from this identification, so that it is straightforward to write down the splitting laws for primes. Note that we now have two kinds of class groups, one traditionally done in algebraic geometry, that is the class group (divisors of degree zero modulo principal) for the complete curve, and the other the class group for the integral closure of A in these cyclotomic fields (which typically has many points at infinity). Hence we give the following definition.

Definition 5.17. Let C (respectively C^+, $\tilde{C}$, $\tilde{C}^+$) denote the p-primary component of the class group of the complete curve for $K(\Lambda_\wp)$ (resp. $K(\Lambda_\wp)^+$, ring of integers of $K(\Lambda_\wp)$, $K(\Lambda_\wp)^+$).

Let W be the ring of Witt vectors of $A/\wp$. Then we have a decomposition into isotypical components $C \otimes_{\mathbb{Z}_p} W = \oplus_{0 \le k < q^d-1} C(w^k)$ according to the characters of $(A/\wp)^*$, where w is the Teichmüller character. (Similarly for C^+, $\tilde{C}$, $\tilde{C}^+$.)

Theorem 5.18 (Okada [Oka91], Goss [G96]). *Let* $A = \mathbb{F}_q[T]$. *For* $0 < k < q^d - 1$ *and* k *'even', if* $\tilde{C}(w^k) \ne 0$ *then* $\wp$ *divides* B_k.

Proof. (Sketch) We define analogs of Kummer homomorphisms $\psi_i \colon \mathcal{O}_F^* \to A/\wp$, $0 < i < q^d - 1$ (note that pth powers map to zero) by $\psi_i(u) = u_{i-1}$, where u_i is defined as follows. Let $u(t) \in A[[t]]$ be such that $u = u(\lambda)$ and define u_i to be $\Pi(i)$ times the coefficient of z^i in the logarithmic derivative of $u(e(z))$. Using the definition of Bernoulli numbers, we calculate that the ith Kummer homomorphism takes the basic cyclotomic unit $\lambda^{\sigma_a - 1}$ to $(a^i - 1)B_i/\Pi(i)$. If $\tilde{C}(w^k) \ne 0$, then by the (component-wise version due to Goss–Sinnott of the) Galovich–Rosen theorem (analog of Kummer's theorem) that the class number of the ring of integers of $K(\Lambda_\wp^+)$ is the index of cyclotomic units in full units of that field, we have $\psi_k(\lambda^{\sum w^{-k}(\sigma)\sigma^{-1}}) = 0$. Hence the calculation above implies that $\wp$ divides B_k. $\square$

The converse is false and Gekeler has suggested the following modification (see also [Ang01]) using the Frobenius action restriction:

Conjecture (Gekeler [Gek90]). *If* $\wp | B_{k'}$ *for all* k' *such that* $k' \equiv p^m k(q^d - 1)$ *with* $0 < k, k' < q^d - 1$, $k \equiv 0 \ (q - 1)$, *then* $\tilde{C}(w^k) \ne 0$.

5.4 First relations with cyclotomic theory: At non-positive integers

For the zeta values at non-positive integers, we have the following story.

Theorem 5.19 (Goss–Sinnott [GS85]). *For* $0 < k < q^d - 1$, *we have* $C(w^{-k}) \ne 0$ *if and only if* p *divides* $L(w^k, 1)$.

Proof. (Sketch) The duality between the Jacobian and the p-adic Tate module T_p transforms the connection between the Jacobian and the class group in Tate's proof of the Stickelberger Theorem [T04, Ch. 1] to $T_p(w^{-k})/(1-F)T_p(w^{-k}) \cong C(w^{-k})$. On the other hand, we have a Weil-type result: $\det(1 - F : T_p(w^{-k})) = L_u(w^k, 1)$. Here L_u is the unit root part of the L-function and hence it has the same pth power divisibility as the complete L-function. Hence $\mathrm{ord}_p(L(w^k, 1))$ is the length of $C(w^{-k})$ as a $\mathbb{Z}_p[G]$-module.   □

For a proof without going through the complex L-function, we refer to the lecture notes of Böckle.

The interpretation above of the L-function as the characteristic polynomial in the above is precisely the result on which Iwasawa's main conjecture is based. Since this is already known, the Gras conjecture giving the component-wise results above, which follows classically from the main conjecture, is known here. This was recognized in [GS85].

Comparison with the corresponding classical result shows that we are looking at divisibility by p, the characteristic, rather than the prime $\wp$ relevant to the cyclotomic field. To bring $\wp$ in, we need to look at the finite characteristic zeta function.

Theorem 5.20 (Goss–Sinnott [GS85]). *We have $C(w^{-k}) \neq 0$ if and only if $\wp | \beta(k)$ for $0 < k < q^d - 1$.*

Proof. The identification $W/pW \cong A/\wp$ provides us with the Teichmüller character $w : (A/\wp)^* \to W^*$ satisfying $w^k(n \bmod \wp) = (n^k \bmod \wp) \bmod p$. Hence the reduction of L value (since it also has 'trivial zero' factors missing) in the theorem above modulo p is $\beta(k) \bmod \wp$.   □

Remarks 5.21.

(1) Recall that for $A = \mathbb{F}_q[t]$ and 'odd' k we have $\beta(k) = \zeta(-k)$, so the result is in analogy with the Herbrand–Ribet theorem; however, for 'even k' we get a new phenomenon. This is connected with the failure of the Spiegelungssatz for Carlitz–Drinfeld cyclotomic theory. Classically the leading terms at even k are conjectured to be transcendental; here they are rational, even integral. On the other hand, for values at positive integers the situation seems to be as expected with naive analogies.

(2) For $A = \mathbb{F}_q[t]$, both analogs of B_n/n mentioned in 5.2 thus connect to arithmetic of related class groups suggesting a stronger connection between the two analogs than what is currently understood.

5.5 Orders of vanishing mystery

Now we turn to the question of the order of vanishing. Classically, the answer is simple: The Euler product representation shows that there are no zeros in the region where it is valid and hence the orders of vanishing of the trivial zeros (namely those at the negative even integers) are easily proved by looking at poles of the gamma function factors in the functional equation. (For the number field situation in general, these are predicted by motives and these orders of vanishing are connected to K-theory and extensions of motives, so many structural issues and clues are at stake in this simple question.) We do not have functional equations. As described below, Goss gave a lower bound for orders of vanishing when L contains H and mentioned as an open question whether they are exact or not. The lower bounds match the naive analogies. But we will see that they are not exact orders and, indeed, that the patterns of extra vanishing are quite surprising in terms of the established analogies. The full situation is still not understood – not even conjecturally.

The main idea of Goss (already mentioned in the proof of Theorem 5.20) is to turn the similarity in the definition of our zeta function with the classical one into a double congruence formula, using the Teichmüller character, and to use the knowledge of classical L-function Euler factors to understand the order of vanishing. This is done as follows.

Let $\wp$ be a prime of A and let W be the Witt ring of $A/\wp$. The identification $W/pW \cong A/\wp$ provides us with the Teichmüller character $w\colon (A/\wp)^* \to W^*$ satisfying $w^k(a \bmod \wp) = (a^k \bmod \wp) \bmod p$.

Now let $\Lambda_\wp$ denote the $\wp$-torsion of the rank one, sign-normalized Drinfeld module ρ of generic characteristic. Let L contain H and let G be the Galois group of $L(\Lambda_\wp)$ over L. Then G can be thought of as a subgroup of $(A/\wp)^*$ and hence w can be thought of as a W-valued character of G. Let $L(w^{-s}, u) \in W(u)$ be the classical L-series of Artin and Weil in $u := q^{-sm}$, where m is the extension degree of the field of constants of L over $\mathbb{F}_q$.

Let $S_\infty := \{\infty_j\}$ denote the set of the infinite places of L and let G_j denote the Galois group of $L_{\infty_j}(\Lambda_\wp)$ over L_{∞_j}. Then $G_j \subset \mathbb{F}_q^*$. Given s, let $S_s \subset S_\infty$ be the subset of the infinite places at which w^{-s} is an unramified character of G. Then S_s does not depend on $\wp$. Put

$$\tilde\zeta_\mathcal{O}(s, X) := \zeta_\mathcal{O}(s, X) \prod_{\infty_j \in S_s} (1 - w^{-s}(\infty_j) X^{\deg(\infty_j)})^{-1}.$$

Theorem 5.22 (Goss [G92]). *Let L contain H and let s be a negative integer. Then $\tilde\zeta_\mathcal{O}(s, X) \in A[X]$.*

Proof. (Sketch) Tracing through the definitions, the property of w mentioned above gives the double congruence formula $L(w^{-s}, X^m) \bmod p = \tilde\zeta_\mathcal{O}(s, X) \bmod \wp$ for infinitely many $\wp$. But as the L-function is known to be a polynomial, the result follows. $\qquad\square$

This gives the following lower bound for the order of vanishing:

Theorem 5.23 (Goss [G92]). *If L contains H and s is a negative integer, then the order of vanishing of $\zeta_\mathcal{O}(s)$ is at least*

$$V_s := \operatorname{ord}_{X=1} \prod_{\infty_j \in S_s} (1 - w^{-s}(\infty_j) X^{\deg(\infty_j)}).$$

Remark 5.24. Note that V_s depends on s only through its value modulo $q-1$ and $V_s \leq [L:K]$. This is analogous to the properties of the exact orders of vanishing in the classical case, but not in our case, as we will see.

Examples 5.25.

(i) $L = K = H$: Then $V_s = 0$ or 1, according to as s is 'odd' or 'even'. We have seen that for $A = \mathcal{O} = \mathbb{F}_q[t]$ the order of vanishing is V_s.

(ii) L is a totally real extension of degree d of K containing H, i.e., ∞ splits completely in L: V_s is 0 or d according to whether s is 'odd' or 'even'. (Note that the bounds V_s in (i) and (ii) are in analogy with the orders of vanishing in the classical case.)

(iii) $L = K(\Lambda_\wp)$: $V_s = (q^{\deg(\wp)} - 1)/(q-1)$ for all s.

(iv) $L = \mathbb{F}_{q^n}(t)$ and $K = \mathbb{F}_q(t)$: $V_s = 0$ when s is 'odd', and for s 'even' we have $V_s = p^k$ if $n = p^k l$, $(l, p) = 1$.

(v) $L = \mathbb{F}_5(\sqrt{-t})$ and $K = \mathbb{F}_5(t)$: V_s is 1 or 0 according to whether 2 (not 4) does or does not divide s.

Let O_s denote the order of vanishing of the relevant zeta function at s. Then simple exact calculations using the vanishing theorem bounds show that $O_s = V_s$ for small s in some examples, such as (i) $A = \mathbb{F}_3[x, y]/y^2 = x^3 - x - 1$; (ii) $A = \mathbb{F}_3[t]$ with $\mathcal{O} = \mathbb{F}_3[t, y]/y^2 = t^3 - t - 1$; (iii) $A = \mathbb{F}_3[t]$ with $\mathcal{O} = \mathbb{F}_9[t]$.

But the following calculation shows that $V_s \neq O_s$ in general.

Examples 5.26. Let (a) $A = \mathbb{F}_2[x, y]/y^2 + y = x^3 + x + 1$; (b) $A = \mathbb{F}_2[x, y]/y^2 + y = x^5 + x^3 + 1$. In both these situations, $hd_\infty = 1$ and hence $K = H$. For small s we compute $\zeta(s, X)$ as follows: The genus of K is 1 and 2 for (a) and (b) respectively. Hence the bounds obtained from the vanishing theorem and the Riemann–Roch theorem show that $\zeta(-1, X) = 1 + 0 * X + (x + (x+1)) * X^2 = 1 + X^2 = (1 + X)^2$ for (a) and (b). Hence the order of vanishing is 2 for $s = -2^n$. A similar simple calculation shows that the order of vanishing is 1 if $2^n \neq -s \leq 9$ for (a), and it is 1 for $s = -7$ and 2 for other s with $-s \leq 9$ for (b).

In fact, more generally, we have the following result:

Theorem 5.27. *If $d_\infty = 1$, $q = 2$ and K is hyper-elliptic, then the order of vanishing of $\zeta(s)$ at a negative integer s is 2 if $l(-s) \leq g$, where g is the genus of K.*

Proof. Let $x \in A$ be an element of degree 2 and let n be the first odd non-gap for A at ∞. Then the genus g of K is seen to be $(n-1)/2$. Let $S_A(i)$ denote the coefficient of X^i in $\zeta_A(s, X)$. A simple application of our general vanishing theorem and the Riemann–Roch theorem shows that $S_A(i) = 0$ for $i > l(-s) + g$. Hence

$$\zeta_A(s, X) = \sum_{i=0}^{(n-1)/2} S_A(2i) X^{2i}$$

since $l(-s) \leq g = (n-1)/2$. On the other hand, as x has degree 1 in $\mathbb{F}_q[x]$, we have $S_A(2i) = S_{\mathbb{F}_q[x]}(i)$ for $0 \leq i \leq (n-1)/2$. Further, by the vanishing theorem, $S_{\mathbb{F}_q[x]}(i) = 0$ if $i > (n-1)/2$, as $(n-1)/2 \geq l(-s)$. Hence, $\zeta_A(s, X) = \zeta_{\mathbb{F}_q[x]}(s, X^2)$. (This works only for s as above, and is not an identity of zeta functions.) Hence, by Example (i), the order of vanishing is $2 = 2 * 1$ as required. $\qquad\square$

Remarks 5.28.

(i) We do not need a restriction on the class number of K, so apart from (a) and (b) this falls outside the scope of the theorem giving the lower bounds. But there are other examples of this phenomenon with $q > 2$ as well as for the full ideal zeta function for higher class numbers. See [T95] and the University of Arizona thesis (1996) of Javier Diaz-Vargas and [DV06].

(ii) Analogies between number fields and function fields are usually the strongest for $A = \mathbb{F}_q[y]$. But even in that case there are examples of extra vanishing. For example, when $A = \mathbb{F}_4[y]$ and $\mathcal{O} = \mathbb{F}_4[x, y]/y^2 + y = x^3 + \zeta_3$, the order of vanishing of $\zeta_\mathcal{O}$ at $s = -1$ is 2, as can be verified by direct computation, whereas the lower bound is 1.

Remark 5.29. These results, when combined with the Goss–Sinnott results above, imply the non-vanishing of relative class group components for all primes and raise important open questions about the meaning of the real leading terms.

Lecture 6

Period Interpretations and Complete Relations Between Values

For $A = \mathbb{F}_q[t]$, we saw in Section 3.1 that a binomial coefficient, also related to terms of a gamma product, was given as (i) a ratio of natural products related to A, as we all as (ii) a coefficient coming from a series involving Drinfeld exponential and logarithms for rank one sign-normalized Drinfeld A-modules. If one generalizes to A, the two properties do not agree and give rise to two binomial coefficient notions, expressed by round and curly brackets respectively. The following result shows that this coincidence of the two notions allows one to get a nice generating function for the terms of the zeta sums for $\zeta(-k)$, namely the power sums $S_d(k) = \sum a^k$, where the sum is over monic elements of degree d in A.

The basic idea is that power sums are symmetric functions of a's and can be calculated via Newton's formulas from the elementary symmetric functions, which are coefficients of $\prod(x - a)$ related to binomial coefficients, where one uses $x \to x - t^d$ to move from the $\mathbb{F}_q$-vector space of all a's of degree less than d to the affine space of monic of degree d. We can use the reciprocal polynomial for $1/a$'s.

6.1 Generating function

Here is [T04, Theorem 5.6.3] and its proof with signs corrected.

Theorem 6.1 (Carlitz). *If $\mathcal{B}_d(x)$ denotes the binomial coefficients $\left\{\begin{smallmatrix} x \\ q^d \end{smallmatrix}\right\} = \left(\begin{smallmatrix} x \\ q^d \end{smallmatrix}\right)$ for $\mathbb{F}_q[t]$, then $-\mathcal{B}'_d(0)/(1 - \mathcal{B}_d(x))$ has Laurent series expansion $\sum_{n=0}^{\infty} S_d(n) x^{-n-1}$ at $x = \infty$ and $-\sum_{n=0}^{\infty} S_d(-n-1) x^n$ at $x = 0$.*

Proof. Since $1 - \mathcal{B}_d(x)$ has all monic polynomials of degree d as its simple zeros, by taking the logarithmic derivative to convert products into sums we see that

$$\sum_{a \in A_{d+}} \frac{1}{x - a} = -\frac{\mathcal{B}'_d(0)}{1 - \mathcal{B}_d(x)}.$$

Developing the left-hand side as a geometric series in two different ways, we see that the Laurent series expansion is $\sum_{n=0}^{\infty}(\sum_{a\in A_{d+}} a^n)x^{-n-1}$ at $x=\infty$ and $-\sum_{n=0}^{\infty}(\sum_{a\in A_{d+}} a^{-n-1})x^n$ at $x=0$.   □

We note that $\exp_1 x \log_1 = \sum \mathcal{B}_d(x)\tau^d$, and $\mathcal{B}'_d(0) = 1/\ell_d$.

In particular, we see that $S_d(-1) = 1/\ell_d$ and so $\zeta(1) = \log(1)$, implying that $\zeta(1)$ is transcendental. When $q > 2$, 1 is an 'odd' value. For the Riemann zeta, the transcendence of only even values is known because of Euler's result. We will show below how the t-motives allow us to handle all values for $A = \mathbb{F}_q[t]$.

Remark 6.2. Let us give a simplification of the argument, avoiding full evaluation of binomial coefficients, to prove directly the simplest case $S_d(-k) = 1/\ell_d^k$ for $1 \le k \le q$. Let a (respectively b) run through polynomials of degree $< d$ (resp. monic of degree d). Then $\prod(x - a)$ has x-coefficient $(-1)^d(D_0 D_1 \cdots D_{d-1})^{q-1}$. Since this polynomial is $\mathbb{F}_q$-linear, $\prod(x - t^d - a) = \prod(x - b)$ has the same x-coefficient (only the constant coefficient is different), which is now the sum over b of products of all monics except b of degree d, that is $(\prod b)(\sum 1/b) = D_d S_d(-1)$. Comparison gives the claim for $k = 1$. The $\mathbb{F}_q$-linearity of the polynomial makes appropriate coefficients zero so that the Newton formula for power sums implies that the kth power sum is kth power of the first, for the given range of k.

In the other direction, relating explicitly the two different notions of binomial coefficients mentioned above, this relation between zeta and logarithms of Drinfeld modules was generalized in [T92b] (see also [T04, 4.15, 5.9] to some higher genus A. For example, for $A = \mathbb{F}_3[x,y]/y^2 = x^3 - x - 1$, we have $\zeta(1) = \log_\rho(y-1)$ for the corresponding sign-normalized (x and y having signs 1) Drinfeld A-module ρ. We omit the details and move back to the $A = \mathbb{F}_q[t]$ case and $\zeta(n)$ for any n.

6.2   *n*th tensor power of the Carlitz module

By definition, the tensor product of t-motives is the tensor product over $F[t]$ on which τ acts diagonally. Thus ranks multiply under tensor product and so the Carlitz–Tate motive $C^{\otimes n}$ is of rank one. It is a uniformizable, simple, pure t-motive of dimension n and it has no complex multiplications. Now the Carlitz module is of rank one over $F[t]$ with basis m, so that $\tau m = (t - \theta)m$. Thus $C^{\otimes n}$ is free of rank one over $F[t]$ with basis $m_1 = m^{\otimes n}$. With $m_i := (t - \theta)^{i-1}m_1$ as its $f[\tau]$-basis, we see that if we write $[a]_n \in \text{End}(\mathbb{G}_a^n)$ for the image of a, then $[t]_n(x_1, \ldots, x_n) = (tx_1 + x_2, \ldots, tx_{n-1} + x_n, tx_n + x_1^q)$. Here, by abuse of notation, we have identified $\theta = t$. If we write $d[a]_n$ for the coefficient of τ^0, then it is the matrix representing the endomorphism of $\text{Lie}(\mathbb{G}_a^n)$ induced by $[a]_n$ and in particular we have $d[a]_n(*, \ldots, *, x) = (*, \ldots, *, ax)$: The largest quotient of $\text{Lie}(\mathbb{G}_a^n)$ on which the derivative and multiplication actions of A coincide is isomorphic to $\mathbb{G}_a$ with the isomorphism $\ell_n : \text{Lie}(\mathbb{G}_a^n) \to \mathbb{G}_a$ given by the last coordinate.

Note that by the definition of diagonal τ-action, the matrix Φ for $C^{\otimes n}$ is just the scalar matrix $(t - \theta)^n$. There is a $\lambda \in C_\infty^n$ with the last entry $\tilde{\pi}^n$ such that the period lattice of $C^{\otimes n}$ is $\{d[a]_n\lambda : a \in A\}$.

By solving the functional equations for the exponential and logarithm, we can find expansions for these series. In particular, the canonical last co-ordinate ℓ_n of $\log_n(X)$ is given by $\sum_{i=0}^{n-1}(-1)^i\Delta^i\mathcal{L}_n(x_{n-i})$, where $\mathcal{L}_n(x) = \sum_{i=0}^{\infty}(-1)^{in}x^{q^i}/L_i^n$ is the naive poly-logarithm corresponding to the Carlitz logarithm and Δ is our analog of the $z\,d/dz$ operator, which is a commutator with t acting on linear functions $f(z)$ giving $f(tz) - tf(z)$. In particular, the last co-ordinate seems to be a good deformation of the naive poly-log $\ell_n(\log_n(0,\ldots,0,x)) = \mathcal{L}_n(x)$. Note that

$$\Delta^k\mathcal{L}_n(x) = \sum \frac{[i]^k x^{q^i}}{\ell_i^n} = \sum \binom{k}{i}t^{k-i}\mathcal{L}_n(t^i x).$$

Also note the differences with the usual complex logarithms and multilogarithms, namely that in our case $\log(1-x) = \log(1) - \log(x)$, $\mathcal{L}_n(x)$ is a naive generalization of the series for $\log(x)$ rather than for $\log(1-x)$, and $\Delta^i\mathcal{L}_n(x)$ is as above compared to the usual $(z\,d/dz)^i\log_n(x) = \log_{n-i}(x)$ for $i < n$.

6.3 Period interpretation for zeta

We now use the logarithmic derivative trick for the generating function with binomial coefficients as explained above to turn the solitons giving terms of gamma products into terms giving zeta sums, and give an explicit algebraic incarnation of the (transcendental) Carlitz zeta value $\zeta(n)$ (and its v-adic counterparts), for n a positive integer, on the Carlitz–Tate motives $C^{\otimes n}$.

For the classical polylog $\log_n$ and Riemann ζ, we have a simple connection $\zeta(n) = \log_n(1)$, which follows directly from the definition and is of not much use to prove irrationality or transcendence results for these zeta values. For the $\mathbb{F}_q[t]$ case, the situation is quite different. The above relation holds for $n = p^r m$ with $m \leq q$, as we have seen in 6.1, but in general the relation is much more complicated. However, because of the direct motivic connection to abelian t-modules, such as $C^{\otimes n}$, we can draw much stronger transcendence and algebraic independence conclusions.

Let us temporarily write V_n for the vector $(0,\ldots,0,1)$ of length n.

Theorem 6.3. *Let $A = \mathbb{F}_q[t]$. There exists an A-valued point Z_n of $C^{\otimes n}$ (constructed explicitly below) such that the canonical last co-ordinate of $\log_n(Z_n)$ is $\Gamma(n)\zeta(n)$.*

If we put $Z_{n,v} := [v^n - 1]_n Z_n$ for a monic irreducible $v \in A$, then the canonical last co-ordinate of $\log_{n,v}(Z_{n,v})$ is $v^n\Gamma(n)\zeta_v(n)$.

The point Z_n is a torsion point if and only if n is 'even'.

Proof. We use the connection between the binomial coefficients and the power sums giving the terms of the zeta functions explained earlier to construct the point Z_n as follows.

Let $G_n(y) := \prod_{i=1}^{n}(t^{q^n} - y^{q^i})$, so that $G_i(t^{q^k}) = (l_k/l_{k-i})^{q^i}$ and so, for the binomial coefficient notation before, $\mathcal{B}_k(x) = \sum_{i=0}^{k} G_i(t^{q^k})/d_i(x/l_k)^{q^i}$. Hence if we define $H_n(y) \in K[y]$ by

$$\sum_{n=0}^{\infty} \frac{H_n(y)}{n!}x^n = \left(1 - \sum_{i=0}^{\infty} \frac{G_i(y)}{d_i}x^{q^i}\right)^{-1},$$

then the fact that $n!m!$ divides $(n+m)!$ for the Carlitz factorial implies first of all that with $H_{n-1}(y) = \sum h_{ni}y^i$ the coefficients h_{ni} belong to A. Furthermore, by the generating function recalled above, we see that

$$H_n(t^{q^k})/l_k^{n+1} = n!S_k(-(n+1)).$$

This connection with multilogarithms is what we need for later applications. We define $Z_n \in C^{\otimes n}(A) = A^n$ by

$$Z_n := \sum [h_{ni}]_n(t^i V_n).$$

Straight degree estimates [AT90] show that $t^i V_n$ is in the region of convergence for $\log_n$ for $i \leq \deg H_{n-1}$. Hence we have

$$\ell_n(\log_n(Z_n)) = \sum \ell_n(d[h_{ni}]_n \log_n(t^i V_n)) = \sum h_{ni}\ell_n(\log_n(t^i V_n))$$
$$= \sum h_{ni}\frac{t^{iq^k}}{l_k^n} = \Gamma(n)\zeta(n).$$

Note that $\log_n$ is a multi-valued function. With $z_n := \sum d[h_{ni}]_n(t^i V_n) \in K_\infty^n$, we have $\exp_n(z_n) = Z_n$.

If Z_n were a torsion point when n is 'odd', then, since $\tilde{\pi}^n$ is the last co-ordinate logarithm of zero, $\zeta(n)$ would be a rational multiple of $\tilde{\pi}^n$. But we know (e.g., from the fractional degree $q/(q-1)$ of $\tilde{\pi}$) that $\zeta(n)/\tilde{\pi}^n$ is not in K_∞ when n is 'odd'. Conversely, if n is 'even' and Z_n is not a torsion point, then by Jing Yu's Hermite–Lindemann type transcendence result for $C^{\otimes n}$, we would have $\zeta(n)/\tilde{\pi}^n$ transcendental, contradicting the Carlitz result mentioned above. (In [ATp] we have a direct algebraic proof of this without appealing to transcendence theory.)

Notice that the calculation proceeds degree by degree, so that instead of dealing with $\zeta(s) \in K_\infty$ we can also work with $\zeta(s, X) \in K[[X]]$; hence, multiplication by $1 - v^n$, which is just removing the Euler factor at $s = -n$, then gives the corresponding sum where the power sums are now over a prime to v, thus giving the v-adic zeta value (see [AT90] for a detailed treatment paying attention to convergence questions). $\qquad\square$

Example 6.4. For $n = rp^k$ with $0 < r < q$, we have $Z_n = [\Gamma(n)]_n(0, \ldots, 0, 1)$ and $Z_{q+1} = (1, 0, \ldots, 0) + [t^q - t]_{q+1}V_{q+1}$.

6.4 Transcendence and algebraic independence results

Proving analogs for $C^{\otimes n}$ of the Hermite–Lindemann theorems (and their v-adic counterparts) about transcendence of values of the logarithm for $C^{\otimes n}$, Jing Yu [Y91] concluded the following from the above theorem.

Theorem 6.5. *Let $A = \mathbb{F}_q[t]$. Then $\zeta(n)$ is transcendental for all positive integers n, and $\zeta(n)/\tilde{\pi}^n$ and $\zeta_v(n)$ are transcendental for 'odd' n.*

More recently, directly using multilogarithms and using the techniques of [ABP04, P08], much stronger results (at least at ∞) were proved by calculations of dimensions of relevant motivic groups. We now turn to explaining these.

Theorem 6.6.

(1) [CY] *The only algebraic relations between $\zeta(n)$'s come from the Carlitz–Euler evaluation at 'even' n, and $\zeta(pn) = \zeta(n)^p$. In particular, for n 'odd', $\zeta(n)$ and $\tilde{\pi}$ are algebraically independent and the transcendence degree of the field $K(\tilde{\pi}, \zeta(1), \dots, \zeta(n))$ is $n + 1 - \lfloor n/p \rfloor - \lfloor n/(q-1) \rfloor + \lfloor n/(p(q-1)) \rfloor$.*

(2) [CPYa] *The only algebraic relations between $\zeta_\ell(n)$'s for all ℓ and n, where ζ_ℓ denotes the Carlitz zeta over $\mathbb{F}_{q^\ell}[t]$, are those as above coming from n 'even' or divisible by p. The periods $\tilde{\pi}_\ell$ of the Carlitz modules for $\mathbb{F}_{q^\ell}[t]$ are all algebraically independent.*

(3) [CPYb] *The only algebraic relations between $\zeta(n)$'s and geometric $\Gamma(z)$'s at proper fractions are those between zeta above and bracket relations for gamma.*

(4) [CPTY] *The only algebraic relations between $\zeta(n)$'s and arithmetic gamma values at proper fractions are those for zeta mentioned above and those for gamma coming from the bracket relations, and thus the transcendence degree of the field*
$$K(\tilde{\pi}, \zeta(1), \dots, \zeta(s), (c/(1 - q^\ell))!)_{1 \le c \le q^\ell - 2}$$
is $s - \lfloor s/p \rfloor - \lfloor s/(q-1) \rfloor + \lfloor s/(p(q-1)) \rfloor + \ell$.

We quickly describe relevant motives and corresponding algebraic groups (which can often be calculated using recipes in [P08] or using Pink's results [Pin97] on images of Galois representations) leaving the details to the references above.

For the Carlitz module M, we have $\Phi = (t - \theta)$, $\Psi = \Omega$ and $\Gamma_M = \mathbb{G}_m$. This follows from the fact that $\text{Hom}(C^{\otimes m}, C^{\otimes n})$ is $\mathbb{F}_q(t)$ or zero, according as $m = n$ or not, and thus that the category generated by M is equivalent to the $\mathbb{Z}$-graded category of vector spaces over $\mathbb{F}_q(\theta)(t)$ with fiber functor to the category of vector spaces over $\mathbb{F}_q(t)$.

For its nth tensor power M (with $n \in \mathbb{Z}$ and with $n = -1$ corresponding to the dual), we have $\Phi = (t - \theta)^n$, $\Psi = \Omega^n$, and $\Gamma_M = \mathbb{G}_m$.

The rank two Drinfeld module M given by $\rho_t = t + g\tau + \tau^2$ is in analogy with elliptic curves. We give the set-up here, the higher rank situation being a straight generalization. It corresponds to the square matrix $\Phi = [0, 1; t - \theta, -g^{(-1)}]$ of size two and the corresponding period matrix $\Psi^{-1}(\theta) = [w_1, w_2; \eta_1, \eta_2]$, where w_i are periods and $\eta_i = F_\tau(w_i)$ are quasi-periods, and where the quasi-periodic function F_τ satisfies $e_\rho(z)^q = F_\tau(\theta z) - \theta F_\tau(z)$. (See [T04, §6.4].) Recall the elliptic curve analog $\eta_i = 2\zeta(w_i/2)$, with Weierstrass ζ related to $\wp$ by $\zeta' = -\wp$. For the non-CM case, we get the group Gl_2.

It is clear how to generalize to define the matrices for higher rank Drinfeld modules, and we can calculate the motivic groups by appealing to Pink's results.

In particular, for getting the Carlitz module over base $\mathbb{F}_{q^\ell}[t]$, which is a rank ℓ Drinfeld module over A, we have the size ℓ matrix $\Phi_\ell := (t - \theta)$ if $\ell = 1$, and otherwise

$$\Phi_\ell := \begin{bmatrix} 0 & 1 & 0 & \cdots & 0 \\ 0 & 0 & 1 & \cdots & 0 \\ \vdots & \vdots & \ddots & \ddots & \vdots \\ 0 & 0 & \cdots & 0 & 1 \\ (t-\theta) & 0 & 0 & \cdots & 0 \end{bmatrix}.$$

Let ξ_ℓ be a primitive element of $\mathbb{F}_{q^\ell}$ and define $\Psi_\ell := \Omega_\ell$ if $\ell = 1$. Otherwise let

$$\Psi_\ell := \begin{bmatrix} \Omega_\ell & \xi_\ell \Omega_\ell & \cdots & \xi_\ell^{\ell-1} \Omega_\ell \\ \Omega_\ell^{(-1)} & (\xi_\ell \Omega_\ell)^{(-1)} & \cdots & (\xi_\ell^{\ell-1} \Omega_\ell)^{(-1)} \\ \vdots & \vdots & \ddots & \vdots \\ \Omega_\ell^{(-(\ell-1))} & (\xi_\ell \Omega_\ell)^{(-1)} & \cdots & (\xi_\ell^{\ell-1} \Omega_\ell)^{(-(\ell-1))} \end{bmatrix},$$

where Ω_ℓ is just obtained from Ω by replacing q^nth powers by $q^{\ell n}$th powers.

The associated motivic Galois group is the Weil restriction of scalars of $\mathbb{G}_m$ from the constant field extension in question, and hence a torus of dimension ℓ (see [CPTY]).

This already allows us to handle the arithmetic gamma values as follows. We have

$$\frac{(\frac{1}{1-q^\ell})!}{(\frac{q^{\ell-1}}{1-q^\ell})!^q} \sim \Omega_\ell(\theta), \qquad \frac{(\frac{q^j}{1-q^\ell})!}{(\frac{q^{j-1}}{1-q^\ell})!^q} \sim \Omega_\ell^{(-(\ell-j))}(\theta) \quad (1 \le j \le \ell - 1),$$

where the first is just the Chowla–Selberg analog that we proved for the arithmetic gamma and the second is its quasi-period analog established similarly.

Now any gamma value with denominator $1-q^\ell$ (note that any denominator is of this form without loss of generality by Fermat's little theorem), by an integral translation resulting in a harmless algebraic factor, is a monomial in the basic values $(q^j/(1 - q^\ell))!$'s $(0 \le j \le \ell - 1)$. Thus the field generated by all these values is ℓ-dimensional, by the dimensional calculation mentioned above. In particular, the $(q^j/(1 - q^\ell))!$'s $(0 \le j \le \ell - 1)$ are algebraically independent.

To handle zeta values, we use the multilogarithm connection mentioned above and first handle a set of multilogarithms $\mathcal{L}_n$ at algebraic α_i's.

For $\alpha \in \bar{k}^\times$ with $|\alpha|_\infty < |\theta|_\infty^{\frac{nq}{q-1}}$, the power series

$$L_{\alpha,n}(t) := \alpha + \sum_{i=1}^{\infty} \frac{\alpha^{q^i}}{(t - \theta^q)^n (t - \theta^{q^2})^n \cdots (t - \theta^{q^i})^n}$$

satisfies $L_{\alpha,n}(\theta) = \mathcal{L}_n(\alpha)$. It is easy to check the functional equation

$$(\Omega^n L_{\alpha,n})^{(-1)} = \alpha^{(-1)}(t - \theta)^n \Omega^n + \Omega^n L_{\alpha,n}. \tag{6.1}$$

Let

$$\Phi_n = \Phi(\alpha_{n0}, \ldots, \alpha_{nk}) := \begin{bmatrix} (t-\theta)^n & 0 & \cdots & 0 \\ \alpha_{n0}^{(-1)}(t-\theta)^n & 1 & \cdots & 0 \\ \vdots & \vdots & \ddots & \vdots \\ \alpha_{nk}^{(-1)}(t-\theta)^n & 0 & \cdots & 1 \end{bmatrix},$$

$$\Psi_n = \Psi(\alpha_{n0}, \ldots, \alpha_{nk}) := \begin{bmatrix} \Omega^n & 0 & \cdots & 0 \\ \Omega^n \mathcal{L}_{n0} & 1 & \cdots & 0 \\ \vdots & \vdots & \ddots & \vdots \\ \Omega^n \mathcal{L}_{nk} & 0 & \cdots & 1 \end{bmatrix}.$$

The functional equation translates to $\Psi_n^{(-1)} = \Phi_n \Psi_n$, and Φ_n defines a t-motive which is an extension of the $(k+1)$-dimensional trivial t-motive $\mathbf{1}^{k+1}$ over $\bar{k}(t)$ by $C^{\otimes n}$ and its motivic Galois group is an extension of $\mathbb{G}_m$ by a vector group (see [CY, Lemma A.1]). Using this, [CY] proves that the linear independence over K implies algebraic independence over $\bar{K}$ for nth multilogarithms at algebraic quantities, generalizing Papanikolas' result [P08] for $n = 1$ by the same method, which generalizes an earlier result of Jing Yu giving an analog of Baker's theorem, yielding linear independence over $\bar{K}$.

To deal with the algebraic independence issue for the zeta values $\zeta(n)$ for $n \le s$, one takes direct sums of such appropriate multilog motives for appropriate 'odd' $n \le s$ not divisible by p and α_i's coming from the relation before, and calculates dimensions [CY].

For the geometric gamma, because of complex multiplications from the cyclotomic function field, the motivic Galois group of the corresponding 'gamma motive' is a torus inside a certain finite product of the Weil restriction of scalars of $\mathbb{G}_m$ from the cyclotomic function field. The direct sum with the 'zeta motive' has motivic Galois group an extension of a torus by a vector group. The dimension can be written down explicitly and gives what we want.

Lecture 7

Cyclotomic and Class Modules, Special Points, Zeros and Other Aspects

In this lecture we review some new developments without proofs, just stressing the new objects, ideas and interconnections. We begin by mentioning new connections of the special zeta value theory with cyclotomic theory.

7.1 Log-algebraicity, cyclotomic module, Vandiver

In our first look at the relations with cyclotomic theory, the p-divisibilities of orders of groups related to the class groups for $\wp$th cyclotomic fields were linked to the $\wp$-divisibilities of the zeta values. This mixture of p and $\wp$ is not satisfactory analog. Also, we see that naive analogs of Kummer's theorem that 'p does not divide $h(\mathbb{Q}(\zeta_p))^-$ implies that p does not divide $h(\mathbb{Q}(\zeta_p)^+)$' and of Vandiver's conjecture that 'p does not divide $h(\mathbb{Q}(\zeta_p)^+)$' both fail for simple examples. The p-divisibilities or even Norm($\wp$)-divisibilities do not work. To get a good analog [T94, p. 163] so that we can talk about $\wp$ divisibilities, we want A-modules and not abelian groups which are just $\mathbb{Z}$-modules.

In a very nice important work [A96], Greg Anderson provided 'cyclotomic A-module, class polynomial', and an analog of the Vandiver conjecture! To do this, he generalized the log-algebraicity phenomena that we have been looking at in the previous lectures, as follows.

The cyclotomic units come from $1 - \zeta = \exp(\log(1 - \zeta)) = \exp(\sum -\zeta^n/n)$. Instead of the logarithm, which is just the inverse function of the exponential, to get more, one should view the series occurring at the right-hand side as the harmonic series or the zeta. For x a fraction, we consider, following Anderson, the triviality $\zeta^n = \exp(2\pi i n m x) = \exp(2\pi i n x)^m$ and look at its analog $e_C(\sum e_C(ax)^m/a)$. By such a mixing of the Carlitz action by $a \in A = \mathbb{F}_q[T]$ and the usual multiplicative action by $m \in \mathbb{Z}$ (thus getting the usual powers and polynomials in the game using functional equations for the exponential), Anderson considered, for $A = \mathbb{F}_q[T]$,

$$S_m(t, z) := \sum_{i=0}^{\infty} \frac{1}{d_i} \sum_{a \in A_+} \left(\frac{C_a(t)^m}{a} \right)^{q^i} z^{q^{i+\deg a}} \in K[[t, z]]$$

and proved the following 'log-algebraicity' theorem:

Theorem 7.1 (Anderson [A96]). *We have $S_m(t, z) \in A[t, z]$.*

Using this theorem, Anderson defined the cyclotomic module $\mathcal{C}$ (an analog of the group of cyclotomic units in $\mathbb{Q}(\zeta_p)$) to be the A-sub-module of $\mathcal{O} := \mathcal{O}_{K(\Lambda_\wp)}$ (under the Carlitz action) generated by 'special points'

$$S_m(e_C(\tilde{\pi}b/\wp), 1) = e_C\left(\sum e_C(ab\tilde{\pi}/\wp)^m/a \right).$$

He proved that the cyclotomic module $\mathcal{C}$ is a Galois stable module of rank equal to $(\mathrm{Norm}(\wp) - 1)(1 - 1/(q - 1))$.

His first proof [A94], which works in more generality for any A with $d_\infty = 1$, uses his soliton ideas of deformation theory to get explicit interpolating functions on powers of the cyclotomic cover similar to what we have seen for solitons, and then showing directly by analysis of divisors and zeros that the coefficients for a large enough power of z are zero. His second proof [A96] is more elementary and uses Dwork's v-adic trick (for all finite v) to deduce integrality of coefficients, and then uses analytic theory to get degree estimates to show that coefficients tend to zero eventually, so that they are zero from some point onwards and the power series is really a polynomial.

We record a family of examples. Let $m = \sum_{e=0}^{k} q^{\mu_e}$ with $k < q$. Then we have [T04, p. 300] $S_i(m - 1) = (\prod [\mu_e]_i)/l_i$, where $[h]_i = d_h/d_{h-i}^{q^i}$, and

$$S_m(t, z) = \sum_{l_1=0}^{\mu_1} \cdots \sum_{l_k=0}^{\mu_k} C_{P_{(\mu_j, l_j)}(T)} \left(z \prod_{e=1}^{k} C_{T^{l_e}}(t) \right),$$

where

$$P_{(\mu_j, l_j)}(T) = (-1)^{\sum (\mu_e - l_e)} \sum T^{q^{j_{1,1}} + \cdots + q^{j_{1,\mu_1 - l_1}} + \cdots + q^{j_{k,1}} + \cdots + q^{j_{k,\mu_k - l_k}}}$$

with $0 \leq j_{e,1} < j_{e,2} < \cdots < j_{e,\mu_e - l_e} \leq \mu_e - 1$, for $1 \leq e \leq k$.

This suggests that there is more underlying structure to these special polynomials and it would be desirable to get such an expression for general m.

Anderson also expresses $L(1, w^i)$ as an algebraic linear combination of logarithms of these 'special points' (and a v-adic analog). (In a recent preprint, Papanikolas has shown, using these facts and ideas from the $C^{\otimes n}$ section, that $L(m, \chi)$ for $m \leq q$ and χ a character of $(A/f)^*$ which is an mth power can be expressed as an explicit algebraic linear combination of mth polylogs at algebraic arguments.)

Now the Kummer–Vandiver conjecture that p does not divide the class number of $\mathbb{Q}(\zeta_p)^+$ can be rephrased as the canonical map taking cyclotomic units mod pth powers to all units modulo pth powers is injective. Replacing $\mathcal{O}^*$, which are integral points for the multiplicative group, by $\mathcal{O}$, which are integral points for the additive group (with Carlitz action), Anderson's analog of the Vandiver conjecture would be that the map $\mathcal{C}/\wp\mathcal{C} \to \mathcal{O}/\wp\mathcal{O}$ is injective. It would be nice to settle this.

As an analog of Kummer's theorem that 'p does not divide h^- implies that p does not divide h^+', Anderson proved that the map above is injective if $\wp$ does not divide $\zeta(1 - i)$ for $1 \le i \le \mathrm{Norm}(\wp)$ and $1 - i$ 'odd'.

If $\sqrt{\mathcal{C}}$ is the divisible closure of $\mathcal{C}$ in $\mathcal{O}$ thought of as $\mathcal{O}$-points of the Carlitz module, then since the units are the divisible closure of cyclotomic units, the analogies suggest that the analog of the class number h^+ is the class polynomial giving the (Fitting) index of the A-module $\sqrt{\mathcal{C}}/\mathcal{C}$.

Using the analogy with cyclotomic units and the relation between class numbers and index of cyclotomic units, we thus get a polynomial as some kind of A-module version of the class number in this special case. How about getting an A-module 'class group' concept in general?

This leads us to the discussion of the recent beautiful work of Taelman.

7.2 Taelman's class number formula conjecture

In a very nice recent work [Ta10], Lenny Taelman gave a good analog of Dirichlet's unit theorem in the context of Drinfeld modules and gave a conjectural 'class number formula' for the Dedekind zeta $\zeta_{\mathcal{O}/A}(1)$ for $A = \mathbb{F}_q[t]$. We now describe some of the ideas.

Let $A = \mathbb{F}_q[t]$. We use the notation of the zeta section, with finite extension L of K and $R = \mathcal{O}$ being the integral closure of A in it. Define $L_\infty := L \otimes K_\infty$ and L_∞^{sep} similarly. Note that these are products of fields. Let $G = \mathrm{Gal}(K_\infty^{\mathrm{sep}}/K_\infty)$.

Let ρ be the Drinfeld A-module with coefficients in R. Taking the G-invariants of the G-equivariant short exact sequence $\Lambda \hookrightarrow \mathrm{Lie}_\rho(L_\infty^{\mathrm{sep}}) \twoheadrightarrow \rho(L_\infty^{\mathrm{sep}})$ of A-modules given by $\exp_\rho$, we get $\Lambda^G \hookrightarrow \mathrm{Lie}(L_\infty) \to \rho(L_\infty) \twoheadrightarrow H^1(G, \Lambda)$ by additive Hilbert 90. Taelman proves the following.

Theorem 7.2. *The cokernel (kernel respectively) of $\rho(R) \to H^1$ is finite (finitely generated respectively) and the inverse image under $\exp_\rho$ of $\rho(R)$ is a discrete and co-compact sub-A-module of $\mathrm{Lie}_\rho(L_\infty)$.*

He defines the cokernel to be the class module H_R and the kernel to be the Mordell–Weil group U_R, and shows that $U_R = \sqrt{\mathcal{C}}$ when L is the cyclotomic extension of K and ρ is the Carlitz module.

By the theorem, the natural map $\ell\colon \exp^{-1}(\rho(R)) \otimes_A K_\infty \to \mathrm{Lie}_\rho(R) \otimes_A K_\infty$ induced by $\exp^{-1}(\rho(R)) \to \mathrm{Lie}_\rho(L_\infty)$ is an isomorphism of K_∞-vector spaces. With respect to any A-basis of $\exp^{-1}(\rho(R))$ and $\mathrm{Lie}_\rho(R)$, the map has a well-defined determinant in K_∞^*/A^*. We call its monic representative in K_∞^* the *regulator* Reg_R. Finally, for an A-module M, let $|M|$ denote a monic generator of its first fitting ideal, i.e., a product of monic polynomials $f_i \in A$ such that $M \equiv \oplus A/(f_i)$.

Now let ρ be the Carlitz module. Then Taelman's conjectural class number formula is

$$\zeta_{R/A}(1) = \mathrm{Reg}_R * |H_R|.$$

He proves it for $R = A = \mathbb{F}_q[t]$ and gives numerical evidence. The author has proved the naive generalization of this for $A = R$ of class number one, having already calculated the left-hand side as a logarithm [T92b] as we mentioned above. For details, we refer to [Tp]. Note that Anderson, in the work already mentioned above, already proved $\zeta_{R/A}(1)$ to be some explicit combination of logarithms of special points for abelian extensions.

Let us now look at the example $A = \mathbb{F}_q[t]$ and $R = \mathbb{F}_{q^2}[t]$. Then $\zeta_{R/A}(1) = \zeta(1)L(1)$, where $L(1) = \sum \chi(a)/a$ with $\chi(a) = (-1)^{\deg(a)}$, so if ζ is a $(q-1)$th root of -1 then $L(1) = \log(\zeta)/\zeta$, as can be verified from the formulas we have seen for logarithms and power sums, and it is obvious for $q = 2$. The class number formula in this case can be verified using this fact.

Remark 7.3. While $\mathbb{C}$ is a local field and an extension of degree 2 of $\mathbb{R}$, C_∞ is not local and it is of infinite degree over K_∞. But the infinitude of the degree has the virtue of allowing lattices, and thus Drinfeld modules, of arbitrarily large ranks. Similarly, Taelman's work above shows another virtue by getting finitely generated quantities by taking invariance with respect to infinite $\mathrm{Gal}(K_\infty^{\mathrm{sep}}/K_\infty)$.

Update. In recent preprints [Tap1, Tap2], Taelman has proved his conjecture and has given several nice interpretations in the Carlitz case.

7.3   Valuations of power sums and zero-distribution for zeta

We will go in reverse chronological order to explain this topic. Let $A = \mathbb{F}_q[t]$ to start with.

For $k \in \mathbb{Z}$ and $d \in \mathbb{Z}_{\geq 0}$, let $S_d(k) = \sum a^{-k}$, where the sum is over $a \in A_{d+}$ (note that we changed sign from the earlier notation) and let $s_d(k)$ be its valuation (i.e., negative of the degree) at the infinite place. While the individual terms have valuations dk, which are monotonic in either d or k, if we fix the other constant (k or d respectively) jumps, then the cancellations in summation makes $s_d(k)$ a very erratic function, not monotonic in k and monotonic in d, but for not so obvious reasons.

For $q = p$ a prime, in [T09a] we found a nice but strange recursion (another proof by Böckle using his cohomological formula can be found in the lecture notes of the parallel lecture series) $s_d(k) = s_{d-1}(s_1(k)) + s_1(k)$, which immediately implies that $s_d(k)$ is monotonic in d and the jumps $s_{d+1}(k) - s_d(k)$ are also strictly monotonic in d, just by straight induction on d, the recursion relation reducing the statement for d to $d - 1$, and the initial case being obvious. (The recursion works for both k positive and negative and so one can approach the p-adic integer y below both from positive and negative integers.)

Now, for a fixed $y \in \mathbb{Z}_p$, the zeros of Goss' zeta function $\zeta(y, X)$ in $X \in C_\infty$ can be calculated by the Newton polygon of this power series, but the above monotonicity in jumps exactly translates to the slopes of the polygon increasing at each vertex, so that the horizontal width of each segment is one, and thus all these zeros, which are a priori 'complex' are simple and 'real', i.e., lie on the 'real line' $K_\infty \subset C_\infty$. (Also, there is at most one zero for each valuation.) This zero distribution result is the Riemann hypothesis for the Goss zeta function, first proved by Daqing Wan in the case $\mathbb{F}_p[t]$. The author noticed that the statement can be quickly reduced to a combinatorial unproved assertion made by Carlitz. This was then proved for $q = p$ by Javier Diaz-Vargas [DV96] in his University of Arizona thesis. But the general non-prime case proved to be much more difficult and after the initial proof by Bjorn Poonen for $q = 4$, Jeff Sheats, then a postdoc in combinatorics at University of Arizona, gave a complete but quite complicated proof of the general case, proving the same statement for $A = \mathbb{F}_q[t]$.

It would be nice if this proof could be simplified in any way, for example by finding some simple recursion similar to that above for any q, which gives the slope statement immediately. For more on power sums, references to the literature, see [T09a].

What happens for general A? Notice that our examples in the extra vanishing phenomena were of the type $\zeta_A(s, X) = \zeta_{\mathbb{F}_q[x]}(s, X^p)$, for some s and some $x \in A$, and since these polynomials are often non-trivial, we get zeros giving inseparable extensions of K_∞. (The simplest example being A of class number one, genus 2 – there is a unique such, given as Example (b) before Theorem 44 – and $s = -3$.) So the naive generalization of the Riemann hypothesis as above does not work. But David Goss [G00, G03] has made some interesting conjectures for the general case of L-functions and also at v-adic interpolations, regarding the finiteness of maximal separable extension when one adjoins zeros, absolute values of the zeros, etc. For precise statements, a discussion and examples, we refer to these papers.

Wan already noticed that the v-adic situation for degree one primes can be handled similarly, at least partially.

The author, in ongoing work, has found interesting patterns (partially proved, partially conjectured) in valuation tables of $S_d(k)$ at finite primes of higher degree. This should lead to some interesting consequences for the zero distribution of the v-adic zeta functions of Goss. We also saw some interesting computationally observed phenomena in the higher genus case in Böckle's lecture series.

7.4   The zeta measure

For $A = \mathbb{F}_q[t]$, polynomials taking integers to integers are integral linear combinations of binomial coefficients, just as in the $\mathbb{Z}$ case, with the binomial coefficients we talked about.

Now Mahler's theorem that continuous functions from $\mathbb{Z}_p$ to itself are exactly the functions of the form $f(x) = \sum f_k \binom{x}{k}$, with $f_k \to 0$ as k tends to infinity, has an exact analog obtained by replacing $\mathbb{Z}_p$ by $A_\wp$ and replacing the usual binomial coefficients by the Carlitz binomial coefficients we introduced above. This was proved by Carlitz' student Wagner.

A $\mathbb{Z}_p$-valued p-adic measure on $\mathbb{Z}_p$ is a $\mathbb{Z}_p$-linear map $\mu : \operatorname{Cont}(\mathbb{Z}_p, \mathbb{Z}_p) \to \mathbb{Z}_p$. One writes $\mu(f)$ symbolically as $\int_{\mathbb{Z}_p} f(x)\, d\mu(x)$.

In the binomial coefficient basis, such a measure μ is uniquely determined by the sequence $\mu_k := \int_{\mathbb{Z}_p} \binom{x}{k} d\mu(x)$ of elements of $\mathbb{Z}_p$. Any sequence μ_k determines a $\mathbb{Z}_p$-valued measure μ by the formula $\int_{\mathbb{Z}_p} f\, d\mu = \sum_{k \geq 0} f_k \mu_k = \sum_{k \geq 0} (\triangle^k f)(0) \mu_k$.

Convolution $*$ of measures μ, μ' is defined by

$$\int f(x)\, d(\mu * \mu')(x) = \int \int f(x+t)\, d\mu(x)\, d\mu'(t).$$

In other words, if one identifies a measure μ with $\sum \mu_k X^k$ then convolution on measures is just multiplication on the corresponding power series. This identification of $\mathbb{Z}_p$-valued measures on $\mathbb{Z}_p$ with power series is called *Iwasawa isomorphism*. In the $\mathbb{F}_q[t]$ case, with function field binomial coefficient analogs, an $A_\wp$-valued measure μ on $A_\wp$ (called just a measure μ for short) is uniquely determined by the sequence $\mu_k = \int_{A_\wp} \binom{x}{k} d\mu$ of elements of $A_\wp$. Any sequence μ_k determines a measure μ by $\int_{A_\wp} f\, d\mu = \sum_{k \geq 0} f_k \mu_k$.

Because of different properties of this analog, if we identify a measure μ with the divided power series $\sum \mu_k (X^k/k!)$ (recall the formal nature of $X^k/k!$ in the context of divided power series), then convolution on measures is multiplication on the corresponding divided power series.

Classically, Nick Katz showed that the measure μ whose moments $\int_{\mathbb{Z}_p} x^k\, d\mu$ are given by $(1 - a^{k+1})\zeta(-k)$ for some $a \geq 2$ with $(a,p) = 1$ has the associated power series $(1+X)/(1-(1+X)) - a(1+X)^a/(1-(1+X)^a)$. We need a twisting factor in front of the zeta values to compensate for the fact that the zeta values are rational rather than integral, in contrast to our case. It would be quite interesting to understand the comparison of this result with the following result [T90]. (This reference contains references and details about all the results mentioned in this section.)

Theorem 7.4. *For $A = \mathbb{F}_q[t]$, the divided power series corresponding to the zeta measure μ, namely the measure whose ith moment $\int_{A_\wp} x^i\, d\mu$ is $\zeta(-i)$, is given by $\sum \mu_k (X^k/k!)$ with $\mu_k = (-1)^m$ if $k = cq^m + (q^m - 1)$, $0 < c < q - 1$, and $\mu_k = 0$ otherwise.*

7.5 Multizeta values, period interpretation and relations

We briefly mention [T04, §5.10] and more recent works [AT09, T09b, T10, Lr09, Lr10, Tp] on multizeta values. For a detailed discussion, a bigger context in the number field and function field situation, full results and proofs, these references should be consulted.

There has been a strong recent interest in multizeta values which are iterated sums

$$\zeta(s_1, \ldots, s_k) := \sum_{n_1 > \cdots > n_k > 0} \frac{1}{n_1^{s_1} \cdots n_k^{s_k}}$$

initially defined by Euler, through their appearance in the Grothendieck–Ihara program [I91] of understanding the absolute Galois group of $\mathbb{Q}$ through the algebraic fundamental group of the projective line minus zero, one and infinity, and related interesting mathematical and mathematical-physics structures in diverse fields. The connection comes through an integral representation from them by an iterated integral of holomorphic differentials dx/x and $dx/(1-x)$ on this space giving a period for its fundamental group. The relations that they satisfy, such as sum shuffle and integral shuffle relations, have many structural implications. We will just mention the simplest sum shuffle

$$\zeta(s_1)\zeta(s_2) = \sum \frac{1}{n_1^{s_1}} \sum \frac{1}{n_2^{s_2}} = \zeta(s_1, s_2) + \zeta(s_2, s_1) + \zeta(s_1 + s_2),$$

which follows from $n_1 > n_2$ or $n_1 < n_2$ or $n_1 = n_2$, the generalization being obtained by shuffling the orders, when one multiplies multizeta values, of terms or of differential forms to write the expression as the sums of some multizeta values. Thus the span of multizeta values is an algebra.

For $s_i \in \mathbb{Z}_+$, we define multizeta values $\zeta(s_1, \ldots, s_r)$ by using the partial order on A_+ given by the degree, and grouping the terms according to it:

$$\zeta(s_1, \ldots, s_r) = \sum_{d_1 > \cdots > d_r \geq 0} S_{d_1}(s_1) \cdots S_{d_r}(s_r) = \sum \frac{1}{a_1^{s_1} \cdots a_r^{s_r}} \in K_\infty,$$

where the second sum is over $a_i \in A_{d_i+}$ satisfying the conditions as in the first sum.

We say that this multizeta value (or rather the tuple $(s_1, \ldots, s_r)$) has depth r and weight $\sum s_i$. Note that we do not need the $s_1 > 1$ condition for convergence as in the classical case.

Since there are many polynomials of a given degree (or norm), the usual proof of sum shuffle relations fails. In fact, it can be seen that naive analogs of sum or integral shuffle relations fail. The Euler identity $\zeta(2, 1) = \zeta(3)$ fails in our case, for the simple reason that degrees on both sides do not match.

On the other hand, we showed that any classical sum-shuffle relation with fixed s_i's works for q large enough.

More interestingly, we can show that a linear span of multizeta values is an algebra, by a completely different type of complicated 'shuffle relations' such as the following. If b is odd,

$$\zeta(2)\zeta(b) = \zeta(2+b) + \sum_{1\le i\le(b-3)/2} \zeta(2i+1, 1+b-2i),$$

and if b is even,

$$\zeta(2)\zeta(b) = \zeta(2+b) + \sum_{1\le i\le b/2-1} \zeta(2i, b+2-2i).$$

There are very interesting recursive recipes [T09b, Lr09, Lr10] which are conjectural. To give a flavor of combinatorics, we will just say that for q prime, given an expression of $\zeta(a)\zeta(b)$ as a sum of multizeta values, one can get an expression for $\zeta(a)\zeta(b + (q-1)p^m)$, where m is the smallest integer such that $a \le p^m$, by adding precisely described new multizeta terms $t_a = \prod(p-j)^{\mu_j}$, where μ_j is the number of j's in the base p expansion of $a-1$.

There are identities as above with $\mathbb{F}_p$ coefficients (understood in some precise sense) and identities (not yet well understood) with $\mathbb{F}_p(t)$ (or K) coefficients, such as $\zeta(1,2) = \zeta(3)/\ell_1 = \zeta(3)/(t-t^3)$ for $q = 3$, as analog of Euler's identity.

In any case, the period interpretation was provided in [AT09].

Theorem 7.5. *Given a multizeta value $\zeta(s_1, \ldots, s_r)$, we can construct explicitly an iterated extension of Carlitz–Tate t-motives over $\mathbb{F}_q[\theta]$ which has as period matrix entry this multizeta value (suitably normalized).*

Recall from Section 6.3 the polynomials $H_n(y) \in K[y] = \mathbb{F}_q(t)[y]$. We write them as $H_n(y,t)$ below.

Let $s = (s_1, \ldots, s_r) \in \mathbb{Z}_+^r$ and consider the matrices (the first one is a diagonal matrix)

$$D = [(t-\theta)^{s_1+\cdots+s_r}, (t-\theta)^{s_2+\cdots+s_r}, \ldots, (t-\theta)^{s_r}, 1],$$

$$Q = \begin{bmatrix} 1 & & & \\ H_{s_1-1}(\theta,t) & 1 & & \\ & & \ddots & \\ & & H_{s_r-1}(\theta,t) & 1 \end{bmatrix}.$$

Then $\Phi = Q^{(-1)}D$ defines the mixed t-motive of rank $r+1$ whose period matrix (obtained by specializing Ψ^{-1} at $t = \theta$) contains the depth r multizeta value $\zeta(s_1, \ldots, s_r)$ for $A = \mathbb{F}_q[\theta]$. For details of the calculation, we refer to the paper.

In this case, the situation is quite complicated and we are far from understanding the motivic Galois group or the corresponding transcendence degrees, unlike in the cases studied in the last lecture.

There is an interesting Galois side to this motivic study. In [ATp] we achieved the first step of constructing analogs of Ihara power series, Deligne–Soulé cocycles which connect zeta, gamma, Gauss sums, cyclotomic modules, etc. in towers and big Galois actions.

Update. In the recent preprint [Lrp] based on his PhD thesis work with the author, Alejandro Lara Rodríguez has proved many of the explicit conjectures [Lr09, Lr10, T09b] that the author and he made on multizeta relations with $\mathbb{F}_p$ coefficients by making the effective recipe of [T10] very explicit.

7.6 Cohomology approach

Taguchi–Wan [TW96, TW97] and Böckle–Pink [BöP, Böc02] have developed a nice cohomological approach getting determinantal formulas (see also [A00] and [T04, §5.11]) for zeta special values, developing much general L-functions, rationality results, meromorphic continuations, a criterion for analyticity, etc., by Dwork style or Crystalline methods respectively. Fortunately, we got a nice exposition about some of this in the parallel lecture series by Böckle. So we can just refer to those lecture notes.

Recently, Vincent Lafforgue [L09] proved some Bloch–Kato type results at positive integers using Fontaine–Perrin–Riou methods, and he also defined a good notion of extensions which inspired Taelman's work mentioned above. Soon after these CRM lecture series, Taelman applied these determinantal formula directly to the special value and proved his class number formula conjecture mentioned in Section 7.1.

7.7 Open questions

We collect some important open questions at the end. Overall the present results bring the special value theory of the arithmetic and geometric gamma functions and the zeta function for $A = \mathbb{F}_q[\theta]$ at the infinite place to a very satisfactory state. However, similar questions ar e still open for (i) v-adic interpolations (see [T04, §11.3] for very partial results about v-adic gamma values using automata methods and [Y91] for transcendence of v-adic zeta values); (ii) generalizations to other rings 'A' in the setting of Drinfeld modules [T04, 4.5, 8.3]; (iii) values of the two-variable gamma function of Goss [T04, 4.12], [T91a, §8], [G88].

Let us gather some other important open questions and projects, apart from generalizations, that we looked at.

(1) What is the arithmetic nature of $\Gamma(0)$ for the gamma function in Section 4.5?

(2) Which general conjectures can be stated for special values of L-functions generalizing Taelman? What are the arithmetic meanings and applications of new class modules, etc.? Which general conjectures are there about orders

of vanishing (good computational data can be easily obtained here with some effort), and what would be the arithmetic significance of new leading terms, or of non-trivial zeros of Goss zeta functions? Is the zeta non-vanishing at 'odd' integers a general phenomenon?

(3) Connection of the zeta values at the positive and negative integers with class groups suggests some connection between them. How can we make it more explicit?

(4) For $d_\infty > 1$, there is a question referred in Section 4.9 on non-vanishing of e_n's and equivalently whether $\xi^{(n)}$ does not belong to the support of the Drinfeld divisor V.

(5) Values of zeta for non-abelian extensions at positive values and whether there is a modular form connection as in the classical case. Better understanding of the zeta measure result is also highly desirable.

(6) Simplification of Sheats' proof of the zero distribution, by possible simple recursive in d relation on the valuations.

Guide to the literature. We give some comments on the references missing in the text. Good references for pre-2003 material are [G96, T04], which contain more details and an extensive bibliography. Except for the comments and developments after 2003 that are mentioned, and except for most of Lectures 3, 6, 7, the other material is based on [T04], which in turn is based on many original papers accessible from the author's home page.

In more detail, most of the material in the first two lectures can be found (except for citations there) in [T87, T88, T91a] or [T04, Ch. 4], that in Lecture 3 in [T99] or [T04, Ch. 7, 8] and that in the fourth lecture in [T96, T98, T93b] or [T04, Ch. 8, 11]. For Lecture 5, see [T95, G96] and [T04, Ch. 5] and references there. For Lecture 6, see [A86, ABP04], [T04, Ch. 7].

Having focused on relatively new results with a particular transcendence method and period connections, the references given here are limited. For surveys with many other references and overlapping results on transcendence, in particular, the Mahler method and the automata method, see [Wal90, Bro98, G96, Pe07], [T04, Ch. 10].

Note added. For many interesting related developments since the lectures were delivered, we refer to papers by Chang and Lara Rodríguez (multizeta relations and independence); the author (multizeta, congruences); Angles and Taelman (cyclotomic and class modules, Herbrand–Ribet, Vandiver analogs, L-values); Papanikolas (L-values); Pellarin and Perkins (deformations of L-values).

Acknowledgment. I thank the Centre de Recerca Matemàtica, Barcelona and its staff for providing nice atmosphere and facilities for the advanced course, Francesc Bars for organization and hospitality, the participants, and in particular Chris Hall and Ignazio Longhi for their comments which helped me improve the draft. I also thank Alejandro Lara Rodríguez for catching several typos.

Bibliography

[ABP04] G.W. Anderson, W.D. Brownawell and M.A. Papanikolas, Determination of the algebraic relations among special Γ-values in positive characteristic, *Ann. of Math.* (2), **160** (2004), 237–313.

[All87] J.-P. Allouche, Automates finis en théorie des nombres. *Exposition. Math.*, **5(3)** (1987), 239–266.

[All96] J.-P. Allouche, Transcendence of the Carlitz–Goss gamma function at rational arguments. *J. Number Theory*, **60(2)** (1996), 318–328.

[A86] G. Anderson, *t*-motives. *Duke Math. J.*, **53(2)** (1986), 457–502.

[A92] G.W. Anderson, A two-dimensional analogue of Stickelberger's theorem. In [G$^+$92], pages 51–73.

[A94] G.W. Anderson, Rank one elliptic A-modules and A-harmonic series. *Duke Math. J.*, **73(3)** (1994), 491–542.

[A96] G.W. Anderson, Log-algebraicity of twisted A-harmonic series and special values of L-series in characteristic p. *J. Number Theory*, **60(1)** (1996), 165–209.

[A00] G.W. Anderson, An elementary approach to L functions mod p. *J. Number Theory*, **80(2)** (2000), 291–303.

[A02] G.W. Anderson, Kronecker–Weber plus epsilon. *Duke Math. J.*, **114(3)** (2002), 439–475.

[A06] G.W. Anderson, A two-variable refinement of the Stark conjecture in the function field case, *Compositio Math.*, **142** (2006), 563–615.

[A07] G.W. Anderson, Digit patterns and the formal additive group, *Israel J. Math.*, **161** (2007), 125–139.

[Ang01] B. Anglès, On Gekeler's conjecture for function fields. *J. Number Theory*, **87(2)** (2001), 242–252.

[AS03] J.-P. Allouche and J. Shallit, *Automatic Sequences*. Cambridge University Press, Cambridge, 2003.

[AT90] G.W. Anderson and D.S. Thakur, Tensor powers of the Carlitz module and zeta values. *Ann. of Math.* (2), **132(1)** (1990), 159–191.

[AT09] G.W. Anderson, D.S. Thakur, Multizeta values for $F_q[t]$, their period interpretation and relations between them. *Internat. Math. Res. Notices*, IMRN, **11** (2009), 2038–2055.

[ATp]   G.W. Anderson and D.S. Thakur, Ihara power series for $\mathbb{F}_q[t]$. Preprint.

[Be06] F. Beukers, A refined version of the Siegel–Shidlovskii theorem, *Ann. of Math.* (2), **163(1)** (2006), 369–379.

[Bha97] M. Bhargava, P-orderings and polynomial functions on arbitrary subsets of Dedekind rings. *J. reine angew. Math.*, **490** (1997), 101–127.

[Bha00] M. Bhargava, The factorial function and generalizations. *Amer. Math. Monthly*, **107(9)** (2000), 783–799.

[Böc02] G. Böckle, Global L-functions over function fields. *Math. Ann.*, **323(4)** (2002), 737–795.

[BöP]   G. Böckle and R. Pink, *Cohomological Theory of Crystals over Function Fields*. European Mathematical Society Tracts in Mathematics, EMS, Zurich, 2009.

[BP02] W.D. Brownawell and M.A. Papanikolas, Linear independence of gamma values in positive characteristic. *J. reine angew. Math.*, **549** (2002), 91–148.

[Bro98] W.D. Brownawell, Transcendence in positive characteristic. In *Number theory* (*Tiruchirapalli*, 1996), vol. 210 of *Contemp. Math.*, pages 317–332. Amer. Math. Soc., Providence, RI, 1998.

[BT98] R.M. Beals and D.S. Thakur, Computational classification of numbers and algebraic properties. *Internat. Math. Res. Notices*, **15** (1998), 799–818.

[Car35] L. Carlitz, On certain functions connected with polynomials in a Galois field. *Duke Math. J.*, **1** (1935), 137–168.

[Ch09] C.-Y. Chang, A note on a refined version of Anderson–Brownawell–Papanikolas criterion, *J. Number Theory*, **129** (2009), 729–738.

[CP08] C.-Y. Chang and M.A. Papanikolas, Algebraic relations among periods and logarithms of rank 2 Drinfeld modules. Preprint (2008), `arXiv:0807.3157`.

[CPTY] C.-Y. Chang, M.A. Papanikolas, D.S. Thakur, and J. Yu, Algebraic independence of arithmetic gamma values and Carlitz zeta values. *Adv. Math.*, **223** (2010), 1137–1154.

[CPYa] C.-Y. Chang, M.A. Papanikolas, and J. Yu, Frobenius difference equations and algebraic independence of zeta values in positive equal characteristic. Preprint `arXiv:0804.0038v2`.

[CPYb] C.-Y. Chang, M.A. Papanikolas, and J. Yu, Geometric gamma values and zeta values in positive characterisitc. *Internat. Math. Res. Notices*, IMRN, **8** (2010), 1432–1455.

[CY]   C.-Y. Chang and J. Yu, Determination of algebraic relations among special zeta values in positive characteristic, *Adv. Math.*, **216** (2007), 321–345.

[Chr79] G. Christol, Ensembles presque periodiques k-reconnaissables. *Theoret. Comput. Sci.*, **9(1)** (1979), 141–145.

[CKMR80] G. Christol, T. Kamae, M. Mendès-France, and G. Rauzy, Suites algébriques, automates et substitutions. *Bull. Soc. Math. France*, **108(4)** (1980), 401–419.

[DMOS82] P. Deligne, J.S. Milne, A. Ogus, and K.-Yen Shih, *Hodge Cycles, Motives, and Shimura Varieties*, vol. 900 of *Lecture Notes in Mathematics*. Springer, Berlin, 1982.

[Dri74] V.G. Drinfel'd, Elliptic modules. *Mat. Sb.*, (N.S.), **94(136)** (1974), 594–627.

[Dri77a] V.G. Drinfel'd, Commutative subrings of certain noncommutative rings. *Funkcional. Anal. i Priložen.*, **11(1)** (1977), 11–14.

[DV96] J. Diaz-Vargas, Riemann hypothesis for $\mathbf{F}_p[T]$. *J. Number Theory*, **59(2)** (1996), 313–318.

[DV06] J. Diaz-Vargas, On zeros of characteristic p zeta functions, *J. Number Theory*, **117** (2006), 241–262.

[Fen97] K. Feng, Anderson's root numbers and Thakur's Gauss sums. *J. Number Theory*, **65(2)** (1997), 279–294.

[G$^+$92] D. Goss et al., eds., *The Arithmetic of Function Fields*, vol. 2 of *Ohio State University Mathematical Research Institute Publications*. Walter de Gruyter & Co., Berlin, 1992.

[G$^+$97] E.-U. Gekeler et al., eds., *Drinfeld Modules, Modular Schemes and Applications*. World Scientific Publishing Co. Inc., River Edge, NJ, 1997.

[Gek89a] E.-U. Gekeler, On the de Rham isomorphism for Drinfel'd modules. *J. reine angew. Math.*, **401** (1989), 188–208.

[Gek89b] E.-U. Gekeler, Quasi-periodic functions and Drinfel'd modular forms. *Compositio Math.*, **69(3)** (1989), 277–293.

[Gek90] E.-U. Gekeler, On regularity of small primes in function fields. *J. Number Theory*, **34(1)** (1990), 114–127.

[GK79] B.H. Gross and N. Koblitz, Gauss sums and the p-adic Γ-function. *Ann. of Math.* (2), **109(3)** (1979), 569–581.

[G79] D. Goss, v-adic zeta functions, L-series and measures for function fields. *Invent. Math.*, **55(2)** (1979), 107–119.

[G87] D. Goss, Analogies between global fields. In *Number Theory*, pages 83–114. Amer. Math. Soc., Providence, RI, 1987.

[G88] D. Goss, The Γ-function in the arithmetic of function fields. *Duke Math. J.*, **56(1)** (1988), 163–191.

[G92] D. Goss, L-series of t-motives and Drinfel'd modules. In [G$^+$92], pages 313–402.

[G94]   D. Goss, Drinfel'd modules: cohomology and special functions. In *Motives* (Seattle, WA, 1991), vol. 55 of *Proc. Sympos. Pure Math.*, pages 309–362. Amer. Math. Soc., Providence, RI, 1994.

[G96]   D. Goss, *Basic Structures of Function Field Arithmetic.* Springer, Berlin, 1996.

[G00]   D. Goss, A Riemann hypothesis for characteristic p L-functions. *J. Number Theory*, **82(2)** (2000), 299–322.

[G03]   D. Goss, The impact of the infinite primes on the Riemann hypothesis for characteristic p valued l-series. In *Algebra, Arithmetic and Geometry with Applications*, pages 357–380. Springer, Berlin, 2003.

[GR81a] S. Galovich and M. Rosen, The class number of cyclotomic function fields. *J. Number Theory*, **13(3)** (1981), 363–375.

[GR82] S. Galovich and M. Rosen, Units and class groups in cyclotomic function fields. *J. Number Theory*, **14(2)** (1982), 156–184.

[GS85] D. Goss and W. Sinnott, Class-groups of function fields. *Duke Math. J.*, **52(2)** (1985), 507–516.

[H74]   D.R. Hayes, Explicit class field theory for rational function fields. *Trans. Amer. Math. Soc.*, **189** (1974), 77–91.

[H79]   D.R. Hayes, Explicit class field theory in global function fields. In *Studies in Algebra and Number Theory*, pages 173–217. Academic Press, New York, 1979.

[H85]   D.R. Hayes, Stickelberger elements in function fields. *Compositio Math.*, **55(2)** (1985), 209–239.

[H92]   D.R. Hayes, A brief introduction to Drinfel'd modules. In [G+92], pages 1–32.

[I91]   Y. Ihara, Braids, Galois groups and some arithmetic functions, *Proceedings of the International Congress of Mathematicians, Kyoto*, 1990, pages 99–120, Math. Soc. Japan, Tokyo, 1991.

[Iwa69] K. Iwasawa, Analogies between number fields and function fields. In *Some Recent Advances in the Basic Sciences, vol. 2*, pages 203–208. Belfer Graduate School of Science, Yeshiva University, New York, 1969.

[J09]   S. Jeong, On a question of Goss, *J. Number Theory*, **129** (2009), 1912–1918.

[Kob80] N. Koblitz, *p-Adic Analysis: A Short Course on Recent Work.* Cambridge University Press, Cambridge, 1980.

[L09]   V. Lafforgue, Valeurs spéciales des fonctions L en caractéristique p, *J. Number Theory*, **129** (2009), 2600–2634.

[Lr09] A. Lara Rodríguez, Some conjectures and results about multizeta values for $F_q[t]$, Master's Thesis for The Autonomous University of Yucatan, May 7, 2009.

[Lr10]   A. Lara Rodríguez, Some conjectures and results about multizeta values for $F_q[t]$, *J. Number Theory*, **130** (2010), 1013–1023.

[Lrp]   A. Lara Rodríguez, Relations between multizeta values in characteristic p. Preprint.

[MFY97] M. Mendès-France and J. Yao, Transcendence and the Carlitz–Goss gamma function. *J. Number Theory*, **63(2)** (1997), 396–402.

[Mor75] Y. Morita, A p-adic analogue of the Γ-function. *J. Fac. Sci. Univ. Tokyo Sect. IA Math.*, **22(2)** (1975), 255–266.

[Mum78] D. Mumford, An algebro-geometric construction of commuting operators and of solutions to the Toda lattice equation, Korteweg deVries equation and related nonlinear equation. In *Proceedings of the International Symposium on Algebraic Geometry*, pages 115–153. Kinokuniya Book Store, Tokyo, 1978.

[Oka91] S. Okada, Kummer's theory for function fields. *J. Number Theory*, **38(2)** (1991), 212–215.

[P08]   M.A. Papanikolas, Tannakian duality for Anderson–Drinfeld motives and algebraic independence of Carlitz logarithms, *Invent. Math.*, **171** (2008), 123–174.

[Pin97] R. Pink, The Mumford–Tate conjecture for Drinfeld-modules. *Publ. Res. Inst. Math. Sci.*, **33(3)** (1997), 393–425.

[Poo95] B. Poonen, Local height functions and Mordell–Weil theorem for Drinfel′d modules. *Compositio Math.*, **97(3)** (1995), 349–368.

[PR03]   M.A. Papanikolas and N. Ramachandran, A Weil–Barsotti formula for Drinfeld modules. *J. Number Theory*, **98(2)** (2003), 407–431.

[Pe07]   F. Pellarin, Aspects d'indépendance algébrique en caractéristique non nulle, d'après Anderson, Brownawell, Denis, Papanikolas, Thakur, and Yu, *Séminaire Bourbaki*, no. 973, March 2007.

[She98] J.T. Sheats, The Riemann hypothesis for the Goss zeta function for $\mathbf{F}_q[T]$. *J. Number Theory*, **71(1)** (1998), 121–157.

[Sin97a] S.K. Sinha, Deligne's reciprocity for function fields. *J. Number Theory*, **63(1)** (1997), 65–88.

[Sin97b] S.K. Sinha, Periods of t-motives and transcendence. *Duke Math. J.*, **88(3)** (1997), 465–535.

[Ta10]   L. Taelman, A Dirichlet unit theorem for Drinfeld modules, *Math. Ann.*, **348(4)** (2010), 899–907.

[Tap1]   L. Taelman, *Special L-values of Drinfeld modules*. Preprint, `arXiv:1004.4304`.

[Tap2]   L. Taelman, *The Carlitz shtuka*. Preprint, `arXiv:1008.4234`.

[Tat84] J. Tate, *Les conjectures de Stark sur les fonctions L d'Artin en $s = 0$*. Birkhäuser Boston Inc., Boston, MA, 1984.

[T87] D.S. Thakur, Gauss sums and gamma functions for function fields and periods of Drinfeld modules. Thesis, Harvard University, 1987.

[T88] D.S. Thakur, Gauss sums for $\mathbf{F}_q[T]$. *Invent. Math.*, **94(1)** (1988), 105–112.

[T90] D.S. Thakur, Zeta measure associated to $\mathbf{F}_q[T]$. *J. Number Theory*, **35(1)** (1990), 1–17.

[T91a] D.S. Thakur, Gamma functions for function fields and Drinfel′d modules. *Ann. of Math.* (2), **134(1)** (1991), 25–64.

[T91b] D.S. Thakur, Gauss sums for function fields. *J. Number Theory*, **37(2)** (1991), 242–252.

[T92a] D.S. Thakur, On gamma functions for function fields. In [G$^+$92], pages 75–86.

[T92b] D.S. Thakur, Drinfel′d modules and arithmetic in the function fields. *Internat. Math. Res. Notices*, **9** (1992), 185–197.

[T93b] D.S. Thakur, Shtukas and Jacobi sums. *Invent. Math.*, **111(3)** (1993), 557–570.

[T94] D.S. Thakur, Iwasawa theory and cyclotomic function fields. In *Arithmetic Geometry (Tempe, AZ, 1993)*, vol. 174 of *Contemp. Math.*, pages 157–165. Amer. Math. Soc., Providence, RI, 1994.

[T95] D.S. Thakur, On characteristic p zeta functions. *Compositio Math.*, **99(3)** (1995), 231–247.

[T96] D.S. Thakur, Transcendence of gamma values for $\mathbf{F}_q[T]$. *Ann. of Math.* (2), **144(1)** (1996), 181–188.

[T98] D.S. Thakur, Automata and transcendence. In *Number Theory (Tiruchirapalli, 1996)*, vol. 210 of *Contemp. Math.*, pages 387–399. Amer. Math. Soc., Providence, RI, 1998.

[T99] D.S. Thakur, An alternate approach to solitons for $\mathbf{F}_q[t]$. *J. Number Theory*, **76(2)** (1999), 301–319.

[T01] D.S. Thakur, Integrable systems and number theory in finite characteristic. *Phys. D*, 152/153:1–8, 2001.

[T04] D.S. Thakur, *Function Field Arithmetic*. World Scientific, NJ, 2004.

[T09a] D.S. Thakur, Power sums with applications to multizeta and zeta zero distribution for $\mathbb{F}_q[t]$. *Finite Fields and their Applications*, **15** (2009), 534–552.

[T09b] D.S. Thakur, Relations between multizeta values for $F_q[t]$. *Internat. Math. Res. Notices*, IMRN, **11** (2009), 2038–2055.

[T10] D.S. Thakur, Shuffle relations for function field multizeta values, *Internat. Math. Res. Notices*, IMRN, to appear.

[Tp]   D.S. Thakur, Multizeta in function field arithmetic, to appear in Proceedings of Banff Conference 2009, to be published by the European Mathematical Society.

[Thi] A. Thiery, Indépendance algébrique des périodes et quasi-périodes d'un module de Drinfel'd. In [G⁺92], pages 265–284.

[TW96] Y. Taguchi and D. Wan, L-functions of ϕ-sheaves and Drinfeld modules. *J. Amer. Math. Soc.*, **9(3)** (1996), 755–781.

[TW97] Y. Taguchi and D. Wan, Entireness of L-functions of ϕ-sheaves on affine complete intersections. *J. Number Theory*, **63(1)** (1997), 170–179.

[Wal90] M. Waldschmidt, Transcendence problems connected with Drinfel'd modules. *İstanbul Üniv. Fen Fak. Mat. Derg.*, **49** (1990), 57–75 (1993).

[Wan93] D. Wan, Newton polygons of zeta functions and L functions. *Ann. of Math.* (2), **137(2)** (1993), 249–293.

[Wan96a] D. Wan, Meromorphic continuation of L-functions of p-adic representations. *Ann. of Math.* (2), **143(3)** (1996), 469–498.

[Y86] J. Yu, Transcendence and Drinfel'd modules. *Invent. Math.*, **83(3)** (1986), 507–517.

[Y89] J. Yu, Transcendence and Drinfel'd modules: several variables. *Duke Math. J.*, **58(3)** (1989), 559–575.

[Y90] J. Yu, On periods and quasi-periods of Drinfel'd modules. *Compositio Math.*, **74(3)** (1990), 235–245.

[Y91] J. Yu, Transcendence and special zeta values in characteristic p. *Ann. of Math.* (2), **134(1)** (1991), 1–23.

[Y92] J. Yu, Transcendence in finite characteristic. In [G⁺92], pages 253–264.

[Y97] J. Yu, Analytic homomorphisms into Drinfeld modules. *Ann. of Math.* (2), **145(2)** (1997), 215–233.

Curves and Jacobians
over Function Fields

Douglas Ulmer

Curves and Jacobians over Function Fields

Douglas Ulmer

Introduction

These notes originated in a 12-hour course of lectures given at the Centre de Recerca Matemàtica in February 2010. The aim of the course was to explain results on curves and their Jacobians over function fields, with emphasis on the group of rational points of the Jacobian, and to explain various constructions of Jacobians with large Mordell–Weil rank.

More so than the lectures, these notes emphasize foundational results on the arithmetic of curves and Jacobians over function fields, most importantly the breakthrough works of Tate, Artin, and Milne on the conjectures of Tate, Artin–Tate, and Birch and Swinnerton-Dyer. We also discuss more recent results such as those of Kato and Trihan. Constructions leading to high ranks are only briefly reviewed, because they are discussed in detail in other recent and forthcoming publications.

These notes may be viewed as a continuation of my Park City notes [70]. In those notes, the focus was on elliptic curves and finite constant fields, whereas here we discuss curves of high genera and results over more general base fields.

It is a pleasure to thank the organizers of the CRM Research Program, especially Francesc Bars, for their efficient organization of an exciting meeting, the audience for their interest and questions, the National Science Foundation for funding to support the travel of junior participants (grant DMS 0968709), and the editors for their patience in the face of many delays. It is also a pleasure to thank Marc Hindry, Nick Katz, and Dino Lorenzini for their help. Finally, thanks to Timo Keller and René Pannekoek for corrections.

I welcome all comments, and I plan to maintain a list of corrections and supplements on my web page. Please check there for updates if you find the material in these notes useful.

1   Dramatis personae

1.1   Notation and standing hypotheses

Unless explicitly stated otherwise, all schemes are assumed to be Noetherian and separated and all morphisms of schemes are assumed to be separated and of finite type.

A *curve* over a field F is a scheme over F that is reduced and purely of dimension 1, and a *surface* is similarly a scheme over F which is reduced and purely of dimension 2. Usually our curves and surfaces will be subject to further hypotheses, like irreducibility, projectivity, or smoothness.

We recall that a scheme Z is *regular* if each of its local rings is regular. This means that for each point $z \in Z$, with local ring $\mathcal{O}_{Z,z}$, maximal ideal $\mathfrak{m}_z \subset \mathcal{O}_{Z,z}$, and residue field $\kappa(z) = \mathcal{O}_{Z,z}/\mathfrak{m}_z$, we have

$$\dim_{\kappa(z)} \mathfrak{m}_z/\mathfrak{m}_z^2 = \dim_z Z.$$

Equivalently, $\mathfrak{m}_z$ should be generated by $\dim_z Z$ elements.

A morphism $f\colon Z \to S$ is *smooth* (of relative dimension n) at $z \in Z$ if there exist affine open neighborhoods U of z and V of $f(z)$ such that $f(U) \subset V$ and a diagram

$$
\begin{array}{ccc}
U & \longrightarrow & \operatorname{Spec} R[x_1,\ldots,x_{n+k}]/(f_1,\ldots,f_k) \\
f\downarrow & & \downarrow \\
V & \longrightarrow & \operatorname{Spec} R
\end{array}
$$

where the horizontal arrows are open immersions and where the Jacobian matrix $(\partial f_i/\partial x_j(z))_{ij}$ has rank k. Also, f is *smooth* if it is smooth at each $z \in Z$.

Smoothness is preserved under arbitrary change of base. If $f\colon Z \to \operatorname{Spec} F$ with F a field, then f smooth at $z \in Z$ implies that z is a regular point of Z. The converse holds if F is perfect, but fails in general. See Section 2.2 below for one example. A lucid discussion of smoothness, regularity, and the relations between them may be found in [45, Ch. V].

If Y is a scheme over a field F, the notation $\overline{Y}$ will mean $Y \times_F \overline{F}$ where $\overline{F}$ is an algebraic closure of F. Letting $F^{\mathrm{sep}} \subset \overline{F}$ be a separable closure of F, we will occasionally want the Galois group $\operatorname{Gal}(F^{\mathrm{sep}}/F)$ to act on objects associated to $\overline{Y}$. To do this, we note that the Galois action on F^{sep} extends uniquely to an action on $\overline{F}$. We also note that the étale cohomology groups $H^i(Y \times_F \overline{F}, \mathbb{Q}_\ell)$ and $H^i(Y \times_F F^{\mathrm{sep}}, \mathbb{Q}_\ell)$ are canonically isomorphic [39, VI.2.6].

1.2   Base fields

Throughout, k is a field, $\mathcal{C}$ is a smooth, projective, absolutely irreducible curve over k, and $K = k(\mathcal{C})$ is the field of rational functions on $\mathcal{C}$. Thus K is a regular extension of k (i.e., K/k is separable and k is algebraically closed in K) and the transcendence degree of K over k is 1. The function fields of our title are of the form $K = k(\mathcal{C})$.

We view the base field K as more-or-less fixed, except that we are willing to make a finite extension of k if it simplifies matters. At a few places below (notably in Sections 2.1 and 2.2), we extend k for convenience, to ensure for example that we have a rational point or smoothness.

We now introduce standing notations related to the field K. Places (equivalence classes of valuations, closed points of $\mathcal{C}$) will be denoted v and for each place v we write K_v, $\mathcal{O}_v$, $\mathfrak{m}_v$, and k_v for the completion of K, its ring of integers, its maximal ideal, and its residue field respectively.

Fix once and for all separable closures K^{sep} of K and K_v^{sep} of each K_v and embeddings $K^{\mathrm{sep}} \hookrightarrow K_v^{\mathrm{sep}}$. Let k^{sep} be the algebraic closure of k in K^{sep}; it is a separable closure of k and the embedding $K^{\mathrm{sep}} \hookrightarrow K_v^{\mathrm{sep}}$ identifies k^{sep} with the residue field of K_v^{sep}.

We write G_K for $\mathrm{Gal}(K^{\mathrm{sep}}/K)$ and similarly for G_{K_v} and G_k. The embeddings $K^{\mathrm{sep}} \hookrightarrow K_v^{\mathrm{sep}}$ identify G_{K_v} with a decomposition group of K at v. We also write $D_v \subset G_K$ for this subgroup, and $I_v \subset D_v$ for the corresponding inertia group.

1.3 The curve X

Throughout, X will be a curve over K which is always assumed to be smooth, projective, and absolutely irreducible. Thus $K(X)$ is a regular extension of K of transcendence degree 1.

The genus of X will be denoted g_X, and since we are mostly interested in the arithmetic of the Jacobian of X, we always assume that $g_X > 0$.

We do not assume that X has a K-rational point. More quantitatively, we let δ denote the *index* of X, i.e., the gcd of the degrees of the residue field extensions $\kappa(x)/K$ as x runs over all closed points of X. Equivalently, δ is the smallest positive degree of a K-rational divisor on X. We write δ' for the *period* of X, i.e., the smallest degree of a K-rational divisor class. (In fancier language, δ' is the order of Pic^1 as an element of the Weil–Châtelet group of J_X. It is clear that δ' divides δ. It is also easy to see (by considering the divisor of a K-rational regular 1-form) that $\delta | (2g_X - 2)$. Lichtenbaum [29, Theorem 8] showed that $\delta | 2\delta'^2$, and if $(2g_X - 2)/\delta$ is even, then $\delta | \delta'^2$.

Similarly, for a closed point v of $\mathcal{C}$, we write δ_v and δ_v' for the index and period of $X \times_K K_v$. It is known that $\delta_v' | (g_X - 1)$, the ratio δ/δ' is either 1 or 2, and it is 1 if $(g_X - 1)/\delta'$ is even. (These facts follow from the arguments in [29, Theorem 7] together with the duality results in [40].)

1.4 The surface $\mathcal{X}$

Given X, there is a unique surface $\mathcal{X}$ equipped with a morphism $\pi: \mathcal{X} \to \mathcal{C}$ with the following properties: $\mathcal{X}$ is irreducible and regular, π is proper and relatively minimal (defined below), and the generic fiber of $\mathcal{X} \to \mathcal{C}$ is $X \to \mathrm{Spec}\, K$.

Provided that we are willing to replace k with a finite extension, we may also insist that $\mathcal{X} \to \operatorname{Spec} k$ be smooth. We generally make this extension, and also if necessary extend k so that $\mathcal{X}$ has a k-rational point.

The construction of $\mathcal{X}$ starting from X and discussion of its properties (smoothness, cohomological flatness, etc.) will be given in Section 2 below.

We note that any smooth projective surface $\mathcal{S}$ over k is closely related to a surface $\mathcal{X}$ of our type. Indeed, after possibly blowing up a finite number of points, $\mathcal{S}$ admits a morphism to $\mathbb{P}^1$ whose generic fiber is a curve of our type (except that it might have genus 0). Thus an alternative point of view would be to start with the surface $\mathcal{X}$ and construct the curve X. We prefer to start with X because specifying a curve over a field can be done succinctly by specifying its field of functions.

1.5   The Jacobian J_X

We write J_X for the Jacobian of X, a g-dimensional, principally polarized abelian variety over K. It represents the relative Picard functor $\underline{\operatorname{Pic}}^0_{X/K}$. The definition of this functor and results on its representability are reviewed in Section 3.

We denote by (B, τ) the K/k-trace of J_X. By definition, this is the final object in the category of pairs (A, σ) where A is an abelian variety over k and $\sigma \colon A \times_k K \to J_X$ is a K-morphism of abelian varieties. We refer to [11] for a very complete account of the K/k-trace in the language of schemes. We will calculate (B, τ) (completely in a special case, and up to inseparable isogeny in general) in Subsection 4.2.

One of the main aims of these notes is to discuss the arithmetic of J_X, especially the Mordell–Weil group $J_X(K)/\tau B(k)$ and the Tate–Shafarevich group $\text{Ш}(J_X/K)$ as well as their connections with analogous invariants of $\mathcal{X}$.

1.6   The Néron model $\mathcal{J}_X$

We denote the Néron model of J_X over $\mathcal{C}$ by $\mathcal{J}_X \to \mathcal{C}$, so that $\mathcal{J}_X \to \mathcal{C}$ is a smooth group scheme satisfying a universal property. More precisely, for every place v of K, every K_v-valued point of J_X should extend uniquely to a section of $\mathcal{J}_X \times_{\mathcal{C}} \operatorname{Spec} \mathcal{O}_v \to \operatorname{Spec} \mathcal{O}_v$. We refer to [3] for a brief overview and [9] for a thorough treatment of Néron models.

There are many interesting results to discuss about $\mathcal{J}_X$, including a fine study of the component groups of its reduction at places of $\mathcal{C}$, monodromy results, etc. Due to constraints of time and space, we will not be able to discuss any of these, and so we will have nothing more to say about $\mathcal{J}_X$ in these notes.

1.7   Our plan

Our goal is to discuss the connections between the objects X, $\mathcal{X}$, and J_X. Specifically, we will study the arithmetic of J_X (its rational points, Tate–Shafarevich group, *BSD* conjecture), the arithmetic of $\mathcal{X}$ (its Néron–Severi group, Brauer group, Tate and Artin–Tate conjectures), and connections between them.

In Sections 2 and 3 we discuss the construction and first properties of $\mathcal{X} \to \mathcal{C}$ and J_X. In the following three sections, we work out the connections between the arithmetic of these objects, mostly in the case when k is finite, with the *BSD* and Tate conjectures as touchstones. In Section 7 we give a few complements on related situations and other ground fields. In Section 8 we review known cases of the Tate conjecture and their consequences for Jacobians, and in Section 9 we review how one may use these results to produce Jacobians with large Mordell–Weil groups.

2 Geometry of $\mathcal{X} \to \mathcal{C}$

2.1 Construction of $\mathcal{X}$

Recall that we have fixed a smooth, projective, absolutely irreducible curve X over $K = k(\mathcal{C})$ of genus $g_X > 0$.

Proposition 2.1. *There exists a regular, irreducible surface $\mathcal{X}$ over k equipped with a morphism $\pi \colon \mathcal{X} \to \mathcal{C}$ which is projective, relatively minimal, and with generic fiber $X \to \operatorname{Spec} K$. The pair $(\mathcal{X}, \pi)$ is uniquely determined by these properties. We have that $\mathcal{X}$ is absolutely irreducible and projective over k and that π is flat, $\pi_*\mathcal{O}_\mathcal{X} = \mathcal{O}_\mathcal{C}$, and π has connected fibers.*

Proof. To show the existence of $\mathcal{X}$, we argue as follows. Choose a non-empty affine open subset $U \subset X$ and an affine model for U:

$$U = \operatorname{Spec} K[x_1, \ldots, x_m]/(f_1, \ldots, f_n).$$

Let $\mathcal{C}^0 \subset \mathcal{C}$ be a non-empty affine open where all of the coefficients of the f_i (which are elements of K) are regular functions. Let $R = \mathcal{O}_\mathcal{C}(\mathcal{C}^0)$, a k-algebra and Dedekind domain. Let $\mathcal{U}$ be the affine k-scheme

$$\mathcal{U} = \operatorname{Spec} R[x_1, \ldots, x_m]/(f_1, \ldots, f_n).$$

Then $\mathcal{U}$ is reduced and irreducible, and the inclusion

$$R \longrightarrow R[x_1, \ldots, x_m]/(f_1, \ldots, f_n)$$

induces a morphism $\mathcal{U} \to \mathcal{C}^0$ whose generic fiber is $U \to \operatorname{Spec} K$. Imbed $\mathcal{U}$ in some affine space $\mathbb{A}_k^N$ and let $\mathcal{X}_0$ be the closure of $\mathcal{U}$ in $\mathbb{P}_k^N$. Thus $\mathcal{X}_0$ is reduced and irreducible, but it may be quite singular.

Lipman's general result on desingularizing two-dimensional schemes (see [32] or [2]) can be used to find a non-singular model of $\mathcal{X}_0$. More precisely, normalizing $\mathcal{X}_0$ results in a scheme with isolated singularities. Let $\mathcal{X}_1$ be the result of blowing up the normalization of $\mathcal{X}_0$ once at each (reduced) closed singular point. Now inductively let $\mathcal{X}_n$ $(n \geq 2)$ be the result of normalizing $\mathcal{X}_{n-1}$ and blowing up each singular point. Lipman's theorem is that the sequence $\mathcal{X}_n$ yields a regular scheme after finitely many steps. The resulting regular projective scheme $\mathcal{X}_n$ is equipped with a rational map to $\mathcal{C}$.

After further blowing up and/or blowing down, we arrive at an irreducible, regular, two-dimensional scheme projective over k with a projective, relatively minimal morphism $\mathcal{X} \to \mathcal{C}$ whose generic fiber is $X \to \operatorname{Spec} K$. Here relatively minimal means that if $\mathcal{X}'$ is regular with a proper morphism $\mathcal{X}' \to \mathcal{C}$, and if there is a factorization

$$\mathcal{X} \xrightarrow{f} \mathcal{X}' \longrightarrow \mathcal{C}$$

with f a birational proper morphism, then f is an isomorphism. This is equivalent to there being no (-1)-curves in the fibers of π. The uniqueness of $\mathcal{X}$ (subject to the properties) follows from [28, Theorem 4.4]. Since $\mathcal{C}$ is a smooth curve, $\mathcal{X}$ is irreducible, and π is dominant, π is flat.

If k' is an extension of k over which $\mathcal{X}$ becomes reducible, then every component of $\mathcal{X} \times_k k'$ dominates $\mathcal{C} \times_k k'$. (This is because π is flat, so $\mathcal{X} \times_k k' \to \mathcal{C} \times_k k'$ is flat.) In this case, X would be reducible over $k'K$. But we assumed that X is absolutely irreducible, so $\mathcal{X}$ must also be absolutely irreducible.

By construction $\mathcal{X}$ is projective over k.

Let $\mathcal{C}'$ be $\operatorname{Spec}_{\mathcal{O}_{\mathcal{C}}} \pi_* \mathcal{O}_{\mathcal{X}}$ (global spec) so that the Stein factorization of π is $\mathcal{X} \to \mathcal{C}' \to \mathcal{C}$. Then $\mathcal{C}' \to \mathcal{C}$ is finite and flat. Let $\mathcal{C}'_\eta = \operatorname{Spec} L \to \eta = \operatorname{Spec} K$ be the generic fiber of $\mathcal{C}' \to \mathcal{C}$. Then the algebra L is finite and thus algebraic over K. On the other hand, L is contained in $k(\mathcal{X})$ and K is algebraically closed in $k(\mathcal{X})$, so we have $L = K$. Thus $\mathcal{C}' \to \mathcal{C}$ is finite flat of degree 1, and, since $\mathcal{C}$ is smooth, it must be an isomorphism. This proves that $\pi_* \mathcal{O}_{\mathcal{X}} = \mathcal{O}_{\mathcal{C}}$. It follows (e.g., by [20, III.11.3]) that the fibers of π are all connected.

This completes the proof of the proposition. $\square$

For the rest of the notes, $\pi \colon \mathcal{X} \to \mathcal{C}$ will be the fibration constructed in the proposition. We will typically extend k if necessary so that $\mathcal{X}$ has a k-rational point.

2.2 Smoothness and regularity

If k is perfect, then, since $\mathcal{X}$ is regular, $\mathcal{X} \to \operatorname{Spec} k$ is automatically smooth. However, it need not be smooth if k is not perfect.

Let us consider an example. Since smoothness and regularity are local properties, we give an affine example and leave to the reader the easy task of making it projective. Let $\mathbb{F}$ be a field of characteristic $p > 0$ and let $\tilde{\mathcal{X}}$ be the closed subset of $\mathbb{A}^4_{\mathbb{F}}$ defined by $y^2 + xy - x^3 - (t^p + u)$. The projection $(x, y, t, u) \mapsto (x, y, t)$ induces an isomorphism $\tilde{\mathcal{X}} \to \mathbb{A}^3$ and so $\tilde{\mathcal{X}}$ is a regular scheme. Let $k = \mathbb{F}(u)$ and let $\mathcal{X} \to \operatorname{Spec} k$ be the generic fiber of the projection $\tilde{\mathcal{X}} \to \mathbb{A}^1$, $(x, y, t, u) \mapsto u$. Since the local rings of $\mathcal{X}$ are also local rings of $\tilde{\mathcal{X}}$, $\mathcal{X}$ is a regular scheme. On the other hand, $\mathcal{X} \to \operatorname{Spec} k$ is not smooth at the point $x = y = t = 0$. Now let X be the generic fiber of the projection $\mathcal{X} \to \mathbb{A}^1$, $(x, y, t) \mapsto t$. Then X is the affine scheme $\operatorname{Spec} K[x, y]/(y^2 + xy - x^3 - (t^p - u))$ and this is easily seen to be smooth over K.

This shows that the minimal regular model $\mathcal{X} \to \mathcal{C}$ of $X \to \operatorname{Spec} K$ need not be smooth over k.

If we are given $X \to \operatorname{Spec} K$ and the regular model $\mathcal{X}$ is not smooth over k, we can remedy the situation by extending k. Indeed, let $\bar{k}$ denote the algebraic closure of k, and let $\overline{\mathcal{X}} \to \overline{\mathcal{C}}$ be the regular model of $X \times_k \bar{k} \to K\bar{k}$. Then since $\bar{k}$ is perfect, $\overline{\mathcal{X}}$ is smooth over $\bar{k}$. It is clear that there is a finite extension k' of k such that $\overline{\mathcal{X}}$ is defined over k' and birational over k' to $\mathcal{X}$. So replacing k by k' we find that the regular model $\mathcal{X} \to \mathcal{C}$ of $X \to \operatorname{Spec} K$ is smooth over k.

We will generally assume below that $\mathcal{X}$ is smooth over k.

2.2.1 Correction to [70]

In [70, Lecture 2, §2, p. 237], speaking of a two-dimensional, separated, reduced scheme of finite type over a field k, we say "Such a scheme is automatically quasi-projective and is projective if and only if it is complete." This is not correct in general – we should also assume that $\mathcal{X}$ is non-singular. In fact, when the ground field is finite, it suffices to assume that $\mathcal{X}$ is normal: see [15].

2.3 Structure of fibers

We write $\mathcal{X}_v$ for the fiber of π over the closed point v of $\mathcal{C}$. We already noted that the fibers $\mathcal{X}_v$ are connected.

We next define certain multiplicities of components, following [9, Ch. 9]. Let $\mathcal{X}_{v,i}$, $i = 1, \ldots, r$ be the reduced irreducible components of $\mathcal{X}_v$ and let $\eta_{v,i}$ be the corresponding generic points of $\mathcal{X}_v$. Let $\bar{k}_v$ be an algebraic closure of the residue field at v and write $\overline{\mathcal{X}}_v$ for $\mathcal{X}_v \times_{k_v} \bar{k}_v$ and $\bar{\eta}_{v,i}$ for a point of $\overline{\mathcal{X}}_v$ over $\eta_{v,i}$. We define the *multiplicity* of $X_{v,i}$ in X_v to be the length of the Artin local ring $\mathcal{O}_{X_v, \eta_{v,i}}$, and the *geometric multiplicity* of $X_{v,i}$ in X_v to be the length of $\mathcal{O}_{\overline{X}_v, \bar{\eta}_{v,i}}$. The geometric multiplicity is equal to the multiplicity when the characteristic of k is zero and it is a power of p times the multiplicity if the characteristic of k is $p > 0$.

We write $\mathcal{X}_v = \sum_i m_{v,i} \mathcal{X}_{v,i}$ where $m_{v,i}$ is the multiplicity of $\mathcal{X}_{v,i}$ in $\mathcal{X}_v$. This is an equality of Cartier divisors on $\mathcal{X}$. We define the *multiplicity* m_v of the fiber $\mathcal{X}_v$ to be the gcd of the multiplicities $m_{v,i}$.

The multiplicity m_v divides the gcd of the geometric multiplicities of the components of $\mathcal{X}_v$ which in turn divides the index δ_v of $\mathcal{X}_v$. In particular, if X has a K-rational point (so that $\mathcal{X} \to \mathcal{C}$ has a section) then for every v we have $m_v = 1$.

We now turn to the combinatorial structure of the fiber $\mathcal{X}_v$. A convenient reference for what follows is [33, Ch. 9].

We write $D.D'$ for the intersection multiplicity of two divisors on $\mathcal{X}$. It is known [33, 9.1.23] that the intersection form restricted to the divisors supported in a single fiber $\mathcal{X}_v$ is negative semi-definite, and that its kernel consists exactly of the divisors which are rational multiples of the entire fiber. (Thus if the multiplicity of the fiber $\mathcal{X}_v$ is m_v, then $(1/m_v)\mathcal{X}_v := \sum_i (m_{v,i}/m_v)\mathcal{X}_{v,i}$ generates the kernel of the pairing.)

It is in principle possible to use this result to give a classification of the possible combinatorial types of fibers (genera and multiplicities of components, intersection numbers) for a fixed value of g_X. Up to a suitable equivalence relation, the set of possibilities for a given value of g_X is finite [5]. When X is an elliptic curve and the residue field is assumed perfect, this is the well-known Kodaira–Néron classification. For higher genus, the situation rapidly becomes intractable. We note that the list of possibilities can be strictly longer when one does not assume that the residue field k_v is perfect. See [59] for a complete analysis of the case where X is an elliptic curve.

2.4   Leray spectral sequence

Fix a prime $\ell \neq \operatorname{Char}(k)$. We consider the Leray spectral sequence for $\pi \colon \overline{\mathcal{X}} \to \overline{\mathcal{C}}$ in ℓ-adic cohomology. The E_2 term

$$E_2^{pq} = H^p(\overline{\mathcal{C}}, R^q \pi_* \mathbb{Q}_\ell)$$

vanishes outside the range $0 \leq p, q \leq 2$ and so the only possibly non-zero differentials are d_2^{01} and d_2^{02}. We will show that these both vanish and so the sequence degenerates at E_2.

The differential d_2^{01} sits in an exact sequence of low degree terms that includes

$$H^0(\overline{\mathcal{C}}, R^1 \pi_* \mathbb{Q}_\ell) \xrightarrow{d_2^{01}} H^2(\overline{\mathcal{C}}, \pi_* \mathbb{Q}_\ell) \longrightarrow H^2(\overline{\mathcal{X}}, \mathbb{Q}_\ell).$$

Now $\pi_* \mathbb{Q}_\ell = \mathbb{Q}_\ell$, and the edge morphism $H^2(\overline{\mathcal{C}}, \mathbb{Q}_\ell) \to H^2(\overline{\mathcal{X}}, \mathbb{Q}_\ell)$ is simply the pull-back π^*. If $i \colon \mathcal{D} \hookrightarrow \mathcal{X}$ is a multisection of degree n and $j = \pi i$, then we have a factorization of j^*:

$$H^2(\overline{\mathcal{C}}, \mathbb{Q}_\ell) \longrightarrow H^2(\overline{\mathcal{X}}, \mathbb{Q}_\ell) \longrightarrow H^2(\overline{\mathcal{D}}, \mathbb{Q}_\ell)$$

as well as a trace map

$$H^2(\overline{\mathcal{D}}, \mathbb{Q}_\ell) \longrightarrow H^2(\overline{\mathcal{C}}, \mathbb{Q}_\ell)$$

for the finite morphism $j \colon \mathcal{D} \to \mathcal{C}$. The composition j^* followed by the trace map is just multiplication by n, which is injective, and therefore $H^2(\overline{\mathcal{C}}, \mathbb{Q}_\ell) \to H^2(\overline{\mathcal{X}}, \mathbb{Q}_\ell)$ is also injective, which implies that $d_2^{01} = 0$.

Now consider d_2^{02}, which sits in an exact sequence

$$H^2(\overline{\mathcal{X}}, \mathbb{Q}_\ell) \longrightarrow H^0(\overline{\mathcal{C}}, R^2 \pi_* \mathbb{Q}_\ell) \xrightarrow{d_2^{02}} H^2(\overline{\mathcal{C}}, R^1 \pi_* \mathbb{Q}_\ell).$$

We can deduce that it too is zero by using duality, or by the following argument, which unfortunately uses terminology not introduced until Section 5. The careful reader will have no trouble checking that there is no circularity.

Here is the argument: Away from the reducible fibers of π, the sheaf $R^2\pi_*\mathbb{Q}_\ell$ is locally constant of rank 1. At a closed point $\overline{v}$ of $\overline{\mathcal{C}}$ where the fiber of π is reducible, the stalk of $R^2\pi_*\mathbb{Q}_\ell$ has rank $f_{\overline{v}}$, the number of components in the fiber. The cycle class of a component of a fiber in $H^2(\overline{\mathcal{X}}, \mathbb{Q}_\ell)$ maps onto the corresponding section in $H^0(\overline{\mathcal{C}}, R^2\pi_*\mathbb{Q}_\ell)$ and so $H^2(\overline{\mathcal{X}}, \mathbb{Q}_\ell) \to H^0(\overline{\mathcal{C}}, R^2\pi_*\mathbb{Q}_\ell)$ is surjective. This implies that $d_2^{02} = 0$, as desired.

For later use, we record the exact sequence of low degree terms (where the zero on the right is because $d_2^{01} = 0$):

$$0 \longrightarrow H^1(\overline{\mathcal{C}}, \mathbb{Q}_\ell) \longrightarrow H^1(\overline{\mathcal{X}}, \mathbb{Q}_\ell) \longrightarrow H^0(\overline{\mathcal{C}}, R^1\pi_*\mathbb{Q}_\ell) \longrightarrow 0. \qquad (2.1)$$

We end this subsection by noting a useful property of $R^1\pi_*\mathbb{Q}_\ell$ when k is finite.

Proposition 2.2. *Suppose that k is finite and $\mathcal{X}$ is a smooth, proper surface over k equipped with a flat, generically smooth, proper morphism $\pi\colon \mathcal{X} \to \mathcal{C}$. Let ℓ be a prime not equal to the characteristic of k and let $\mathcal{F} = R^1\pi_*\mathbb{Q}_\ell$, a constructible ℓ=adic sheaf on $\mathcal{C}$. If $j\colon \eta \hookrightarrow \mathcal{C}$ is the inclusion of the generic point, then the canonical morphism $\mathcal{F} \to j_*j^*\mathcal{F}$ is an isomorphism.*

Remark 2.3. The same proposition holds if we let $j\colon U \hookrightarrow \mathcal{C}$ be the inclusion of a non-empty open subset over which π is smooth. Sheaves with this property are sometimes called "middle extension" sheaves. It is also useful to note that for $\mathcal{F}$ and j as above, we have an isomorphism

$$H^1(\overline{\mathcal{C}}, \mathcal{F}) \cong \operatorname{Im}\left(H^1_c(\overline{U}, j^*\mathcal{F}) \longrightarrow H^1(\overline{U}, j^*\mathcal{F})\right)$$

(image of compactly supported cohomology in usual cohomology). This is "well known" but the only reference I know is [67, 7.1.6].

Proof of Proposition 2.2. We will show that for every geometric point $\overline{x}$ over a closed point x of $\mathcal{C}$, the stalks of $\mathcal{F}$ and $j_*j^*\mathcal{F}$ are isomorphic. (Note that the latter is the group of invariants of inertia at x in the Galois module $\mathcal{F}_{\overline{\eta}}$.)

The local invariant cycle theorem (Theorem 3.6.1 of [13]) says that the map of stalks $\mathcal{F}_{\overline{x}} \to (j_*j^*\mathcal{F})_{\overline{x}}$ is surjective. A closer examination of the proof shows that it is also injective in our situation. Indeed, in the diagram (8) on p. 214 of [13], the group on the left $H^0(X_{\overline{\eta}})_I(-1)$ is pure of weight 2, whereas $H^1(X_s)$ is mixed of weight ≤ 1; also the preceding term in the vertical sequence is (with $\mathbb{Q}_\ell$ coefficients) dual to $H^3(X_s)(2)$ and this vanishes for dimension reasons. Thus the diagonal map sp^* is also injective.

(Note the paragraph after the display (8): this argument only works with $\mathbb{Q}_\ell$ coefficients, not necessarily with $\mathbb{Z}_\ell$ coefficients.) $\qquad\square$

See [25, Proposition 7.5.2] for a more general result with a similar proof.

We have a slightly weaker result in integral cohomology.

Corollary 2.4. *Under the hypotheses of the proposition, the natural map*

$$R^1\pi_*\mathbb{Z}_\ell \longrightarrow j_*j^*R^1\pi_*\mathbb{Z}_\ell$$

has kernel and cokernel supported at finitely many closed points of $\mathcal{C}$ and with finite stalks. Thus the induced map

$$H^i\big(\overline{C}, R^1\pi_*\mathbb{Z}_\ell\big) \longrightarrow H^i\big(\overline{C}, j_*j^*R^1\pi_*\mathbb{Z}_\ell\big)$$

is an isomorphism for $i > 1$ and surjective with finite kernel for $i = 1$.

Proof. The proof of the Proposition shows that the kernel and cokernel are supported at points where π is not smooth, a finite set of closed points. Also, the Proposition shows that the stalks are torsion. Since they are also finitely generated, they must be finite. $\qquad\square$

2.5   Cohomological flatness

Since $\pi\colon \mathcal{X} \to \mathcal{C}$ is flat, general results on cohomology and base change (e.g., [20, III.12]) imply that the coherent Euler characteristic of the fibers of π is constant, i.e., the function $v \mapsto \chi(\mathcal{X}_v, \mathcal{O}_{\mathcal{X}_v})$ is constant on $\mathcal{C}$. Moreover, the dimensions of the individual cohomology groups $h^i(\mathcal{X}_v, \mathcal{O}_{\mathcal{X}_v})$ are upper semi-continuous. They are not in general locally constant.

To make this more precise, we recall a standard exact sequence from the theory of cohomology and base change:

$$0 \longrightarrow \big(R^i\pi_*\mathcal{O}_\mathcal{X}\big) \otimes_{\mathcal{O}_\mathcal{C}} \kappa(v) \longrightarrow H^i\big(\mathcal{X}_v, \mathcal{O}_{\mathcal{X}_v}\big) \longrightarrow \big(R^{i+1}\pi_*\mathcal{O}_\mathcal{X}\big)\big[\varpi_v\big] \longrightarrow 0.$$

Here $\mathcal{X}_v$ is the fiber of π at v, the left-hand group is the fiber of the coherent sheaf $R^i\pi_*\mathcal{O}_\mathcal{X}$ at v, and the right-hand group is the ϖ_v-torsion in $R^{i+1}\pi_*\mathcal{O}_\mathcal{X}$ where ϖ_v is a generator of $\mathfrak{m}_v$. Since $R^i\pi_*\mathcal{O}_\mathcal{X}$ is coherent, the function

$$v \longmapsto \dim_{\kappa(v)} \big(R^i\pi_*\mathcal{O}_\mathcal{X}\big) \otimes_{\mathcal{O}_\mathcal{C}} \kappa(v)$$

is upper semi-continuous, and it is locally constant if and only if $R^i\pi_*\mathcal{O}_\mathcal{X}$ is locally free. Thus the obstruction to $v \mapsto h^i(\mathcal{X}_v, \mathcal{O}_{\mathcal{X}_v})$ being locally constant is controlled by torsion in $R^i\pi_*\mathcal{O}_\mathcal{X}$ and $R^{i+1}\pi_*\mathcal{O}_\mathcal{X}$.

We say that π is *cohomologically flat in dimension i* if formation of $R^i\pi_*\mathcal{O}_\mathcal{X}$ commutes with arbitrary change of base, i.e., if for all $\phi\colon T \to \mathcal{C}$ the base change morphism

$$\phi^*R^i\pi_*\mathcal{O}_\mathcal{X} \longrightarrow R^i\pi_T\mathcal{O}_{T\times_\mathcal{C}\mathcal{X}}$$

is an isomorphism. Because the base $\mathcal{C}$ is a smooth curve, this is equivalent to the same condition where $T \to \mathcal{C}$ runs through inclusions of closed point, i.e., to the condition that

$$\big(R^i\pi_*\mathcal{O}_\mathcal{X}\big) \otimes_{\mathcal{O}_\mathcal{C}} \kappa(v) \longrightarrow H^i\big(\mathcal{X}_v, \mathcal{O}_{\mathcal{X}_v}\big)$$

be an isomorphism for all closed points $v \in \mathcal{C}$. By the exact sequence above, this is equivalent to $R^{i+1}\pi_*\mathcal{O}_\mathcal{X}$ being torsion free, and thus locally free.

Since $\pi\colon \mathcal{X} \to \mathcal{C}$ has relative dimension 1, $R^{i+1}\pi_*\mathcal{O}_\mathcal{X} = 0$ for $i \geq 1$ and π is automatically cohomologically flat in dimension $i \geq 1$. It is cohomologically flat in dimension 0 if and only if $R^1\pi_*\mathcal{O}_\mathcal{X}$ is locally free. To lighten terminology, in this case we simply say that π is cohomologically flat.

Since $\pi_*\mathcal{O}_\mathcal{X} = \mathcal{O}_\mathcal{C}$ is free of rank 1, we have that π is cohomologically flat if and only if $v \mapsto h^i(\mathcal{X}_v, \mathcal{O}_{\mathcal{X}_v})$ is locally constant (and thus constant since $\mathcal{C}$ is connected) for $i = 0, 1$. Obviously the common value is 1 for $i = 0$ and $g_\mathcal{X}$ for $i = 1$.

Raynaud gave a criterion for cohomological flatness in [50, 7.2.1]. Under our hypotheses ($\mathcal{X}$ regular, $\mathcal{C}$ a smooth curve over k, and $\pi\colon \mathcal{X} \to \mathcal{C}$ proper and flat with $\pi_*\mathcal{O}_\mathcal{X} = \mathcal{O}_\mathcal{C}$), π is cohomologically flat if $\mathrm{Char}(k) = 0$ or if k is perfect and the following condition holds: For each closed point v of $\mathcal{C}$, let d_v be the gcd of the geometric multiplicities of the components of $\mathcal{X}_v$. The condition is that d_v is prime to $p = \mathrm{Char}(k)$ for all v.

Raynaud also gave an example of non-cohomological flatness which we will make more explicit below. Namely, let S be a complete DVR with fraction field F and algebraically closed residue field of characteristic $p > 0$, and let $\mathcal{E}$ be an elliptic curve over S with either multiplicative or good supersingular reduction. Let Y be a principal homogeneous space for $E = \mathcal{E} \times \mathrm{Spec}\, F$ of order p^e ($e > 0$) and let $\mathcal{Y}$ be a minimal regular model for Y over S. Then $\mathcal{Y} \to \mathrm{Spec}\, S$ is not cohomologically flat. Moreover, the invariant δ defined as above is p^e. It follows from later work [34, Theorem 6.6] that the special fiber of $\mathcal{Y}$ is like that of $\mathcal{E}$, but with multiplicity p^e. The explicit example below should serve to make the meaning of this clear.

2.6 Example

Let $k = \mathbb{F}_2$ and $\mathcal{C} = \mathbb{P}^1$, so that $K = \mathbb{F}_2(t)$. We give an example of a curve of genus 1 over $\mathbb{A}^1_K$ which is not cohomologically flat at $t = 0$.

Consider the elliptic curve E over K given by

$$y^2 + xy = x^3 + tx.$$

The point $P = (0,0)$ is 2-torsion. The discriminant of this model is t^2 so E has good reduction away from the place $t = 0$ of K. At $t = 0$, E has split multiplicative reduction with minimal regular model of type I_2.

The quotient of E by the subgroup generated by P is $\phi\colon E \to E'$, where E' is given by

$$s^2 + rs = r^3 + t.$$

One checks that the degree of the conductor of E' is 4 and so, by [70, Lecture 1, Theorem 9.3 and Theorem 12.1(1)], the rank of $E'(K)$ is zero. Also, $E'(K)$ has no 2-torsion.

Therefore, taking cohomology of the sequence

$$0 \longrightarrow E[\phi] \longrightarrow E \longrightarrow E' \longrightarrow 0,$$

we find that the map

$$K/\wp(K) \cong H^1(K, \mathbb{Z}/2\mathbb{Z}) \cong H^1(E, E[\phi]) \longrightarrow H^1(K, E)$$

is injective. For $f \in K$, we write X_f for the torsor for E obtained from the class of f in $K/\wp(K)$ via this map.

Let L be the quadratic extension of K determined by f, i.e.,

$$L = K[u]/(\wp(u) - f).$$

The action of $G = \mathrm{Gal}(L/K) = \mathbb{Z}/2\mathbb{Z}$ on L is $u \mapsto u + 1$. Let G act on $K(E)$ by translation by P. Explicitly, one finds

$$(x, y) \longmapsto (t/x, t(x+y)/x^2)$$

and so $y/x \mapsto y/x + 1$.

The function field of X_f is the field of G-invariants in $L(E)$ where G acts as above. One finds that

$$K(X_f) = \frac{K(r)[s, z]}{(s^2 + rs + r^3 + t, z^2 + z + r + f)}$$

which presents X_f as a double cover of E'.

Let us now specialize to $f = t^{-1}$ and find a minimal regular model of X_f over $R = \mathbb{F}_2[t]$. Let

$$\mathcal{U} = \mathrm{Spec} \frac{R[r, s, w]}{(s^2 + rs + r^3 + t, w^2 + tw + t^2r + t)}.$$

Then $\mathcal{U}$ is regular away from $t = r = s = w = 0$ and its generic fiber is isomorphic (via $w = tz$) to an open subset of X_f. Let

$$\mathcal{V} = \mathrm{Spec} \frac{R[r', s', w']}{(s' + r's' + r'^3 + ts'^3, w'^2 + tw' + t^3r's' + t)},$$

which is a regular scheme whose generic fiber is another open subset of X_f. Let $\mathcal{Y}_f$ be the result of gluing $\mathcal{U}$ and $\mathcal{V}$ via $(r, s, w) = (r'/s', 1/s', w' + tr'^2/s')$. The generic fiber of $\mathcal{Y}_f$ is X_f and $\mathcal{Y}_f$ is regular away from $r = s = w = t = 0$. Note that the special fiber of $\mathcal{Y}_f$ is isomorphic to the product of the doubled point $\mathrm{Spec}\, \mathbb{F}_2[w]/(w^2)$ with the projective nodal plane cubic

$$\mathrm{Proj} \frac{\mathbb{F}_2[r, s, v]}{(s^2v + rsv + r^3)}.$$

In particular, the special fiber over $t = 0$, call it $\mathcal{Y}_{f,0}$, satisfies $H^0(\mathcal{Y}_{f,0}, \mathcal{O}_{\mathcal{Y}_{f,0}}) = \mathbb{F}_2[w]/(w^2)$. This shows that $\mathcal{Y}_f$ is not cohomologically flat over $\mathcal{C}$ at $t = 0$.

To finish the example, we should blow up $\mathcal{Y}_f$ at its unique non-regular point. The resulting scheme $\mathcal{X}_f$ is regular and flat over R, but it is not cohomologically flat at $t = 0$. The fiber over $t = 0$ is the product of a double point and a Néron configuration of type I_2, and its global regular functions are $\mathbb{F}_2[w]/(w^2)$. This is in agreement with [34, 6.6].

We will re-use parts of this example below.

Exercise 2.5. By the earlier discussion, $R^1\pi_*\mathcal{O}_{\mathcal{X}_f}$ has torsion at $t = 0$. Make this explicit.

3 Properties of J_X

3.1 Review of Picard functors

We quickly review basic material on the relative Picard functor. The original sources [18] and [50] are still very much worth reading. Two excellent modern references with more details and historical comments are [9] and [26].

3.1.1 The relative Picard functor

For any scheme $\mathcal{Y}$, we write $\mathrm{Pic}(\mathcal{Y})$ for the *Picard group* of $\mathcal{Y}$, i.e., for the group of isomorphism classes of invertible sheaves on $\mathcal{Y}$. This group can be calculated cohomologically: $\mathrm{Pic}(\mathcal{Y}) \cong H^1(\mathcal{Y}, \mathcal{O}_{\mathcal{Y}}^{\times})$ (cohomology computed in the Zariski, étale, or finer topologies).

Now fix a morphism of schemes $f\colon \mathcal{Y} \to \mathcal{S}$ (separated and of finite type, as always). If $\mathcal{T} \to \mathcal{S}$ is a morphism of schemes, we write $\mathcal{Y}_{\mathcal{T}}$ for $\mathcal{Y} \times_{\mathcal{S}} \mathcal{T}$ and $f_{\mathcal{T}}$ for the projection $\mathcal{Y}_{\mathcal{T}} \to \mathcal{T}$. Define a functor $P_{\mathcal{Y}/\mathcal{S}}$ from schemes over $\mathcal{S}$ to abelian groups by the rule

$$\mathcal{T} \longmapsto P_{\mathcal{Y}/\mathcal{S}}(\mathcal{T}) = \frac{\mathrm{Pic}(\mathcal{Y}_{\mathcal{T}})}{f_{\mathcal{T}}^* \mathrm{Pic}(\mathcal{T})}.$$

We define the *relative Picard functor* $\underline{\mathrm{Pic}}_{\mathcal{Y}/\mathcal{S}}$ to be the fppf sheaf associated to $P_{\mathcal{Y}/\mathcal{S}}$. (Here "fppf" means "faithfully flat and finitely presented". See [9, 8.1] for details on the process of sheafification.) Explicitly, if $\mathcal{T}$ is affine, an element of $\underline{\mathrm{Pic}}_{\mathcal{Y}/\mathcal{S}}(\mathcal{T})$ is represented by a line bundle $\xi' \in \mathrm{Pic}(\mathcal{Y} \times_{\mathcal{S}} \mathcal{T}')$ where $\mathcal{T}' \to \mathcal{T}$ is fppf, subject to the condition that there should exist an fppf morphism $\tilde{\mathcal{T}} \to \mathcal{T}' \times_{\mathcal{T}} \mathcal{T}'$ such that the pull backs of ξ' via the two projections

$$\tilde{\mathcal{T}} \longrightarrow \mathcal{T}' \times_{\mathcal{T}} \mathcal{T}' \rightrightarrows \mathcal{T}'$$

are isomorphic. Two such elements $\xi_i \in \mathrm{Pic}(\mathcal{Y} \times_{\mathcal{S}} \mathcal{T}_i')$ represent the same element of $\underline{\mathrm{Pic}}_{\mathcal{Y}/\mathcal{S}}(\mathcal{T})$ if and only if there is an fppf morphism $\tilde{\mathcal{T}} \to \mathcal{T}_1' \times_{\mathcal{T}} \mathcal{T}_2'$ such that the pull-backs of the ξ_i to $\tilde{\mathcal{T}}$ via the two projections are isomorphic. Fortunately, under mild hypotheses, this can be simplified quite a bit!

Assume that $f_* \mathcal{O}_{\mathcal{Y}} = \mathcal{O}_{\mathcal{S}}$ universally. This means that for all $\mathcal{T} \to \mathcal{S}$ we have $f_{\mathcal{T}*} \mathcal{O}_{\mathcal{Y}_{\mathcal{T}}} = \mathcal{O}_{\mathcal{T}}$. Equivalently, $f_* \mathcal{O}_{\mathcal{Y}} = \mathcal{O}_{\mathcal{S}}$ and f is cohomologically flat in dimension 0. In this case, for all $\mathcal{T} \to \mathcal{S}$, we have an exact sequence

$$0 \longrightarrow \mathrm{Pic}\,(\mathcal{T}) \longrightarrow \mathrm{Pic}\,(\mathcal{Y}_{\mathcal{T}}) \longrightarrow \underline{\mathrm{Pic}}_{\mathcal{Y}/\mathcal{S}}\,(\mathcal{T}) \longrightarrow \mathrm{Br}\,(\mathcal{T}) \longrightarrow \mathrm{Br}\,(\mathcal{Y}_{\mathcal{T}}). \qquad (3.1)$$

(This is the exact sequence of low-degree terms in the Leray spectral sequence for $f\colon \mathcal{Y}_{\mathcal{T}} \to \mathcal{T}$, computed with respect to the fppf topology.) Here the groups $\mathrm{Br}(\mathcal{T}) = H^2(\mathcal{T}, \mathcal{O}_{\mathcal{T}}^{\times})$ and $\mathrm{Br}(\mathcal{Y}_{\mathcal{T}}) = H^2(\mathcal{Y}_{\mathcal{T}}, \mathcal{O}_{\mathcal{Y}_{\mathcal{T}}}^{\times})$ are the cohomological Brauer groups of $\mathcal{T}$ and $\mathcal{Y}_{\mathcal{T}}$, again computed with the fppf topology. (It is known that the étale topology gives the same groups.) See [9, 8.1] for the assertions in this paragraph and the next.

In case f has a section, we get a short exact sequence

$$0 \longrightarrow \mathrm{Pic}(\mathcal{T}) \longrightarrow \mathrm{Pic}(\mathcal{Y}_{\mathcal{T}}) \longrightarrow \underline{\mathrm{Pic}}_{\mathcal{Y}/\mathcal{S}}(\mathcal{T}) \longrightarrow 0$$

and so in this case

$$\underline{\mathrm{Pic}}_{\mathcal{Y}/\mathcal{S}}\,(\mathcal{T}) = P_{\mathcal{Y}/\mathcal{S}}(\mathcal{T}) = \frac{\mathrm{Pic}(\mathcal{Y}_{\mathcal{T}})}{f_{\mathcal{T}}^* \, \mathrm{Pic}(\mathcal{T})}.$$

3.1.2   Representability and Pic^0 over a field

The simplest representability results will be sufficient for many of our purposes.

To say that $\underline{\mathrm{Pic}}_{\mathcal{Y}/\mathcal{S}}$ is represented by a scheme $\mathrm{Pic}_{\mathcal{Y}/\mathcal{S}}$ means that, for all $\mathcal{S}$-schemes $\mathcal{T}$,

$$\underline{\mathrm{Pic}}_{\mathcal{Y}/\mathcal{S}}\,(\mathcal{T}) = \mathrm{Pic}_{\mathcal{Y}/\mathcal{S}}\,(\mathcal{T}) = \mathrm{Mor}_{\mathcal{S}}\,(\mathcal{T}, \mathrm{Pic}_{\mathcal{Y}/\mathcal{S}}\,).$$

Suppose $\mathcal{S}$ is the spectrum of a field and $\mathcal{Y} \to \mathcal{S}$ is proper. Then $\underline{\mathrm{Pic}}_{\mathcal{Y}/\mathcal{S}}$ is represented by a scheme $\mathrm{Pic}_{\mathcal{Y}/\mathcal{S}}$ which is locally of finite type over $\mathcal{S}$ [9, 8.2, Theorem 3]. The connected component of this group scheme will be denoted by $\mathrm{Pic}^0_{\mathcal{Y}/\mathcal{S}}$. If $\mathcal{Y} \to \mathcal{S}$ is smooth and geometrically irreducible, then $\mathrm{Pic}^0_{\mathcal{Y}/\mathcal{S}}$ is proper [9, 8.4, Theorem 3].

The results of the previous paragraph apply in particular to X/K and $\mathcal{X}/k$. Moreover, since X/K is a curve, $H^2(X, \mathcal{O}_X) = 0$ and so $\mathrm{Pic}^0_{X/K}$ is smooth and hence an abelian variety [9, 8.4, Proposition 2] (plus our assumption that morphisms are of finite type to convert formal smoothness into smoothness) or [26, 9.5.19].

In general $\mathrm{Pic}_{\mathcal{X}/k}$ need not be reduced and so need not be smooth over k. If k has characteristic zero or $H^2(\mathcal{X}, \mathcal{O}_{\mathcal{X}}) = 0$, then $\mathrm{Pic}^0_{\mathcal{X}/k}$ is again an abelian variety. We define

$$\mathrm{PicVar}_{\mathcal{X}/k} = \left(\mathrm{Pic}^0_{\mathcal{X}/k} \right)_{\mathrm{red}},$$

the *Picard variety* of $\mathcal{X}$, which is an abelian variety. See [53] and [44] for an analysis of non-reduced Picard schemes and [31] for more in the case of surfaces.

Since we have assumed that k is large enough so that X has a rational point, for all k-schemes $\mathcal{T}$ we have

$$\mathrm{Pic}^0_{X/k}(\mathcal{T}) = \frac{\mathrm{Pic}^0(\mathcal{X}_{\mathcal{T}})}{\pi_{\mathcal{T}}^* \mathrm{Pic}^0(\mathcal{T})}.$$

3.1.3 More general bases

For any $\mathcal{Y} \to \mathcal{S}$ which is proper, define $\underline{\mathrm{Pic}}^0_{\mathcal{Y}/\mathcal{S}}$ to be the subfunctor of $\underline{\mathrm{Pic}}_{\mathcal{Y}/\mathcal{S}}$ consisting of elements whose restrictions to fibers $\mathcal{Y}_s$, $s \in \mathcal{S}$, lie in $\mathrm{Pic}^0_{\mathcal{Y}_s/\kappa(s)}$.

We need a deeper result to handle $X \to C$. Namely, assume that $\mathcal{Y}$ is regular, $\mathcal{S}$ is one-dimensional and regular, $f: \mathcal{Y} \to \mathcal{S}$ is flat, and projective of relative dimension 1 such that $f_* \mathcal{O}_{\mathcal{S}} = \mathcal{O}_C$. Over the open subset of $\mathcal{S}$ where f is smooth, $\underline{\mathrm{Pic}}^0_{\mathcal{Y}/\mathcal{S}}$ is represented by an abelian scheme [9, 9.4, Proposition 4].

Over all of $\mathcal{S}$, we cannot hope for reasonable representability results unless we make further hypotheses on f. If we assume that each fiber of f has the property that the gcd of the geometric multiplicities of its irreducible components is 1, then $\underline{\mathrm{Pic}}^0_{\mathcal{Y}/\mathcal{S}}$ is represented by a separated $\mathcal{S}$-scheme [9, 9.4, Theorem 2].

3.1.4 Example

Here is an explicit example where an element of $\mathrm{Pic}_{X/K}(K)$ is not represented by an element of $\mathrm{Pic}(X)$.

Let E/K be the elliptic curve of Section 2.6 and let X_f be the homogeneous space for E as above, where $f \in K$ will be selected later. Then E is the Jacobian of X and so, by the results quoted in Subsection 3.1.2, E represents the functor $\underline{\mathrm{Pic}}^0_{X/K}$. Let us consider the point $P = (0,0)$ above in $E(K) = \mathrm{Pic}^0_{X/K}(K)$ and show that, for many choices of f, this class is not represented by an invertible sheaf on $X = X_f$. Equivalently, we want to show that P does not go to 0 in $\mathrm{Br}(K)$ in the sequence (3.1) above.

The image of P in $\mathrm{Br}(K)$ is given by a pairing studied by Lichtenbaum [29]. More precisely, consider the image of P under the coboundary

$$E(K) \longrightarrow H^1\big(K, E'[\phi^\vee]\big) \cong H^1\big(K, \mu_2\big) \cong K^\times / K^{\times 2}.$$

By Kramer's results on 2-descent in characteristic 2 [27, 1.1b], the image of P is the class of t. On the other hand, suppose that $X = X_f$ is the torsor for E corresponding to $f \in K/\wp(K)$. Then the local invariant at v of the image of P in $\mathrm{Br}(K)$ is given by

$$\frac{1}{2} \mathrm{Res}_v \left(f \frac{dt}{t} \right) \in \frac{1}{2} \mathbb{Z}/\mathbb{Z} \subset \mathrm{Br}\big(K_v\big).$$

Thus, for example, if we take $f = 1/(t-1)$, then we get an element of the Brauer group ramified at 1 and ∞ and so the class of P does not come from $\mathrm{Pic}(X_f)$.

3.2   The Jacobian

By definition, the Jacobian J_X of X is the group scheme $\mathrm{Pic}^0_{X/K}$. Because X is a smooth, projective curve, J_X is an abelian variety of dimension g_X where g_X is the genus of X, and it is equipped with a canonical principal polarization given by the theta divisor. We refer to [41] for more on the basic properties of J_X, including the Albanese property and autoduality.

3.3   The K/k trace of J_X

Recall that (B, τ) denotes the K/k trace of J_X, which by definition is a final object in the category of pairs (A, σ) where A is a k-abelian variety and $\sigma \colon A \times_k K \to J_X$ is a K-morphism of abelian varieties.

We refer to [11] for a discussion in modern language of the existence and basic properties of (B, τ). In particular, Conrad proves that (B, τ) exists, it has good base change properties, and (when K/k is regular, as it is in our case) τ has finite connected kernel with connected Cartier dual. In particular, τ is purely inseparable.

3.4   The Lang–Néron theorem

Define the Mordell–Weil group $\mathrm{MW}(J_X)$ as

$$\mathrm{MW}\left(J_X\right) = \frac{J_X(K)}{\tau B(k)},$$

where as usual (B, τ) is the K/k trace of J_X. (In view of the theorem below, perhaps this would be better called the Lang–Néron group.)

Generalizing the classical Mordell–Weil theorem, we have the following finiteness result, proven independently by Lang and Néron.

Theorem 3.1. $\mathrm{MW}(J_X)$ *is a finitely generated abelian group.*

Note that when k is finitely generated, $B(k)$ is finitely generated as well, and so $J_X(K)$ is itself a finitely generated abelian group. For large fields k, $B(k)$ may not be finitely generated, and so it really is necessary to quotient by $\tau B(k)$ in order to get a finitely generated abelian group.

We refer to [11] for a proof of the theorem in modern language. Roughly speaking, the proof there follows the general lines of the usual proof of the Mordell–Weil theorem for an abelian variety over a global field: One shows that $\mathrm{MW}(J_X)/n$ is finite by embedding it in a suitable cohomology group, and then uses a theory of heights to deduce finite generation of $\mathrm{MW}(J_X)$.

Another proof proceeds by relating $\mathrm{MW}(J_X)$ to the Néron–Severi group $\mathrm{NS}(\mathcal{X})$ (this is the Shioda–Tate isomorphism discussed in the next section) and then proving that $\mathrm{NS}(\mathcal{X})$ is finitely generated. The latter was proven by Kleiman in [55, XIII].

4 Shioda–Tate and heights

4.1 Points and curves

We write $\mathrm{Div}(\mathcal{X})$ for the group of (Cartier or Weil) divisors on $\mathcal{X}$ and similarly with $\mathrm{Div}(X)$. A prime divisor on $\mathcal{X}$ is *horizontal* if it is flat over $\mathcal{C}$ and *vertical* if it is contained in a fiber of π. The group $\mathrm{Div}(\mathcal{X})$ is the direct sum of its subgroups $\mathrm{Div}^{\mathrm{hor}}(\mathcal{X})$ and $\mathrm{Div}^{\mathrm{vert}}(\mathcal{X})$ generated respectively by horizontal and vertical prime divisors.

Restriction of divisors to the generic fiber of π induces a homomorphism

$$\mathrm{Div}(\mathcal{X}) \longrightarrow \mathrm{Div}(X)$$

whose kernel is $\mathrm{Div}^{\mathrm{vert}}(\mathcal{X})$ and which induces an isomorphism

$$\mathrm{Div}^{\mathrm{hor}}(\mathcal{X}) \cong \mathrm{Div}(X).$$

The inverse of this isomorphism sends a closed point of X to its scheme-theoretic closure in $\mathcal{X}$.

We define a filtration of $\mathrm{Div}(\mathcal{X})$ by declaring that $L^1 \mathrm{Div}(\mathcal{X})$ be the subgroup of $\mathrm{Div}(\mathcal{X})$ consisting of divisors whose restriction to X has degree 0, and by declaring that $L^2 \mathrm{Div}(\mathcal{X}) = \mathrm{Div}^{\mathrm{vert}}(\mathcal{X})$.

We define $L^i \mathrm{Pic}(\mathcal{X})$ to be the image of $L^i \mathrm{Div}(\mathcal{X})$ in $\mathrm{Pic}(\mathcal{X})$. Also, recall that $\mathrm{Pic}^0(\mathcal{X}) = \mathrm{Pic}^0_{\mathcal{X}/k}(k)$ is the group of invertible sheaves which are algebraically equivalent to zero (equivalence over $\overline{k}$ as usual).

Recall that the Néron–Severi group $\mathrm{NS}(\overline{\mathcal{X}})$ is $\mathrm{Pic}(\overline{\mathcal{X}})/\mathrm{Pic}^0(\overline{\mathcal{X}})$ and $\mathrm{NS}(\mathcal{X})$ is by definition the image of $\mathrm{Pic}(\mathcal{X})$ in $\mathrm{NS}(\overline{\mathcal{X}})$. We define $L^i \mathrm{NS}(\mathcal{X})$ as the image of $L^i \mathrm{Pic}(\mathcal{X})$ in $\mathrm{NS}(\mathcal{X})$.

It is obvious that $\mathrm{NS}(\mathcal{X})/L^1 \mathrm{NS}(\mathcal{X})$ is an infinite cyclic group; as generator we may take the class of a horizontal divisor of total degree δ over $\mathcal{C}$, where δ is the index of X.

Recall that the intersection pairing restricted to the group generated by the components of a fiber $\mathcal{X}_v$ is negative semi-definite, with kernel $(1/m_v)\mathcal{X}_v$. From this one deduces that $L^2 \mathrm{NS}(\mathcal{X})$ is the group generated by the irreducible components of fibers, with relations $\mathcal{X}_v = \mathcal{X}_{v'}$ for any two closed points v and v'. Note that if for some $v \neq v'$ we have $m = \gcd(m_v, m_{v'}) > 1$, then $L^2 \mathrm{NS}(\mathcal{X})$ has non-trivial torsion: $m((1/m)\mathcal{X}_v - (1/m)\mathcal{X}_{v'}) = 0$ in $\mathrm{NS}(\mathcal{X})$.

Summarizing the above, we have that $\mathrm{NS}(\mathcal{X})/L^1 \mathrm{NS}(\mathcal{X})$ is infinite cyclic, and $L^2 \mathrm{NS}(\mathcal{X})$ is finitely generated of rank $1 + \sum_v (f_v - 1)$, where f_v is the number of irreducible components of $\mathcal{X}_v$. Also, $L^2 \mathrm{NS}(\mathcal{X})$ is torsion free if and only if $\gcd(m_v, m_{v'}) = 1$ for all $v \neq v'$.

The interesting part of $\mathrm{NS}(\mathcal{X})$, namely $L^1 \mathrm{NS}(\mathcal{X})/L^2 \mathrm{NS}(\mathcal{X})$, is the subject of the Shioda–Tate theorem.

4.2   Shioda–Tate theorem

Write $\mathrm{Div}^0(X)$ for the group of divisors on X of degree 0. Restriction to the generic fiber gives a homomorphism $L^1 \mathrm{Div}(\mathcal{X}) \to \mathrm{Div}^0(X)$ which descends to a homomorphism $L^1 \mathrm{Pic}(\mathcal{X}) \to J_X(K) = \mathrm{Pic}^0_{X/K}(K)$. The Shioda–Tate theorem uses this map to describe $L^1 \mathrm{NS}(\mathcal{X})/L^2 \mathrm{NS}(\mathcal{X})$.

Recall that (B, τ) is the K/k-trace of J_X.

Proposition 4.1 (Shioda–Tate). *The map above induces a homomorphism*

$$\frac{L^1 \mathrm{NS}(\mathcal{X})}{L^2 \mathrm{NS}(\mathcal{X})} \longrightarrow \mathrm{MW}(J_X) = \frac{J_X(K)}{\tau B(k)}$$

with finite kernel and cokernel. In particular the two sides have the same rank as finitely generated abelian groups. This homomorphism is an isomorphism if X has a K-rational point and k is either finite or algebraically closed

Taking into account what we know about $\mathrm{NS}(\mathcal{X})/L^1 \mathrm{NS}(\mathcal{X})$ and $L^2 \mathrm{NS}(\mathcal{X})$, we have a formula relating the ranks of $\mathrm{NS}(\mathcal{X})$ and $\mathrm{MW}(J_X)$.

Corollary 4.2. *We have*

$$\mathrm{Rank}\,\mathrm{NS}(\mathcal{X}) = \mathrm{Rank}\,\mathrm{MW}(J_X) + 2 + \sum_v (f_v - 1).$$

Various versions of Proposition 4.1 appear in the literature, notably in [16], [21], [58], and [62], and, as Shioda notes [58, p. 359], it was surely known to the ancients.

Proof of Proposition 4.1. Specialized to X/K, the exact sequence of low degree terms (3.1) gives an exact sequence

$$0 \longrightarrow \mathrm{Pic}(X) \longrightarrow \mathrm{Pic}_{X/K}(K) \longrightarrow \mathrm{Br}(K).$$

If X has a K-rational point, then $\mathrm{Pic}(X) \to \mathrm{Pic}_{X/K}(K)$ is an isomorphism. In any case, X has a rational point over a finite extension K' of K of degree δ (the index of X), and over K' the coboundary in the analogous sequence

$$\mathrm{Pic}_{X/K'}(K') \longrightarrow \mathrm{Br}(K')$$

is zero. This implies that the cokernel of $\mathrm{Pic}(X) \to \mathrm{Pic}_{X/K}(K)$ maps to the kernel of $\mathrm{Br}(K) \to \mathrm{Br}(K')$ and therefore lies in the δ-torsion subgroup of $\mathrm{Br}(K)$ [54, p. 157]. The upshot is that the cokernel of $\mathrm{Pic}(X) \to \mathrm{Pic}_{X/K}(K)$ has exponent dividing δ.

A simple geometric argument as in Shioda [58, p. 363] shows that the kernel of the homomorphism $L^1 \mathrm{Pic}(\mathcal{X}) \to \mathrm{Pic}^0(X)$ is exactly $L^2 \mathrm{Pic}(\mathcal{X})$, so we have an isomorphism

$$\frac{L^1 \mathrm{Pic}(\mathcal{X})}{L^2 \mathrm{Pic}(\mathcal{X})} \cong \mathrm{Pic}^0(X).$$

Now consider the composite homomorphism

$$\mathrm{Pic}^0(\mathcal{X}) = \mathrm{Pic}^0_{\mathcal{X}/k}(k) \longrightarrow \mathrm{Pic}^0(X) \longrightarrow \mathrm{Pic}^0_{X/K}(K).$$

There is an underlying homomorphism of algebraic groups

$$\mathrm{Pic}^0_{\mathcal{X}/k} \times_k K \longrightarrow \mathrm{Pic}^0_{X/K}$$

inducing the homomorphism above on points. By the definition of the K/k-trace, this morphism must factor through B, i.e., we have a morphism of algebraic groups over k,

$$\mathrm{Pic}^0_{\mathcal{X}} \longrightarrow B.$$

We are going to argue that this last morphism is surjective. To do so, first note that a similar discussion applies over $\bar{k}$ and yields the following diagram:

$$\begin{array}{ccccccc}
\dfrac{L^1 \mathrm{Pic}(\mathcal{X} \times \bar{k})}{L^2 \mathrm{Pic}(\mathcal{X} \times \bar{k})} & \longrightarrow & \mathrm{Pic}^0\left(X \times K\bar{k}\right) & \longrightarrow & \mathrm{Pic}\left(X \times K\bar{k}\right) & \longrightarrow & \mathrm{Pic}_{X/K\bar{k}}\left(K\bar{k}\right) \\
\uparrow & & & & & & \uparrow \\
\mathrm{Pic}^0\left(\mathcal{X} \times \bar{k}\right) & & \longrightarrow & & & & B\left(\bar{k}\right).
\end{array}$$

Now the cokernel of the left vertical map is a subquotient of $\mathrm{NS}(\mathcal{X})$, so finitely generated, and the cokernels of the horizontal maps across the top are trivial, $\mathbb{Z}$, and of finite exponent respectively. On the other hand, $B(\bar{k})$ is a divisible group. This implies that the image of $B(\bar{k})$ in $\mathrm{Pic}_{X/K\bar{k}}(K\bar{k})$ is equal to the image of $\mathrm{Pic}^0(\mathcal{X} \times \bar{k})$ in that same group. Since the kernel of $B(\bar{k}) \to \mathrm{Pic}_{X/K\bar{k}}(K\bar{k})$ is finite, this in turn implies that the morphism $\mathrm{Pic}^0(\mathcal{X}) \to B$ is surjective.

An alternative proof of the surjectivity is to use the exact sequence (2.1). The middle term is $V_\ell \mathrm{Pic}^0(\mathcal{X} \times \bar{k})$ whereas the right hand term is $V_\ell B(\bar{k})$. Thus the morphism $\mathrm{Pic}^0_{\mathcal{X}} \to B$ is surjective.

Next we argue that the map of k-points $\mathrm{Pic}(\mathcal{X}) = \mathrm{Pic}_{\mathcal{X}/k}(k) \to B(k)$ has finite cokernel. More generally, if $\phi \colon A \to A'$ is a surjective morphism of abelian varieties over a field k, then we claim that the map of points $A(k) \to A'(k)$ has finite cokernel. If ϕ is an isogeny, then considering the dual isogeny $\phi^\vee$ and the composition $\phi\phi^\vee$ shows that the cokernel is killed by $\deg \phi$, so it is finite. For a general surjection, if $A'' \subset A$ is a complement (up to a finite group) of $\ker \phi$, then by the above $A''(k) \to A'(k)$ has finite cokernel, and *a fortiori* so does $A(k) \to A'(k)$. (Thanks to Marc Hindry for suggesting this argument.)

When k is algebraically closed, $\mathrm{Pic}^0_{\mathcal{X}}(k) \to B(k)$ is obviously surjective. It is also surjective when k is finite and X has a K-rational point. This follows from Lang's theorem because, according to Proposition 4.4 below, the kernel of $\mathrm{Pic}^0_{\mathcal{X}} \to B$ is an abelian variety.

Summing up, the map

$$\frac{L^1 \operatorname{Pic}(\mathcal{X})}{L^2 \operatorname{Pic}(\mathcal{X})} \longrightarrow J_X(K)$$

has cokernel of finite exponent, and

$$\operatorname{Pic}^0(\mathcal{X}) \longrightarrow B(k)$$

has finite cokernel. Thus the map

$$\frac{L^1 \operatorname{NS}(\mathcal{X})}{L^2 \operatorname{NS}(\mathcal{X})} = \frac{L^1 \operatorname{Pic}(\mathcal{X})}{L^2 \operatorname{Pic}(\mathcal{X}) + \operatorname{Pic}^0(\mathcal{X})} \longrightarrow \operatorname{MW}(J_X) = \frac{J_X(K)}{\tau B(k)}$$

has finite kernel and its cokernel has finite exponent. Since the target group is finitely generated (Lang–Néron), the cokernel must be finite. When X has a rational point the first map displayed above is surjective, and when k is finite or algebraically closed, the second map is surjective. Thus under both of these hypotheses, the third map is an isomorphism.

This completes the proof of the theorem. $\square$

Remark 4.3. It is also possible to deduce the theorem in the case when k is finite from the case when k is algebraically closed by taking Galois invariants.

It will be convenient to have an explicit description of B.

Proposition 4.4. *If X has a K-rational point and k is perfect, then we have an exact sequence*

$$0 \longrightarrow \operatorname{Pic}^0_{\mathcal{C}/k} \longrightarrow \operatorname{PicVar}_{\mathcal{X}/k} \longrightarrow B \longrightarrow 0.$$

Sketch of proof. Obviously there are morphisms $\operatorname{Pic}^0_{\mathcal{C}/k} \to \operatorname{PicVar}_{\mathcal{X}/k} \to B$. The first is injective because $\pi \colon \mathcal{X} \to \mathcal{C}$ has a section, and the second was seen to be surjective in the proof of the Shioda–Tate theorem. It is also clear that the composed map is zero.

Again because π has a section, the argument in Section 2.4 shows that the integral ℓ-adic Leray spectral sequence degenerates and we have exact sequences

$$0 \longrightarrow H^1\big(\overline{\mathcal{C}}, \mathbb{Z}_\ell(1)\big) \longrightarrow H^1\big(\overline{\mathcal{X}}, \mathbb{Z}_\ell(1)\big) \longrightarrow H^0\big(\overline{\mathcal{C}}, R^1\pi_*\mathbb{Z}_\ell(1)\big) \longrightarrow 0$$

for all ℓ. These cohomology groups can be identified with the Tate modules of $\operatorname{Pic}^0_{\mathcal{C}/k}$, $\operatorname{PicVar}_{\mathcal{X}/k}$, and B respectively. (For $\ell = p$ we should use flat cohomology.) This shows that $\operatorname{PicVar}_{\mathcal{X}/k} / \operatorname{Pic}^0_{\mathcal{C}/k} \to B$ is purely inseparable. But we also have

$$0 \longrightarrow H^1\big(\mathcal{C}, \mathcal{O}_{\mathcal{C}}\big) \longrightarrow H^1\big(\mathcal{X}, \mathcal{O}_{\mathcal{X}}\big) \longrightarrow H^0\big(\mathcal{C}, R^1\pi_*\mathcal{O}_{\mathcal{X}}\big) \longrightarrow 0$$

and (using the existence of a section and thus cohomological flatness) these three groups can be identified with the tangent spaces of $\operatorname{Pic}^0_{\mathcal{C}/k}$, $\operatorname{PicVar}_{\mathcal{X}/k}$, and B respectively. This shows that $\operatorname{PicVar}_{\mathcal{X}/k} / \operatorname{Pic}^0_{\mathcal{C}/k} \to B$ is separable. Thus it must be an isomorphism. $\square$

Remark 4.5. If we no longer assume that k is perfect, but continue to assume that X has a rational point, then the conclusion of the proposition still holds. Indeed, we have a morphism $\mathrm{PicVar}_{\mathcal{X}/k} / \mathrm{Pic}^0_{\mathcal{C}/k} \to B$ which becomes an isomorphism over $\bar{k}$. (Here we use that K/k is regular, so that formation of B commutes with extension of k [11, 6.8] and similarly for the Picard varieties.) It must therefore have already been an isomorphism over k. In general, the morphism $\mathrm{PicVar}_{\mathcal{X}/k} / \mathrm{Pic}^0_{\mathcal{C}/k} \to B$ may be purely inseparable.

4.3 Heights

We use the Shioda–Tate theorem to define a non-degenerate bilinear pairing on $\mathrm{MW}(J_X)$ which is closely related to the Néron–Tate canonical height when k is finite.

To simplify the notation we write

$$L^1 = (L^1 \, \mathrm{NS}(\mathcal{X})) \otimes \mathbb{Q} \quad \text{and} \quad L^2 = (L^2 \, \mathrm{NS}(\mathcal{X})) \otimes \mathbb{Q},$$

so that $\mathrm{MW}(J_X) \otimes \mathbb{Q} \cong L^1/L^2$.

It is well known that the intersection form on $\mathrm{NS}(\mathcal{X}) \otimes \mathbb{Q}$ is non-degenerate (cf. [70, p. 29]). Since L^1 has codimension 1 in $\mathrm{NS}(\mathcal{X}) \otimes \mathbb{Q}$, the kernel of the intersection form restricted to L^1 has dimension at most 1, and it is easy to see that it is in fact one-dimensional, generated by F, the class of a fiber of π. Also, our discussion of the structure of the fibers of π shows that the kernel of the intersection form restricted to L^2 is also one-dimensional, generated by F.

It follows that for each class in $D \in L^1/L^2$, there is a representative $\tilde{D} \in L^1$ such that $\tilde{D}$ is orthogonal to all of L^2; moreover, $\tilde{D}$ is determined by D up to the addition of a multiple of F. We thus have a homomorphism $\phi \colon L^1/L^2 \to L^1/\mathbb{Q}F$ defined by $\phi(D) = \tilde{D}$. We define a bilinear pairing on $\mathrm{MW}(J_X) \otimes \mathbb{Q} = L^1/L^2$ by the rule

$$\bigl(D, D'\bigr) = -\phi\bigl(D\bigr) \cdot \phi\bigl(D'\bigr)$$

where the dot signifies the intersection pairing on $\mathcal{X}$, extended by $\mathbb{Q}$-linearity. The right-hand side is well defined because $F \in L^2$.

Proposition 4.6. *The formula above defines a symmetric, $\mathbb{Q}$-valued bilinear form on $\mathrm{MW}(J_X) \otimes \mathbb{Q}$ which is positive definite on $\mathrm{MW}(J_X) \otimes \mathbb{R}$. If k is a finite field of cardinality q, then $(-,-) \log q$ is the Néron–Tate canonical height on $\mathrm{MW}(J_X) \cong J_X(K) \otimes \mathbb{Q}$*

Sketch of proof. It is clear that the pairing is bilinear, symmetric, and $\mathbb{Q}$-valued.

To see that it is positive definite on $\mathrm{MW}(J_X) \otimes \mathbb{Q}$, we use the Hodge index theorem. Given $D \in L^1/L^2$, choose an irreducible multisection P and a representative $\tilde{\phi}(D) \in L^1$ for $\phi(D)$. The intersection number $P \cdot D$ is in $\mathbb{Q}$ and replacing D with nD for suitable n we can assume it is in $\mathbb{Z}$. Then adding a multiple of F to $\tilde{\phi}(D)$ and calling the result again $\tilde{\phi}(D)$, we get a new representative of $\phi(D)$

such that $P \cdot \tilde{\phi}(D) = 0$. On the other hand $(P + mF)^2 > 0$ if m is large enough and $(P+mF) \cdot \tilde{\phi}(D) = 0$. The Hodge index theorem (e.g., [6, 2.4]) then implies that $\tilde{\phi}(D)^2 < 0$ and so $(D, D) > 0$.

Since the pairing is positive definite and $\mathbb{Q}$-valued on $\mathrm{MW}(J_X) \otimes \mathbb{Q}$, it is also positive definite on $\mathrm{MW}(J_X) \otimes \mathbb{R}$.

The connection with Néron–Tate canonical height can be seen by using the local canonical heights. These were constructed purely on the relative curve $\mathcal{X} \to \mathcal{C}$ (as opposed to on the Néron model $\mathcal{J}_X \to \mathcal{C}$) by Gross in [17]. An inspection of his construction shows that summing the local heights over all places of K gives exactly our definition above times $\log q$. $\qquad \square$

Exercise 4.7. When X is an elliptic curve, give a global proof of the connection between the Shioda–Tate height and the canonical height as follows: Check from the definition that $\log q$ times the Shioda–Tate height differs from the naive height associated to the origin of X (as a divisor on X) by a bounded amount. Since the Shioda–Tate height is already bilinear, it must thus be the canonical height as constructed by Tate.

5   Cohomology and cycles

5.1   Tate's conjecture for $\mathcal{X}$

Throughout this section, k is a finite field of characteristic p and $\mathcal{X}$ is a smooth and projective surface over k. We write $\overline{\mathcal{X}}$ for $\mathcal{X} \times_k \overline{k}$. Our goal is to discuss Tate's first conjecture on cycles and cohomology classes for $\mathcal{X}$. Since this was already discussed in [70, Lecture 2], we will be brief.

5.1.1   Basic exact sequences

Let $\mathrm{Br}(\mathcal{X}) = H^2(\mathcal{X}, \mathcal{O}_X^\times) = H^2(\mathcal{X}, \mathbb{G}_m)$ be the (cohomological) Brauer group of $\mathcal{X}$. Here we use the étale or flat topologies. It is known that $\mathrm{Br}(\mathcal{X})$ is a torsion group and it is conjectured to be finite in our situation.

For any positive integer n, consider the Kummer sequence on $\overline{\mathcal{X}}$,

$$0 \longrightarrow \mu_n \longrightarrow \mathbb{G}_m \overset{n}{\longrightarrow} \mathbb{G}_m \longrightarrow 0.$$

Here if $p|n$ the sequence is exact in the flat topology, but not in the étale topology.

Let ℓ be a prime (with $\ell = p$ allowed). Taking cohomology and noting that $\mathrm{Pic}(\overline{\mathcal{X}})/\ell^n = \mathrm{NS}(\overline{\mathcal{X}})/\ell^n$ (because $\mathrm{Pic}^0(\overline{\mathcal{X}})$ is divisible), we find an exact sequence

$$0 \longrightarrow \mathrm{NS}\left(\overline{\mathcal{X}}\right)/\ell^n \longrightarrow H^2\left(\overline{\mathcal{X}}, \mu_{\ell^n}\right) \longrightarrow \mathrm{Br}\left(\overline{\mathcal{X}}\right)[\ell^n] \longrightarrow 0.$$

Taking the inverse limit over n yields an exact sequence

$$0 \longrightarrow \mathrm{NS}\left(\overline{\mathcal{X}}\right) \otimes \mathbb{Z}_\ell \longrightarrow H^2\left(\overline{\mathcal{X}}, \mathbb{Z}_\ell(1)\right) \longrightarrow T_\ell \,\mathrm{Br}\left(\overline{\mathcal{X}}\right) \longrightarrow 0.$$

Now $H^1(G_k, \mathrm{NS}(\overline{\mathcal{X}}) \otimes \mathbb{Z}_\ell)$ is a finite group. On the other hand, if H is a group with finite ℓ-torsion, $T_\ell H$ is finitely generated and torsion-free over $\mathbb{Z}_\ell$. This applies in particular to $T_\ell \mathrm{Br}(\overline{\mathcal{X}})$. Thus taking G_k invariants, we get an exact sequence

$$0 \longrightarrow \left(\mathrm{NS}\left(\overline{\mathcal{X}}\right) \otimes \mathbb{Z}_\ell\right)^{G_k} \longrightarrow H^2\left(\overline{\mathcal{X}}, \mathbb{Z}_\ell(1)\right)^{G_k} \longrightarrow \left(T_\ell \mathrm{Br}\left(\overline{\mathcal{X}}\right)\right)^{G_k} \longrightarrow 0.$$

Next note that, since k is finite, $H^1(G_k, \mathrm{Pic}^0(\overline{\mathcal{X}})) = 0$ and so

$$\left(\mathrm{NS}\left(\overline{\mathcal{X}}\right) \otimes \mathbb{Z}_\ell\right)^{G_k} = \mathrm{NS}\left(\mathcal{X}\right) \otimes \mathbb{Z}_\ell$$

since by definition $\mathrm{NS}(\mathcal{X})$ is the image of $\mathrm{Pic}(\mathcal{X})$ in $\mathrm{NS}(\overline{\mathcal{X}})$.

Now, using the Hochschild–Serre spectral sequence, we have a homomorphism $H^2(\mathcal{X}, \mathbb{Z}_\ell(1)) \to H^2(\overline{\mathcal{X}}, \mathbb{Z}_\ell(1))^{G_k}$ which is surjective with finite kernel. Since $T_\ell \mathrm{Br}(\mathcal{X})$ is torsion-free, the commutative diagram

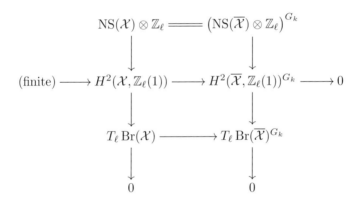

induces an isomorphism $T_\ell \mathrm{Br}(\mathcal{X}) \xrightarrow{\sim} T_\ell \mathrm{Br}(\overline{\mathcal{X}})^{G_k}$.

Putting everything together, we get an exact sequence

$$0 \longrightarrow \mathrm{NS}\left(\mathcal{X}\right) \otimes \mathbb{Z}_\ell \longrightarrow H^2\left(\overline{\mathcal{X}}, \mathbb{Z}_\ell(1)\right)^{G_k} \longrightarrow T_\ell \mathrm{Br}\left(\mathcal{X}\right) \longrightarrow 0. \qquad (5.1)$$

5.1.2 First Tate conjecture

Tate's first conjecture for a prime ℓ, which we denote $T_1(\mathcal{X}, \ell)$, says that the cycle class map induces an isomorphism

$$\mathrm{NS}\left(\mathcal{X}\right) \otimes \mathbb{Q}_\ell \cong H^2\left(\overline{\mathcal{X}}, \mathbb{Q}_\ell(1)\right)^{G_k}.$$

The exact sequence (5.1) leads immediately to several equivalent forms of this conjecture.

Proposition 5.1. *The following are equivalent:*

1. $T_1(\mathcal{X}, \ell)$.

2. *The cycle class map induces an isomorphism*

$$\mathrm{NS}\,(\mathcal{X}) \otimes \mathbb{Z}_\ell \cong H^2\big(\overline{\mathcal{X}}, \mathbb{Z}_\ell(1)\big)^{G_k}.$$

3. $T_\ell \,\mathrm{Br}(\mathcal{X}) = 0$.

4. *The ℓ-primary component of $\mathrm{Br}(\mathcal{X})$ is finite.*

Proof. Tensoring (5.1) with $\mathbb{Q}_\ell$ shows that $T_1(\mathcal{X}, \ell)$ is equivalent to $V_\ell \,\mathrm{Br}(\mathcal{X}) := T_\ell \,\mathrm{Br}(\mathcal{X}) \otimes \mathbb{Q}_\ell = 0$ and therefore to $T_\ell \,\mathrm{Br}(\mathcal{X}) = 0$, which is in turn equivalent to the injection

$$\mathrm{NS}\,(\mathcal{X}) \otimes \mathbb{Z}_\ell \longrightarrow H^2\big(\overline{\mathcal{X}}, \mathbb{Z}_\ell(1)\big)$$

being an isomorphism. This establishes the equivalence of 1–3. Since $\mathrm{Br}(\mathcal{X})$ is a torsion group and $\mathrm{Br}(\mathcal{X})[\ell]$ is finite, the ℓ-primary component of $\mathrm{Br}(\mathcal{X})$ is finite if and only if $T_\ell Br(\mathcal{X})$ is finite. This establishes the equivalence of 3 and 4. $\qquad\square$

We will see below that $T_1(\mathcal{X}, \ell)$ is equivalent to $T_1(\mathcal{X}, \ell')$ for any two primes ℓ and ℓ'.

We note also that the flat cohomology group $H^2(\overline{\mathcal{X}}, \mathbb{Q}_p(1))$ is isomorphic to the crystalline cohomology group $(H^2(\overline{\mathcal{X}}/W(\overline{k})) \otimes_{\mathbb{Z}_p} \mathbb{Q}_p)^{\mathrm{Fr} = p}$, where the exponent indicates the subspace upon which the absolute Frobenius Fr acts by multiplication by p. One can thus reformulate the $\ell = p$ part of the Tate conjecture in terms of crystalline cohomology.

5.2   Selmer group conjecture for J_X

We assume throughout this subsection that the ground field k is finite.

By the Lang–Néron theorem discussed in Section 3.4 above, the Mordell–Weil group $J(K)/\tau B(k)$ is finitely generated, and since $B(k)$ is obviously finite, we have that $J_X(K)$ is also finitely generated. In this subsection we quickly review the standard mechanism to bound (and conjecturally capture) the rank of this group via descent.

5.2.1   Selmer and Tate–Shafarevich groups

For each positive integer n, consider the exact sequence of sheaves for the flat topology on $\mathrm{Spec}\, K$,

$$0 \longrightarrow J_X[n] \longrightarrow J_X \xrightarrow{\ n\ } J_X \longrightarrow 0,$$

where $J_X[n]$ denotes the kernel of multiplication by n on J_X. Taking cohomology yields

$$0 \longrightarrow J_X(K)/nJ_X(K) \longrightarrow H^1(K, J_X[n]) \longrightarrow H^1(K, J_X)[n] \longrightarrow 0.$$

Similarly, for every completion K_v of K we have a similar sequence and restriction maps:

$$
\begin{array}{ccccccccc}
0 & \longrightarrow & J_X(K)/nJ_X(K) & \longrightarrow & H^1(K, J_X[n]) & \longrightarrow & H^1(K, J_X)[n] & \longrightarrow & 0 \\
& & \downarrow & & \downarrow & & \downarrow & & \\
0 & \longrightarrow & J_X(K_v)/nJ_X(K_v) & \longrightarrow & H^1(K_v, J_X[n]) & \longrightarrow & H^1(K_v, J_X)[n] & \longrightarrow & 0.
\end{array}
$$

We define $\mathrm{Sel}(J_X, n)$ to be

$$
\ker\left(H^1(K, J_X[n]) \longrightarrow \prod_v H^1(K_v, J_X) \right),
$$

where the product is over the places of K. Also, for any prime ℓ, we define $\mathrm{Sel}(J_X, \mathbb{Z}_\ell) = \varprojlim_n \mathrm{Sel}(J_X, \ell^n)$. We define the Tate–Shafarevich group by

$$
\mathrm{III}(J_X) = \ker\left(H^1(K, J_X) \longrightarrow \prod_v H^1(K_v, J_X) \right).
$$

There are exact sequences

$$
0 \longrightarrow J_X(K)/nJ_X(K) \longrightarrow \mathrm{Sel}(J_X, n) \longrightarrow \mathrm{III}(J_X)[n] \longrightarrow 0
$$

for each n. All the groups appearing here are finite. (For n prime to p, the classical proof of finiteness of the Selmer group – reducing it to the finiteness of the class group and finite generation of the unit group of Dedekind domains with fraction field K – works in our context. The p part was first proven by Milne [37].) So taking the inverse limit over powers of a prime ℓ yields an exact sequence

$$
0 \longrightarrow J_X(K) \otimes \mathbb{Z}_\ell \longrightarrow \mathrm{Sel}(J_X, \mathbb{Z}_\ell) \longrightarrow T_\ell \mathrm{III}(J_X) \longrightarrow 0. \tag{5.2}
$$

5.2.2 Tate–Shafarevich conjecture

Tate and Shafarevich conjectured (independently) that $\mathrm{III}(J_X)$ is a finite group. We write $TS(J_X)$ for this conjecture, and $TS(J_X, \ell)$ for the *a priori* weaker conjecture that the ℓ-primary part of $\mathrm{III}(J_X)$ is finite.

For each prime ℓ, we refer to the statement "the homomorphism

$$
J_X(K) \otimes \mathbb{Z}_\ell \longrightarrow \mathrm{Sel}(J_X/K, \mathbb{Z}_\ell)
$$

is an isomorphism" as the *Selmer group conjecture* $S(J_X, \ell)$. (This is perhaps nonstandard and is meant to suggest that the Selmer group captures rational points.)

The following is obvious from the exact sequence (5.2) above:

Proposition 5.2. *For each prime ℓ, the Selmer group conjecture $S(J_X, \ell)$ holds if and only if the Tate–Shafarevich conjecture $TS(J_X, \ell)$ holds.*

Note that, if the ℓ-primary component of $\mathrm{III}(J_X)$ is finite, then knowing $\mathrm{Sel}(J_X, \ell^n)$ for sufficiently large n determines the rank of $J_X(K)$. This observation and some input from L-functions leads (conjecturally) to an effective algorithm for computing generators of $J_X(K)$; see [36].

5.3   Comparison of cohomology groups

There is an obvious parallel between the conjectures $T_1(\mathcal{X}, \ell)$ and finiteness of $\mathrm{Br}(\mathcal{X})[\ell^\infty]$ on the one hand, and $S(J_X, \ell)$ and $TS(J_X, \ell)$ on the other. The Shioda–Tate isomorphism gives a precise connection between the groups of cycles and points involved. In this subsection, we give connections between the cohomology groups involved. We restrict to the simplest cases here – they are already complicated enough. In Subsection 6.3 we will give a comparison for all ℓ and the most general hypotheses on $\mathcal{X}$ and X using analytic methods.

     Thus, for the rest of this subsection, the following hypotheses are in force (in addition to the standing hypotheses): k is finite of characteristic p, ℓ is a prime distinct from p, and X has a K-rational point. This implies that $\mathcal{X} \to \mathcal{C}$ has a section, and so the multiplicities m_v of the fibers of π are all equal to 1 and π is cohomologically flat.

5.3.1   Comparison of $\mathrm{Br}(\mathcal{X})$ and $\mathrm{III}(J_X)$

These groups are closely related – as we will see later, under the standing hypotheses and assuming k is finite, they differ by a finite group. Assuming also that X has a K-rational point, even more is true.

Proposition 5.3 (Grothendieck). *In addition to the standing hypotheses, assume that k is finite and X has a K-rational point. Then we have a canonical isomorphism* $\mathrm{Br}(\mathcal{X}) \cong \mathrm{III}(J_X)$.

     This is proven in detail in [19, §4] by an argument "assez long et technique". We sketch the main points in the rest of this subsection.

     First of all, since π is cohomologically flat, we have $\pi_* \mathbb{G}_m = \mathbb{G}_m$. Also, we have a vanishing/cohomological dimension result of Artin: $R^q \pi_* \mathbb{G}_m = 0$ for $q > 1$. Thus the Leray spectral sequence for π and $\mathbb{G}_m$ gives a long exact sequence

$$H^2(\mathcal{C}, \mathbb{G}_m) \longrightarrow H^2(\mathcal{X}, \mathbb{G}_m) \longrightarrow H^1(\mathcal{C}, R^1 \pi_* \mathbb{G}_m) \longrightarrow H^3(\mathcal{C}, \mathbb{G}_m) \longrightarrow H^3(\mathcal{X}, \mathbb{G}_m).$$

Now $H^2(\mathcal{C}, \mathbb{G}_m) = \mathrm{Br}(\mathcal{C}) = 0$ since $\mathcal{C}$ is a smooth complete curve over a finite field. Also, since π has a section, $H^3(\mathcal{C}, \mathbb{G}_m) \to H^3(\mathcal{X}, \mathbb{G}_m)$ is injective. Thus we have an isomorphism

$$\mathrm{Br}(\mathcal{X}) = H^2(\mathcal{X}, \mathbb{G}_m) \cong H^1(\mathcal{C}, R^1 \pi_* \mathbb{G}_m).$$

     Now write $\mathcal{F} = R^1 \pi_* \mathbb{G}_m$ and consider the inclusion $\eta = \mathrm{Spec}\, K \hookrightarrow \mathcal{C}$. Let $\mathcal{G} = j_* j^* \mathcal{F}$, so that we have a canonical morphism $\mathcal{F} \longrightarrow \mathcal{G}$. If $\bar{x}$ is a geometric point over a closed point of $\mathcal{C}$, we write $\tilde{\mathcal{X}}_{\bar{x}}$, $\tilde{\eta}_{\bar{x}}$, and $\tilde{\mathcal{C}}_{\bar{x}}$ for the strict henselizations at $\bar{x}$. Then the stalk of $\mathcal{F} \to \mathcal{G}$ at $\bar{x}$ is

$$\mathrm{Pic}\left(\tilde{\mathcal{X}}_{\bar{x}}/\tilde{\mathcal{C}}_{\bar{x}}\right) \longrightarrow \mathrm{Pic}\left(\tilde{\mathcal{X}}_{\tilde{\eta}_{\bar{x}}}/\tilde{\eta}_{\bar{x}}\right).$$

Since the Brauer group of $\tilde{\eta}_{\bar{x}}$ vanishes, the displayed map is surjective (cf. the exact sequence (3.1)). The kernel is zero at all $\bar{x}$ where the fiber of π is reduced and irreducible. Thus we have an exact sequence

$$0 \longrightarrow \mathcal{K} \longrightarrow \mathcal{F} \longrightarrow \mathcal{G} \longrightarrow 0$$

where $\mathcal{K}$ is a skyscraper sheaf supported on the points of $\mathcal{C}$ where the fibers of π are reducible. (See [19, pp. 115–118] for a more complete description of this sheaf and its cohomology.) Under our hypotheses (k finite and X with a K-rational point), every fiber of π has a component of multiplicity 1, and this is enough to ensure that $H^1(\mathcal{C}, \mathcal{K}) = 0$.

Now write $\mathcal{X}_x$ and $\mathcal{C}_x$ for the ordinary henselizations of $\mathcal{X}$ and $\mathcal{C}$ at a closed point x of $\mathcal{C}$. Then we have a diagram

$$\begin{array}{ccccccc}
0 & \longrightarrow & H^1(\mathcal{C}, \mathcal{F}) & \longrightarrow & H^1(\mathcal{C}, \mathcal{G}) & \longrightarrow & \prod_x H^2(x, \mathcal{K}_x) \\
& & \downarrow & & \downarrow & & \| \\
0 & \longrightarrow & \prod_x H^1(\mathcal{C}_x, \mathcal{F}) & \longrightarrow & \prod_x H^1(\mathcal{C}_x, \mathcal{G}) & \longrightarrow & \prod_x H^2(x, \mathcal{K}_x).
\end{array}$$

Lang's theorem implies that $H^1(\mathcal{C}_x, \mathcal{F}) = 0$ for all x. Thus $\mathrm{Br}(\mathcal{X}) = H^1(\mathcal{C}, \mathcal{F})$ is identified with the subgroup of $H^1(\mathcal{C}, \mathcal{G})$ consisting of elements which go to zero in $H^1(\mathcal{C}_x, \mathcal{G})$ for all x. Call this group $\mathrm{III}(\mathcal{C}, \mathcal{G})$.

Finally, the Leray spectral sequence for $\eta \hookrightarrow \mathcal{C}$ and the definition of $\mathcal{G}$ leads to an exact sequence

$$0 \longrightarrow \mathrm{III}(\mathcal{C}, \mathcal{G}) \longrightarrow H^1(\eta, j^*\mathcal{F}) \longrightarrow \coprod_x H^1(\kappa(\mathcal{C}_x), j^*\mathcal{F})$$

where $\kappa(\mathcal{C}_x)$ is the field of fractions of the henselization of $\mathcal{C}$ at x. This identifies $\mathrm{III}(\mathcal{C}, \mathcal{G})$ with $\mathrm{III}(J_X)$ as defined in Subsection 5.2.1.

This completes our sketch of Grothendieck's theorem. The reader is encouraged to consult [19] for a much more general discussion delivered in the inimitable style of the master.

One consequence of the proposition is that (under the hypotheses there) the conjectures on the finiteness of $\mathrm{Br}(\mathcal{X})$ and $\mathrm{III}(J_X)$ are equivalent, and thus so are $T_1(\mathcal{X}, \ell)$ and $S(J_X, \ell)$ for all ℓ. We will see below that this holds even without the supplementary hypothesis on X.

5.3.2 Comparison of the Selmer group and $H^1(\overline{\mathcal{C}}, R^1\pi_*\mathbb{Z}_\ell(1))^{G_k}$

The arguments in Subsection 2.4 show that under our hypotheses the Leray spectral sequence degenerates at the integral level as well, and so $H^2(\overline{\mathcal{X}}, \mathbb{Z}_\ell(1))$ has a filtration whose graded pieces are the groups $H^i(\overline{\mathcal{C}}, R^j\pi_*\mathbb{Z}_\ell(1))$ where $i + j = 2$.

As we will see below, from the point of view of the Tate conjecture, the interesting part of $H^2(\overline{\mathcal{X}}, \mathbb{Z}_\ell(1))$ is $H^1(\overline{\mathcal{C}}, R^1\pi_*\mathbb{Z}_\ell(1))$. It turns out that its G_k-invariant part is closely related to the ℓ-adic Selmer group of J_X – they differ by a finite group.

Before being more precise, we make the groups $H^i(\overline{C}, \mathbb{Z}_\ell(1))$ more explicit. Since the fibers of π are connected, $\pi_* \mathbb{Z}_\ell(1) = \mathbb{Z}_\ell(1)$ and so $H^2(\overline{C}, \pi_* \mathbb{Z}_\ell(1)) = \mathbb{Z}_\ell$. It is easy to see that, under the map $H^2(\overline{X}, \mathbb{Z}_\ell(1)) \to H^2(\overline{C}, \pi_* \mathbb{Z}_\ell(1))$, the cycle class of a section maps to a generator of $H^2(\overline{C}, \pi_* \mathbb{Z}_\ell(1))$.

The stalk of $R^2\pi_* \mathbb{Z}_\ell(1)$ at a geometric point $\overline{x}$ over a closed point x of $\mathcal{C}$ is $\mathbb{Z}_\ell^{f_{\overline{x}}}$ where $f_{\overline{x}}$ is the number of irreducible components in the geometric fiber. The action of G_k permutes the factors as it permutes the components. If we write the fiber as $\sum_i m_{\overline{x},i} \mathcal{X}_{\overline{x},i}$ as in Subsection 2.3, then the specialization map

$$\mathbb{Z}_\ell^{f_{\overline{x}}} = \left(R^2\pi_* \mathbb{Z}_\ell(1)\right)_{\overline{x}} \longrightarrow \left(R^2\pi_* \mathbb{Z}_\ell(1)\right)_{\overline{\eta}} = \mathbb{Z}_\ell$$

is $(c_i) \mapsto \sum_i m_{\overline{x},i} c_i$. It follows that $H^0(\overline{C}, R^2\pi_* \mathbb{Z}_\ell(1))^{G_k}$ has rank $1 + \sum_v (f_v - 1)$ where f_v is the number of irreducible components of the fiber $\mathcal{X}_v$ (as a scheme over the residue field k_v). Also, the cycle classes of components of fibers lie in

$$H^0\left(\overline{C}, R^2\pi_* \mathbb{Z}_\ell(1)\right)^{G_k} \subset H^2\left(\overline{X}, \mathbb{Z}_\ell(1)\right)^{G_k}$$

and span this subgroup.

Thus we see that the cycle class map induces a well-defined homomorphism

$$\left(L^1 \operatorname{NS}(\mathcal{X})/L^2 \operatorname{NS}(\mathcal{X})\right) \otimes \mathbb{Z}_\ell \longrightarrow H^1(\overline{C}, R^1\pi_* \mathbb{Z}_\ell(1))^{G_k}$$

which is surjective if and only if $T_1(\mathcal{X}, \ell)$ holds. Since the source of this map is closely related to $J_X(K)$ – it is exactly $(J_X(K)/\tau B(k)) \otimes \mathbb{Z}_\ell$ by the Shioda–Tate theorem and $B(k)$ is finite – we must have a close connection between the groups $H^1(\overline{C}, R^1\pi_* \mathbb{Z}_\ell(1))^{G_k}$ and $\operatorname{Sel}(J_X, \mathbb{Z}_\ell)$. We will prove directly that they differ by a finite group.

We will only sketch a proof, since a complete treatment requires more details on bad reduction and Néron models than we have at our disposal. However, the main ideas should be clear.

To state the result, let $K' = K\overline{k}$ and define another Selmer group as follows:

$$\operatorname{Sel}\left(J_x/K', \mathbb{Z}_\ell\right) = \varprojlim_n \ker\left(H^1\left(K', J_X[\ell^n]\right) \longrightarrow \prod_v H^1\left(K'_v, J_X\right)\right),$$

where the product is over the places of K'.

Proposition 5.4. *There are homomorphisms* $\operatorname{Sel}(J_X, \mathbb{Z}_\ell) \to \operatorname{Sel}(J_X/K', \mathbb{Z}_\ell)^{G_k}$ *and* $H^1(\overline{C}, R^1\pi_* \mathbb{Z}_\ell(1))^{G_k} \to \operatorname{Sel}(J_X/K', \mathbb{Z}_\ell)^{G_k}$ *with finite kernels and cokernels.*

Sketch of proof of Proposition 5.4. Inflation-restriction for $G_{K'} \subset G_K$ yields sequences

$$0 \longrightarrow H^1\left(G_k, J_X[\ell^n](K')\right) \longrightarrow H^1\left(G_K, J_X[\ell^n]\right) \longrightarrow H^1\left(G_{K'}, J[\ell^n]\right)$$
$$\longrightarrow H^2\left(G_k, J_X[\ell^n](K')\right).$$

Now $J[\ell^n](K')$ is an extension of (finite) by $B[\ell^n](\overline{k})$ where (finite) is bounded independently of n, and B is the K/k-trace of J_X. Using Lang's theorem, we have that the kernel and cokernel of $H^1(G_K, T_\ell J) \to H^1(G_{K'}, T_\ell J)$ are finite and therefore so is the kernel of $\mathrm{Sel}(J_X, \mathbb{Z}_\ell) \to \mathrm{Sel}(J_X/K', \mathbb{Z}_\ell)^{G_k}$.

To control the cokernel, consider the following commutative diagram with exact rows and columns:

$$
\begin{array}{ccc}
0 & & 0 \\
\downarrow & & \downarrow \\
\mathrm{Sel}\left(J_X, \ell^n\right) & \longrightarrow & \mathrm{Sel}\left(J_X/K', \ell^n\right)^{G_k} \\
\downarrow & & \downarrow \\
H^1\left(G_K, J_X[\ell^n]\right) & \longrightarrow & H^1\left(G_{K'}, J[\ell^n]\right)^{G_k} \\
\downarrow & & \downarrow \\
\prod_v H^1\left(G_{k_v}, J_X(\overline{k}K_v)\right) & \longrightarrow \prod_v H^1\left(G_{K_v}, J\right) \longrightarrow \prod_v \left(\prod_{\overline{v}|v} H^1\left(G_{K'_{\overline{v}}}, J\right)\right)^{G_k}.
\end{array}
$$

The columns define the Selmer groups and the bottom row is a product of inflation-restriction sequences. By [40, I.3.8], the group on the left is zero for all but finitely many v and it is finite at all v. A diagram chase then shows that $\mathrm{Sel}(J_X, \mathbb{Z}_\ell) \to \mathrm{Sel}(J_X/K', \mathbb{Z}_\ell)^{G_k}$ has finite cokernel.

Now let $\mathcal{F} = j_* j^* \mathbb{Z}_\ell(1)$ where $j \colon \eta \hookrightarrow \mathcal{C}$ is the inclusion of the generic point. By Corollary 2.4, we may replace $H^1(\overline{\mathcal{C}}, R^1\pi_* \mathbb{Z}_\ell(1))$ with $H^1(\overline{\mathcal{C}}, \mathcal{F})$. The fact that $\mathcal{F}$ is a middle extension sheaf gives a III-like description of $H^1(\overline{\mathcal{C}}, \mathcal{F})$: there is an exact sequence

$$
0 \longrightarrow H^1\left(\overline{\mathcal{C}}, \mathcal{F}\right) \longrightarrow H^1\left(K', T_\ell J_X\right) \longrightarrow \prod_{\overline{v}} H^1\left(K'_{\overline{v}}, T_\ell J_X\right).
$$

On the other hand, $\mathrm{Sel}(J_X/K', \mathbb{Z}_\ell)$ is defined by a similar sequence, except that instead of vanishing in $H^1(K'_{\overline{v}}, T_\ell J_X)$, a class in the Selmer group should land in the image of $J_X(K'_{\overline{v}}) \hat{\otimes} \mathbb{Z}_\ell \to H^1(K'_{\overline{v}}, T_\ell J_X)$. But $J(K'_{\overline{v}})$ is an extension of a finite group by an ℓ-divisible group, and for all but finitely many places $\overline{v}$ the finite group is trivial. (This follows from the structure of the Néron model of J_X.) Thus we get a inclusion $H^1(\overline{\mathcal{C}}, \mathcal{F}) \subset \mathrm{Sel}(J_X/K', \mathbb{Z}_\ell)$ with finite cokernel. Taking G_k-invariants finishes the proof. $\qquad\square$

6 Zeta and *L*-functions

6.1 Zeta functions and T_2 for $\mathcal{X}$

We quickly review the zeta function and the second Tate conjecture for $\mathcal{X}$. Most of this material was covered in [70] and so we will be very brief.

Throughout this section, $k = \mathbb{F}_q$ is the finite field with q elements, p is its characteristic, and $\mathcal{X}$ is a surface satisfying our usual hypotheses (as in Subsection 1.4).

6.1.1 Zetas

Let N_n be the number of points on $\mathcal{X}$ rational over $\mathbb{F}_{q^n}$ and form the Z-function

$$Z(\mathcal{X}, T) = \prod_{x \in \mathcal{X}^0} \left(1 - T^{\deg x}\right)^{-1} = \exp\left(\sum_{n \geq 1} N_n \frac{T^n}{n}\right)$$

and the zeta function $\zeta(\mathcal{X}, s) = Z(\mathcal{X}, q^{-s})$.

We choose an auxiliary prime $\ell \neq p$. Then the Grothendieck–Lefschetz trace formula

$$N_n = \sum_{i=0}^{4} (-1)^i \operatorname{Tr}\left(\operatorname{Fr}_q^n | H^i(\overline{\mathcal{X}}, \mathbb{Q}_\ell)\right)$$

leads to an expression for $Z(\mathcal{X}, T)$ as a rational function of T:

$$Z(\mathcal{X}, T) = \prod_{i=0}^{4} P_i(\mathcal{X}, T)^{(-1)^{i+1}}$$

where

$$P_i(\mathcal{X}, T) = \det\left(1 - T \operatorname{Fr}_q | H^i(\overline{\mathcal{X}}, \mathbb{Q}_\ell)\right) = \prod_j \left(1 - \alpha_{ij} T\right).$$

By the Poincaré duality theorem, there is a functional equation relating $\zeta(\mathcal{X}, 1 - s)$ and $\zeta(\mathcal{X}, s)$, and therefore relating $P_i(\mathcal{X}, T)$ with $P_{4-i}(\mathcal{X}, q/T)$.

By Deligne's theorem, the eigenvalues of Frobenius α_{ij} are Weil numbers of size $q^{i/2}$. It follows that the zeroes and poles of $\zeta(\mathcal{X}, s)$ have real parts in the sets $\{1/2, 3/2\}$ and $\{0, 1, 2\}$ respectively, and that the order of pole of $\zeta(\mathcal{X}, s)$ at $s = 1$ is equal to the multiplicity of q as an eigenvalue of Fr_q on $H^2(\overline{\mathcal{X}}, \mathbb{Q}_\ell)$.

6.1.2 Tate's second conjecture

The second main conjecture in Tate's article [60] relates the ζ function to the Néron–Severi group.

Conjecture 6.1 $(T_2(\mathcal{X}))$. *We have*

$$\operatorname{Rank} \operatorname{NS}(\mathcal{X}) = -\operatorname{ord}_{s=1} \zeta(\mathcal{X}, s).$$

Recall from (5.1) that the cycle class map induces an injection

$$\operatorname{NS}(\mathcal{X}) \otimes \mathbb{Z}_\ell \longrightarrow H^2(\overline{\mathcal{X}}, \mathbb{Z}_\ell(1))^{G_k}$$

and Conjecture $T_1(\mathcal{X}, \ell)$ was the statement that this map is an isomorphism. When $\ell \neq p$, we have $H^2(\overline{\mathcal{X}}, \mathbb{Z}_\ell(1))^{G_k} \cong H^2(\overline{\mathcal{X}}, \mathbb{Z}_\ell)^{\operatorname{Fr}_q = q}$, where the latter is the subspace where the q-power Frobenius acts by multiplication by q. When $\ell = p$, $H^2(\overline{\mathcal{X}}, \mathbb{Z}_\ell(1))^{G_k}$ is isomorphic to $H^2_{\mathrm{cris}}(\mathcal{X}/W(k))^{\mathrm{Fr}=p}$, the subgroup of crystalline cohomology where the absolute Frobenius acts by multiplication by p. Thus, in either case the $\mathbb{Z}_\ell$-rank of the target of the cycle class map (an eigenspace) is bounded above by the order of vanishing of the ζ function (which is the multiplicity of the corresponding eigenvalue). This proves the first two parts of the following.

Theorem 6.2 (Artin–Tate, Milne). *We have:*

1. $\mathrm{Rank}\, \mathrm{NS}(\mathcal{X}) \leq -\mathrm{ord}_{s=1}\, \zeta(\mathcal{X}, s).$
2. $T_2(\mathcal{X})$ *(i.e., equality in 1) implies* $T_1(\mathcal{X}, \ell)$ *and thus finiteness of* $\mathrm{Br}(\mathcal{X})[\ell^\infty]$ *for all* ℓ.
3. $T_1(\mathcal{X}, \ell)$ *for any* ℓ *(the case* $\ell = p$ *is allowed) implies* $T_2(\mathcal{X})$.

We sketch the proof of the last part in the next subsection. The prime-to-p was discussed/sketched in [70, Lecture 2, 10.2] and the general case is similar, although it uses more sophisticated cohomology.

6.1.3 Artin–Tate conjecture

We define two more invariants which enter into the Artin–Tate conjecture below.

Recall that there is a symmetric, integral intersection pairing on the Néron–Severi group which gives $\mathrm{NS}(\mathcal{X})/\mathrm{tor}$ the structure of a lattice. We define the regulator $R(\mathcal{X})$ to be the discriminant of this lattice. More precisely,

$$R = R(\mathcal{X}) = \left| \det\left(D_i . D_j \right)_{i,j=1,\ldots,\rho} \right|,$$

where D_i ($i = 1, \ldots, \rho$) is a basis of $\mathrm{NS}(\mathcal{X})/\mathrm{tor}$ and the dot denotes the intersection product.

We also define a Tamagawa-like factor $q^{-\alpha(\mathcal{X})}$ where

$$\alpha = \alpha(\mathcal{X}) = \chi(\mathcal{X}, \mathcal{O}_{\mathcal{X}}) - 1 + \dim \mathrm{PicVar}\,(\mathcal{X}).$$

The following was inspired by the *BSD* conjecture for the Jacobian of X.

Conjecture 6.3 ($AT(\mathcal{X})$). $T_2(\mathcal{X})$ *holds,* $\mathrm{Br}(\mathcal{X})$ *is finite, and we have the asymptotic*

$$\zeta(\mathcal{X}, s) \sim \frac{R\, |\mathrm{Br}(\mathcal{X})|\, q^{-\alpha}}{|\mathrm{NS}(\mathcal{X})_{tor}|^2}\left(1 - q^{1-s}\right)^{\rho(\mathcal{X})}$$

as $s \to 1$.

The analysis showing that $T_1(\mathcal{X}, \ell) \Rightarrow T_2(\mathcal{X})$ can be pushed further to show that $T_2(\mathcal{X}) \Rightarrow AT(\mathcal{X})$. Indeed, we have the following spectacular result:

Theorem 6.4 (Artin–Tate, Milne). *If* $T_2(\mathcal{X})$ *holds, then so does* $AT(\mathcal{X})$.

To prove this, we must show that for all but finitely many ℓ, $\mathrm{Br}(\mathcal{X})[\ell]$ is trivial, and also relate the order of $\mathrm{Br}(\mathcal{X})$ and other invariants to the leading term of the zeta function.

The proofs of the third part of Theorem 6.2 and Theorem 6.4 proceed via a careful consideration of the following big commutative diagram:

$$
\begin{array}{ccc}
\mathrm{NS}\,(\mathcal{X}) \otimes \hat{\mathbb{Z}} \xrightarrow{\ e\ } \mathrm{Hom}\,(\,\mathrm{NS}\,(\mathcal{X}) \otimes \hat{\mathbb{Z}}, \mathbb{Z}_\ell) == \mathrm{Hom}\,(\,\mathrm{NS}\,(\mathcal{X}) \otimes \mathbb{Q}/\mathbb{Z}, \mathbb{Q}/\mathbb{Z}) \\
\left\downarrow h \right. \qquad\qquad\qquad\qquad\qquad\qquad\qquad\qquad\qquad \left\uparrow g^* \right. \\
H^2\big(\overline{\mathcal{X}}, T\mu\big)^{G_k} \xrightarrow{\ f\ } H^2\big(\overline{\mathcal{X}}, T\mu\big)_{G_k} \xrightarrow{\ j\ } \mathrm{Hom}\,\big(H^2(\mathcal{X}, \mu(\infty))\big)^{G_k}, \mathbb{Q}/\mathbb{Z}\big).
\end{array}
$$

Here $H^2(\overline{\mathcal{X}}, T\mu)$ means the inverse limit over n of the flat cohomology groups $H^2(\overline{\mathcal{X}}, \mu_n)$, and $H^2(\mathcal{X}, \mu(\infty))$ is the direct limit over n of the flat cohomology groups $H^2(\mathcal{X}, \mu_n)$. The map e is induced by the intersection form, h is the cycle class map, f is induced by the identity map of $H^2(\overline{\mathcal{X}}, T\mu)$, g^* is the transpose of a map

$$
g \colon \mathrm{NS}(\mathcal{X}) \otimes \mathbb{Q}/\mathbb{Z} \longrightarrow H^2(\mathcal{X}, \mu(\infty))
$$

obtained by taking the direct limit over positive integers n of the Kummer map $\mathrm{NS}(\mathcal{X}) \otimes \mathbb{Z}/n\mathbb{Z} \to H^2(\mathcal{X}, \mu_n)$, and j is induced by the Hochschild–Serre spectral sequence and Poincaré duality.

We say that a homomorphism $\phi \colon A \to B$ of abelian groups is a *quasi-isomorphism* if it has finite kernel and cokernel. In this case, we define

$$
z(\phi) = \frac{\#\ker(\phi)}{\#\mathrm{coker}(\phi)}.
$$

It is easy to check that if $\phi_3 = \phi_2\phi_1$ (composition) and if two of the maps ϕ_1, ϕ_2, ϕ_3 are quasi-isomorphisms, then so is the third and we have $z(\phi_3) = z(\phi_2)z(\phi_1)$.

In the diagram above, the map e is a quasi-isomorphism and $z(e)$ is the order of the torsion subgroup of $\mathrm{NS}(\mathcal{X})$ divided by the discriminant of the intersection form. The map j is also a quasi-isomorphism and $z(j)$ is the order of the torsion subgroup of Néron–Severi times a power of q determined by a certain p-adic cohomology group.

Now assume $T_1(\mathcal{X}, \ell)$ for one ℓ and consider the ℓ-primary part of the diagram above. The assumption $T_1(\mathcal{X}, \ell)$ implies that the ℓ parts of the maps h and g^* are quasi-isomorphisms. Thus the ℓ part of f must also be a quasi-isomorphism. This means that Fr_q acts semisimply on the generalized eigenspace of $H^2(\overline{\mathcal{X}}, \mathbb{Q}_\ell(1))$ corresponding to the eigenvalue 1, and this implies $T_2(\mathcal{X})$.

Now assume $T_2(\mathcal{X})$, so we have that $T_1(\mathcal{X}, \ell)$ holds for all ℓ and that h is an isomorphism. A fairly intricate analysis leads to a calculation of $z(f)$ in terms of the zeta function of $\mathcal{X}$ and shows that $z(g^*)$ (as a "supernatural number") is the order of the Brauer group. Since all the other z's in the diagram are rational numbers, so is $z(g^*)$ and we get finiteness of $\mathrm{Br}(\mathcal{X})$. The product formula

$$z(e) = z(h)\, z(f)\, z(j)\, z(g^*)$$

leads, after some delicate work in p-adic cohomology, to the leading coefficient formula in the conjecture $AT(\mathcal{X})$. We refer to Milne's paper [38] for the details.

6.1.4 The case $p = 2$ and de Rham–Witt cohomology

In Milne's paper [38], the case $\ell = p = 2$ had to be excluded due to the state of p-adic cohomology at the time. More complete results are now available, and so this restriction is no longer needed. This is pointed out by Milne on his web site and was implicit in the more general results in [42].

In slightly more detail, what is needed is a bridge between crystalline co-homology (which calculates the zeta function and receives cycle classes) and flat cohomology (which is closely related to the Brauer group). This bridge is provided by the cohomology of the de Rham–Witt complex, generalizing Serre's Witt vec-tor cohomology. Such a theory was initiated by Bloch, whose approach required $p > 2$, and this is the source of the original restriction in [38]. Soon afterward, Illusie developed a different approach without any restriction on the characteristic, and important further developments were made by Illusie–Raynaud, Ekedahl, and Milne.

Briefly, the theory provides groups $H^j(\mathcal{X}, W\Omega^i)$ with operators F and V and a spectral sequence

$$E_1^{ij} = H^j(\mathcal{X}, W\Omega^i) \implies H_{\mathrm{cris}}^{i+j}(\mathcal{X}/W)$$

which degenerates at E_1 modulo torsion. There are also variants involving sheaves of cycles $ZW\Omega^i$, boundaries $BW\Omega^i$, and logarithmic differentials $W\Omega_{\log}^i$. The var-ious cohomology groups (modulo torsion) are related by the spectral sequence to pieces of crystalline cohomology defined by "slope" conditions, i.e., by the valua-tions of eigenvalues of Frobenius. The cohomology of the logarithmic differentials for $i = 1$ is closely related to the flat cohomology

$$H^j(\mathcal{X}, \mathbb{Z}_p(1)) = \varprojlim_n H^j(\mathcal{X}, \mu_{p^n}).$$

The torsion in the de Rham–Witt cohomology can be "large" and it provides further interesting invariants.

I recommend [22] and [23] for a thorough overview of the theory and [42] for further important developments.

6.2 *L*-functions and the *BSD* conjecture for J_X

In this subsection we again assume that $k = \mathbb{F}_q$ and we consider X and J_X satisfying the usual hypotheses. The discussion will be parallel to that of the preceding subsection.

6.2.1 *L*-functions

Choose an auxiliary prime $\ell \neq p$ and consider the ℓ-adic Tate module

$$V_\ell J_X = \left(\varprojlim_n J_X[\ell^n](\overline{K}) \right) \otimes_{\mathbb{Z}_\ell} \mathbb{Q}_\ell$$

equipped with the natural action of G_K. There are canonical isomorphisms

$$V_\ell J_X \cong H^1(\overline{J}_X, \mathbb{Q}_\ell)^* \cong H^1(\overline{X}, \mathbb{Q}_\ell)^* = H^1(\overline{X}, \mathbb{Q}_\ell(1))$$

where the $*$ indicates the dual vector space.

We define

$$L(J_X, T) = \prod_v \det\left(1 - T \, \mathrm{Fr}_v \, | V_\ell^{I_v}\right)$$

where the product runs over the places of K, I_v denotes an inertia group at v, and Fr_v denotes the *arithmetic* Frobenius at v. (Alternatively, we could use $H^1(\overline{X}, \mathbb{Q}_\ell)$ and the geometric Frobenius.) Also, we write $L(J_X, s)$ for the L-function above with $T = q^{-s}$. The product defining $L(J_X, s)$ converges in a half-plane and the cohomological description below shows that it is a rational function of q^{-s} and so it extends meromorphically to all s.

Let $j \colon U \hookrightarrow \mathcal{C}$ be a non-empty open subset over which X has good reduction. Then the representation $G_K \to \mathrm{Aut}(H^1(\overline{X}, \mathbb{Q}_\ell))$ is unramified at all places $v \in U$ and so it defines a lisse sheaf $\mathcal{F}_U$ over U. We set $\mathcal{F} = j_* \mathcal{F}_U$. The resulting constructible sheaf on $\mathcal{C}$ is independent of the choice of U. Its stalk $\mathcal{F}_{\overline{x}}$ at a geometric point $\overline{x}$ over closed point $x \in \mathcal{C}$ is the group of inertial invariants $H^1(\overline{X}, \mathbb{Q}_\ell)^{I_x}$.

The Grothendieck–Lefschetz trace formula for $\mathcal{F}$ reads

$$\sum_{x \in \mathcal{C}(\mathbb{F}_{q^n})} \mathrm{Tr}\left(\mathrm{Fr}_x \, | \mathcal{F}_{\overline{x}} \right) = \sum_i (-1)^i \, \mathrm{Tr}\left(\mathrm{Fr}_{q^n} \, | H^i(\overline{\mathcal{C}}, \mathcal{F}) \right)$$

and this leads to a cohomological expression for the L-function as a rational function in T:

$$L(J_X, T) = \prod_{i=0}^{2} Q_i(T)^{(-1)^{i+1}} \tag{6.1}$$

where

$$Q_i(J_X, T) = \det\left(1 - T \, \mathrm{Fr}_q \, | H^i(\overline{\mathcal{C}}, \mathcal{F})\right) = \prod_j (1 - \beta_{ij} T).$$

By Deligne's theorem, the β_{ij} are Weil integers of size $q^{(i+1)/2}$.

It is convenient to have a criterion for $L(J_X, T)$ to be a polynomial.

Lemma 6.5. *$L(J_X, T)$ is a polynomial in T if and only if $H^0(\overline{\mathcal{C}}, \mathcal{F}) = H^2(\overline{\mathcal{C}}, \mathcal{F}) = 0$ if and only if the K/k-trace $B = \mathrm{Tr}_{K/k} J_X$ is zero.*

Proof. The first equivalence is immediate from the cohomological formula (6.1). For the second equivalence, we use the Lang–Néron theorem which says that if B is zero, then $J_X(\bar{k}K)$ is finitely generated. In this case, its torsion subgroup is finite and so

$$H^0(\bar{\mathcal{C}}, \mathcal{F}) = \varprojlim_n \left(J_X[\ell^n](\bar{k}K) \right) = 0.$$

Conversely, if $B \neq 0$, then $V_\ell B \subset V_\ell J_X$ and we see that $H^0(\bar{\mathcal{C}}, \mathcal{F})$ contains a subspace of dimension $2 \dim B > 0$. Thus $B = 0$ is equivalent to $H^0(\bar{\mathcal{C}}, \mathcal{F}) = 0$. That these statements are equivalent to $H^2(\bar{\mathcal{C}}, \mathcal{F}) = 0$ follows from duality (using autoduality up to a Tate twist of $\mathcal{F}$). □

Whether or not $L(J_X, s)$ is a polynomial, its zeroes lie on the line $\operatorname{Re} s = 1$, and the order of zero at $s = 1$ is equal to the multiplicity of q as an eigenvalue of Fr_q on $H^1(\bar{\mathcal{C}}, \mathcal{F})$.

6.2.2 The basic *BSD* conjecture

Conjecture 6.6 (*BSD*)**.** *We have*

$$\operatorname{Rank} J_X(K) = \operatorname{ord}_{s=1} L(J_X, s).$$

Parallel to the Tate conjecture case, we have an inequality and implications among the conjectures.

Proposition 6.7. *We have*

1. $\operatorname{Rank} J_X(K) \leq \operatorname{ord}_{s=1} L(J_X, s)$.
2. *BSD (i.e., equality in 1) implies the Tate–Shafarevich conjecture $TS(J_X, \ell)$ (and thus also the Selmer group conjecture $S(J_X, \ell)$) for all ℓ.*
3. *$TS(J_X, \ell)$ for any ℓ implies BSD.*

Proof. The discussion in Subsection 5.3.2 shows that

$$-\operatorname{ord}_{s=1} \zeta(\mathcal{X}, s) - \operatorname{ord}_{s=1} L(J_X, s)$$
$$= \dim H^2(\overline{\mathcal{X}}, \mathbb{Q}_\ell(1))^{G_k} - \dim H^1(\bar{\mathcal{C}}, \mathbb{R}^1 \pi_* \mathbb{Q}_\ell(1))^{G_k}$$
$$= 2 + \sum_v (f_v - 1).$$

By the Shioda–Tate theorem,

$$\operatorname{Rank} \operatorname{NS}(\mathcal{X}) - \operatorname{Rank} J_X(K) = 2 + \sum_v (f_v - 1).$$

Thus

$$\operatorname{Rank} J_X(K) - \operatorname{ord}_{s=1} L(J_X, s) = \operatorname{Rank} \operatorname{NS}(\mathcal{X}) + \operatorname{ord}_{s=1} \zeta(\mathcal{X}, s)$$

and so Rank $J_X(K) \le \mathrm{ord}_{s=1} L(J_X(K), s)$ with equality if and only if $T_2(\mathcal{X})$ holds. But $T_2(\mathcal{X})$ implies the finiteness of the ℓ-primary part of $\mathrm{Br}(\mathcal{X})$ for all ℓ and thus finiteness of the ℓ-primary part of $\mathrm{III}(J_X)$ for all ℓ. By Proposition 5.2, this proves that BSD implies the Selmer group conjecture $S(J_X, \ell)$ for all ℓ.

Conversely, if $S(J_X, \ell)$ holds for some ℓ, then the ℓ-primary part of $\mathrm{III}(J_X)$ is finite and so is the ℓ-primary part of $\mathrm{Br}(\mathcal{X})$. By Proposition 5.1 this implies $T_1(\mathcal{X}, \ell)$ and therefore $T_2(\mathcal{X})$. By the previous paragraph, this implies BSD.   □

6.2.3   The refined BSD conjecture

We need two more invariants. Recall that we have the canonical height pairing

$$J_X(K) \times J_X(K) \longrightarrow \mathbb{R}$$
$$(P, Q) \longmapsto \langle P, Q \rangle,$$

which is bilinear and symmetric, and which takes values in $\mathbb{Q} \cdot \log q$. It makes $J_X(K)/\mathrm{tor}$ into a Euclidean lattice. We define the regulator of J_X to be the discriminant of the height pairing,

$$R = R(J_X) = \left| \det \left(\langle P_i, P_j \rangle \right)_{i,j=1,\ldots,r} \right|,$$

where $P_1, \ldots, P_r$ is a basis of $J_X(K)/\mathrm{tor}$. The regulator is $(\log q)^r$ times a rational number (with an *a priori* bounded denominator).

We also define the Tamagawa number of J_X as follows. Choose a non-zero, top-degree differential form ω on J_X over K. At each place v, let a_v be the integer such that $\pi_v^{a_v} \omega$ is a Néron differential at v. (Here π_v is a generator of the maximal ideal $\mathfrak{m}_v$ at v.) Also, let c_v be the number of connected components in the special fiber of the Néron model at v. Then we set

$$\tau = \tau(J_X) = q^{g_X(1-g_C)} \prod_v q^{a_v} c_v.$$

(See [62] for a less *ad hoc* version of this definition.)

The refined BSD conjecture ($rBSD$ for short) relates the leading coefficient of the L-function at $s = 1$ to the other invariants attached to J_X.

Conjecture 6.8 ($rBSD$). *BSD holds, $\mathrm{III}(J_X)$ is finite, and we have the asymptotic*

$$L(J_X, s) \sim \frac{R|\mathrm{III}(J_X)|\tau}{|J_X(K)_{tor}|^2} (s-1)^r$$

as $s \to 1$.

As with the Tate and Artin–Tate conjectures, the basic conjecture implies the refined conjecture.

Theorem 6.9 (Kato–Trihan). *If J_X satisfies the BSD conjecture, then it also satisfies the $rBSD$ conjecture.*

(Kato and Trihan [24] actually treat the general case of abelian varieties, not just Jacobians.)

This is a difficult theorem which completes a long line of research by many authors, including Artin, Tate, Milne, Gordon, Schneider, and Bauer. Its proof is well beyond what can be covered in these lectures, but we sketch a few of the main ideas.

In [51], Schneider generalized the results of Artin–Tate to abelian varieties: $|\mathrm{III}(A)[\ell^\infty]| < \infty$ for every $\ell \neq p$ implies the prime-to-p part of $rBSD$. In [7], Bauer was able to handle the p-part for abelian varieties with everywhere good reduction. In all of the above, the main thing is to compare a "geometric cohomology" (such as $H^i(\overline{\mathcal{X}}, \mathbb{Z}_\ell)$ or crystalline cohomology) that computes the L-function to an "arithmetic cohomology" (such as the flat cohomology of μ_n) that relates to $A(K)$ and $\mathrm{III}(A)$. Ultimately, on the geometric side one needs an integral theory which is supple enough to handle degenerating coefficients. Log crystalline cohomology does the job in the context of semistable reduction. The Kato–Trihan paper [24] handles the general case by an elaborate argument involving passing to an extension of K where A achieves semistable reduction, using log-syntomic cohomology upstairs to make a comparison with flat cohomology, and showing that the results can be brought back down to K by a Galois argument. It is a *tour de force* of p-adic cohomology theory.

6.3 Summary of implications

The results of this and the preceding two subsections show that many statements related to the BSD and Tate conjectures are equivalent. In fact, there are many redundancies, so the reader is invited to choose his or her favorite way to organize the implications. The net result is the following.

Theorem 6.10. *Let $\mathcal{X}$ be a smooth, proper, geometrically irreducible surface over a finite field k, equipped with a generically smooth morphism $\mathcal{X} \to \mathcal{C}$ to a smooth, proper, geometrically irreducible curve $\mathcal{C}$. Let K be the function field $k(\mathcal{C})$, let $X \to \operatorname{Spec} K$ be the generic fiber of π, and let J_X be the Jacobian of X. Then*

$$\operatorname{ord}_{s=1} L\big(J_X, s\big) - \operatorname{Rank} J_X\big(K\big) = -\operatorname{ord}_{s=1} \zeta\big(\mathcal{X}, s\big) - \operatorname{Rank} \operatorname{NS}\big(\mathcal{X}\big) \geq 0 \quad (6.2)$$

and the following are equivalent:

- *Equality holds in (6.2).*
- $\mathrm{III}(J_X)[\ell^\infty]$ *is finite for every prime number ℓ.*
- $\operatorname{Br}(\mathcal{X})[\ell^\infty]$ *is finite for every prime number ℓ.*
- $\mathrm{III}(J_X)$ *is finite.*
- $\operatorname{Br}(\mathcal{X})$ *is finite.*

If these conditions hold, then the refined BSD conjecture $rBSD(J_X)$ and the Artin–Tate conjecture $AT(\mathcal{X})$ hold as well.

To end the section, we quote one more precise result [35] on the connection between $\mathrm{III}(J_X)$ and $\operatorname{Br}(\mathcal{X})$:

Proposition 6.11 (Liu–Lorenzini–Raynaud). *Assume that the equivalent conditions of Theorem 6.10 hold. Then the order of* $\mathrm{Br}(\mathcal{X})$ *is a square and we have*

$$\left|\text{Ш}(J_X)\right| \prod_v \delta_v \delta_v' = \left| \mathrm{Br}\left(\mathcal{X}\right)\right| \delta^2.$$

Here δ is the index of X, and δ_v and δ_v' are the index and period of $X \times_K K_v$.

We refer to [35] for an interesting history of misconceptions about when the order of the Brauer group or Tate–Shafarevich group is a square.

7   Complements

7.1   Abelian varieties over K

Assume that k is finite and let A be an abelian variety over K. Then the *BSD* and refined *BSD* conjectures make sense for A. (See [24] for the statement, which involves also the dual abelian variety.) Kato–Trihan and predecessors proved in this case too that Rank $A(K) \leq \mathrm{ord}_{s=1} L(A, s)$ with equality if and only if $\text{Ш}(A)[\ell^\infty]$ is finite for one ℓ if and only if $\text{Ш}(A)$ is finite, and that, when these conditions hold, the refined *BSD* conjecture does too.

We add one simple observation to this story:

Lemma 7.1. *If the BSD conjecture holds for Jacobians over K, then it also holds for abelian varieties over K.*

Proof. It is well known that given an abelian variety A over K, there is another abelian variety A' over K and a Jacobian J over K with an isogeny $J \to A \oplus A'$. If *BSD* holds for Jacobians, then it also holds for $A \oplus A'$. But since we have an inequality "rank $\leq$ ord" for abelian varieties, equality for the direct sum implies equality for the factors. Thus *BSD* holds for A as well. $\square$

7.2   Finite k, other special value formulas

Special value conjectures for varieties over finite fields can be made more stream-lined by stating them in terms of Euler characteristics of cohomology of suitable complexes of sheaves (the so-called "motivic cohomology") and the whole zeta function (as opposed to just the piece corresponding to one part of cohomology) – in contrast with the Artin–Tate conjecture, which relates to H^2. See for example [30], [42], and [51]. See also the note [42] on Milne's web site for a particularly streamlined statement using Weil-étale cohomology.

7.3 Finitely generated k

The analytic and algebraic conjectures discussed in the preceding sections have generalizations to the case where k is finitely generated, rather than finite. In this subsection we give a quick overview of the statements and a few words about some of the comparisons.

Assume that k is finitely generated over its prime field. We assume as always that $\mathcal{X}$ is a smooth, proper, geometrically irreducible surface over k equipped with a flat, projective, relatively minimal morphism $\pi\colon \mathcal{X} \to \mathcal{C}$ where $\mathcal{C}$ is a smooth, geometrically irreducible curve over k. As usual, $K = k(\mathcal{C})$ is the function field of $\mathcal{C}$ and $X \to \operatorname{Spec} K$ is the generic fiber of π.

7.3.1 Algebraic conjectures

Fix a prime ℓ (with $\ell = p$ allowed). The conjectures $T_1(\mathcal{X}, \ell)$ and $S(J_X, \ell)$ have straightforward generalizations to the case where k is finitely generated.

As before, we have an exact sequence

$$0 \longrightarrow \operatorname{NS}\left(\overline{\mathcal{X}}\right) \otimes \mathbb{Q}_\ell \longrightarrow H^2\left(\overline{\mathcal{X}}, \mathbb{Q}_\ell(1)\right) \longrightarrow V_\ell \operatorname{Br}\left(\overline{\mathcal{X}}\right) \longrightarrow 0$$

induced from the Kummer sequence as in Subsection 5.1.1.

Taking G_k invariants, we have an exact sequence

$$0 \longrightarrow \operatorname{NS}\left(\mathcal{X}\right) \otimes \mathbb{Q}_\ell \longrightarrow H^2\left(\overline{\mathcal{X}}, \mathbb{Q}_\ell(1)\right)^{G_k} \longrightarrow \left(V_\ell \operatorname{Br}\left(\overline{\mathcal{X}}\right)\right)^{G_K} \longrightarrow 0.$$

There is a 0 on the right since $H^1(k, \operatorname{NS}(\overline{\mathcal{X}}) \otimes \mathbb{Q}_\ell)$ is both a torsion group and a $\mathbb{Q}_\ell$-vector space. Conjecture $T_1(\mathcal{X}, \ell)$ is the statement that the first map above is an isomorphism.

The exact sequence above shows that the vanishing of a certain Brauer group (namely $(V_\ell \operatorname{Br}(\overline{\mathcal{X}}))^{G_K}$) is equivalent to $T_1(\mathcal{X}, \ell)$. Note that one cannot expect that $\operatorname{Br}(\mathcal{X})$ is finite in general, as it may contain copies of $\operatorname{Br}(k)$.

It is also straightforward to generalize the Selmer group and Tate–Shafarevich conjectures. Define groups $\operatorname{Sel}(J_X, \mathbb{Z}_\ell)$ and $\text{Ш}(J_X)$ exactly as in Subsection 5.2.1. Then we have an exact sequence

$$0 \longrightarrow J_X\left(K\right) \otimes \mathbb{Z}_\ell \longrightarrow \operatorname{Sel}\left(J_X, \mathbb{Z}_\ell\right) \longrightarrow T_\ell \text{Ш}\left(J_X\right) \longrightarrow 0.$$

The Selmer group conjecture is that the first map is an isomorphism, and this is obviously equivalent to the vanishing of $T_\ell \text{Ш}(J_X)$.

7.3.2 Setup for analytic conjectures

Recall that k is assumed to be finitely generated over its prime field. This means there is an irreducible, regular scheme Z of finite type over $\operatorname{Spec}\mathbb{Z}$ whose function field is k. Of course Z is only determined up to birational isomorphism and we may shrink it in the course of the discussion. We write d for the dimension of Z.

We now choose models of the data over Z. That is, we choose schemes $\tilde{\mathcal{X}}$ and $\tilde{\mathcal{C}}$ with morphisms

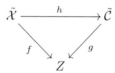

such that the generic fiber is

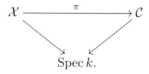

After shrinking Z if necessary, we can and will assume that f, g, and h are smooth and proper. For a closed point $z \in Z$, the residue field at z is finite, and we assume that the fiber over z of the diagram above satisfies our usual hypotheses (in particular h_z should be relatively minimal). Shrinking Z further, we may assume that h is "equisingular" in the sense that as $\bar{z}$ varies through geometric points over closed points $z \in Z$, the fibers $h_{\bar{z}} \colon \tilde{\mathcal{X}}_{\bar{z}} \to \tilde{\mathcal{C}}_{\bar{z}}$ have the same number of singular fibers with the same configurations of components. (More precisely, we may assume that the set of critical values of h is étale over Z.)

7.3.3   Analytic conjectures

For the rest of this subsection, ℓ is a prime $\neq \operatorname{Char}(k)$. For each smooth, projective scheme V of dimension n over a finite field of cardinality q, we write the zeta function of V as

$$\zeta(V, s) = \frac{P_1(V, s) \cdots P_{2n-1}(V, s)}{P_0(V, s) \cdots P_{2n}(V, s)},$$

where the P_i are the characteristic polynomials of Frobenius at q^{-s},

$$P_i(V, s) = \det\left(1 - q^{-s} \operatorname{Fr}_q | H^i(\overline{V}, \mathbb{Q}_\ell)\right).$$

Now define

$$\Phi_2(\tilde{\mathcal{X}}/k, s) = \prod_{z \in Z} P_2(\tilde{\mathcal{X}}_z, s)^{-1}.$$

Here the product is over closed points z of Z and $\tilde{\mathcal{X}}_z$ is the fiber of f over z, which by our hypotheses is a smooth projective surface over the residue field at z.

Similarly, define

$$\Phi_1(X/K, s) = \prod_{c \in \tilde{\mathcal{C}}} P_1(\tilde{\mathcal{X}}_c, s)^{-1};$$

here the product is now over the closed points of $\mathcal{C}$. In this case, the fibers are projective curves, but in general they will not be smooth or irreducible, and so we need a slight extension of our definition of the polynomials P_1. (For the order of vanishing conjecture to follow, we could shrink $\tilde{\mathcal{C}}$ to avoid this problem, but for the comparison between conjectures to follow, it is more convenient to set up the definitions as we have done.)

With these definitions, we have the following conjectures:

$$T_2(\mathcal{X}):\qquad -\operatorname{ord}_{s=d+1}\Phi_2(\mathcal{X}/k,s)=\operatorname{Rank}\operatorname{NS}(\mathcal{X})$$

and

$$BSD(X):\qquad \operatorname{ord}_{s=d+1}\Phi_1(X/K,s)=\operatorname{Rank}J_X(K).$$

7.3.4 Comparison of conjectures

Comparing the various conjectures for finitely generated k seems to be much more complicated than for finite k.

The Shioda–Tate isomorphism (4.1) works for general k and gives a connection between the finitely generated groups $\operatorname{NS}(\mathcal{X})$, $J_X(K)$, and $B(k)$.

To discuss the zeta function side of the analytic conjectures, we make the assumption that $\operatorname{Char}(k)=p>0$. (In characteristic 0, very little can be said about zeta functions without automorphic techniques which are far outside the scope of these notes.) So let us assume that Z has characteristic p and that its field of constants has cardinality q.

In this case, the Grothendieck–Lefschetz trace formula shows that the order of the pole of $\Phi_2(\mathcal{X}/k,s)$ at $s=d+1$ is equal to the multiplicity of q^{d+1} as an eigenvalue of Fr_q on $H_c^{2d}(\overline{Z},R^2f_*\mathbb{Q}_\ell)$ for any $\ell\neq p$. The Leray spectral sequence for $f=gh$ relates $R^2f_*\mathbb{Q}_\ell$ to $R^2g_*(h_*\mathbb{Q}_\ell)$, $R^1g_*(R^1h_*\mathbb{Q}_\ell)$, and $g_*(R^2h_*\mathbb{Q}_\ell)$. It is not hard to work out $H_c^{2d}(\overline{Z},R^2g_*(h_*\mathbb{Q}_\ell))$ and $H_c^{2d}(\overline{Z},g_*(R^2h_*\mathbb{Q}_\ell))$, and one finds that they contribute exactly $1+\sum_v(f_v-1)$ (notation as in Corollary 4.2) to the order of the pole.

It remains to consider $H_c^{2d}(\overline{Z},R^1g_*(R^1h_*\mathbb{Q}_\ell))$. The Leray spectral sequence for h leads to an exact sequence

$$0\longrightarrow H_c^{2d}\big(\overline{Z},R^1g_*\big(R^1h_*\mathbb{Q}_\ell\big)\big)\longrightarrow H_c^{2d+1}\big(\overline{\mathcal{C}},R^1h_*\mathbb{Q}_\ell\big)$$
$$\longrightarrow H_c^{2d-1}\big(\overline{Z},R^2g_*\big(R^1h_*\mathbb{Q}_\ell\big)\big)\longrightarrow 0.$$

The middle group is exactly what controls $\Phi_1(X/K,s)$ (in other words, its order of vanishing at $s=d+1$ is the multiplicity of q^{d+1} as eigenvalue of Fr_q on this group). On the other hand, after some unwinding, one sees that the group $H_c^{2d-1}(\overline{Z},R^2g_*(R^1h_*\mathbb{Q}_\ell))$ is related to B via a BSD conjecture.

Thus it seems possible (after filling in many details) to show that $T_2(\mathcal{X})$ and $BSD(J_X)+BSD(B)$ should be equivalent in positive characteristic. (We remark that Tate's article [60] gives a different analytic comparison, namely between $T_2(\mathcal{X})$ and a BSD conjecture related to the rank of PicVar_X, for general finitely generated k.)

A comparison between the algebraic conjectures along the lines of Proposition 5.4 looks like an interesting project. We remark that it will certainly be more complicated than over a finite k. For example, if $K' = K\overline{k}$, then in the limit

$$\varprojlim_{n} J_X (K')/\ell^n$$

the subgroup $B(\overline{k})$ dies as it is ℓ-divisible. Thus the kernel of a map $\mathrm{Sel}(J_X, \mathbb{Z}_\ell) \to \mathrm{Sel}(J_X/K', \mathbb{Z}_\ell)^{G_k}$ cannot be finite in general.

Finally, a comparison of analytic and algebraic conjectures, along the lines of the results of Artin–Tate and Milne (as always, assuming that k has positive characteristic), also seems to be within the realm of current technology (although the author does not pretend to have worked out the details).

I know of no strong evidence in favor of the conjectures in this subsection beyond the case where k is a global field.

7.4   Large fields k

We simply note that if k is "large" then one cannot expect finiteness of $\mathrm{Br}(\mathcal{X})$ or $\mathrm{III}(J_X)$. For example, the exact sequence

$$0 \longrightarrow \mathrm{NS}\left(\overline{\mathcal{X}}\right) \otimes \mathbb{Z}_\ell \longrightarrow H^2\left(\overline{\mathcal{X}}, \mathbb{Z}_\ell(1)\right) \longrightarrow T_\ell \, \mathrm{Br}\left(\overline{\mathcal{X}}\right) \longrightarrow 0$$

shows that over a separably closed field, when the rank of H^2 is larger than the Picard number, Br has divisible elements.

It may also happen that $\mathrm{Br}(\mathcal{X})$ or $\mathrm{III}(J_X)$ has infinite p-torsion. For examples in the case of III (already for elliptic curves), see [64, 7.12b], where there is a Selmer group which is in a suitable sense linear as a function of the finite ground field, and is thus infinite when $k = \overline{\mathbb{F}}_p$. This phenomenon is closely related to torsion in the de Rham–Witt cohomology groups. See [65] for some related issues.

8   Known cases of the Tate conjecture and consequences

8.1   Homomorphisms of abelian varieties and T_1 for products

Let k be a field finitely generated over its prime field. In the article where he first conjectured $T_1(\mathcal{X}, \ell)$ for smooth projective varieties $\mathcal{X}$ over k, Tate explained that the case of abelian varieties is equivalent to an attractive statement on homomorphisms of abelian varieties. Namely, if A and B are abelian varieties over k, then for any $\ell \neq \mathrm{Char}(k)$ the natural homomorphism

$$\mathrm{Hom}_k\left(A, B\right) \otimes \mathbb{Z}_\ell \longrightarrow \mathrm{Hom}_{G_k}\left(T_\ell A, T_\ell B\right) \tag{8.1}$$

should be an isomorphism.

In [61], Tate gave an axiomatic framework showing that isomorphism in (8.1) follows from a fundamental finiteness statement for abelian varieties over k, which is essentially trivial for finite k. The finiteness statement (and thus isomorphism in (8.1)) was later proven for all finitely generated fields of characteristic $p > 0$ by Zarhin [73] ($p > 2$) and Mori ($p = 2$) – see [43] – and for finitely generated fields of characteristic zero by Faltings [14].

Isomorphism in (8.1) in turn implies $T_1(\mathcal{X}, \ell)$ for products of curves and abelian varieties (and more generally for products of varieties for which T_1 is known). This is explained, for example, in [61] or [70].

Summarizing the part of this most relevant for us, we have the following statement.

Theorem 8.1. *If k is a finitely generated field and $\mathcal{X}$ is a product of curves over k, then for all $\ell \neq \mathrm{Char}(k)$ there is an isomorphism*

$$\mathrm{NS}\left(\mathcal{X}\right) \otimes \mathbb{Z}_\ell \xrightarrow{\sim} H^2\left(\overline{\mathcal{X}}, \mathbb{Z}_\ell(1)\right)^{G_k}.$$

8.2 Descent property of T_1 and domination by a product of curves

The property in question is the following.

Lemma 8.2. *Suppose k is a finitely generated field and $\mathcal{X}$ is smooth, proper variety over k satisfying T_1. If $\mathcal{X} \dashrightarrow \mathcal{Y}$ is a dominant rational map, then $T_1(\mathcal{Y})$ holds as well.*

This is explained in [63] and, with an unnecessary hypothesis on resolution of singularities, in [52]. The latter article points out the utility of combining Theorem 8.1 and Lemma 8.2 to prove the following result.

Corollary 8.3. *If k is a finitely generated field and $\mathcal{X}$ is a variety admitting a dominant rational map from a product of curves $\prod_i C_i \dashrightarrow \mathcal{X}$, then T_1 holds for $\mathcal{X}$.*

The corollary implies most of the known cases of T_1 for surfaces (some of which were originally proven by other methods). Namely, we have T_1 for: rational surfaces, unirational surfaces, Fermat surfaces, and more generally hypersurfaces of dimension 2 defined by an equation in 4 monomials (with mild hypotheses). We refer to [56] for these last surfaces, which Shioda calls "Delsarte surfaces". See also [66] and [68].

Of course, for finite k, surfaces which satisfy T_1 lead to curves whose Jacobians satisfy BSD. We will explain in the next subsection how this leads to Jacobians of large rank over global function fields.

Remark 8.4. In a letter to Grothendieck [10] dated March 31, 1964, Serre constructed an example of a surface which is not dominated by a product of curves. In a note to this letter, Serre says that Grothendieck had hoped to prove the Weil conjectures by showing that every variety is dominated by a product of curves,

thus reducing the problem to the known case of curves. We are thankful to Bruno
Kahn for pointing out this letter. See also [52] for other examples of varieties not
dominated by products of curves.

8.3   Other known cases of T_1 for surfaces

Assume that k is a finite field. The other main systematic cases of T_1 for surfaces
over k are for $K3$ surfaces. Namely, Artin and Swinnerton-Dyer showed in [4] that
$\text{Ш}(E)$ is finite for an elliptic curve E over $k(t)$ when the corresponding elliptic
surface $\mathcal{E} \to \mathbb{P}^1_k$ is a K3 surface. Also, Nygaard and Ogus showed in [46] that if $\mathcal{X}$
is a $K3$ surface of finite height and $\text{Char}(k) \geq 5$, then T_1 holds for $\mathcal{X}$.

  In a recent preprint, Lieblich and Maulik show that the Tate conjecture holds
for $K3$ surfaces over finite fields of characteristic $p \geq 5$ if and only if there are only
finitely many isomorphism classes of $K3$s over each finite field of characteristic p.
This is reminiscent of Tate's axiomatization of T_1 for abelian varieties.

  It was conjectured by Artin in [1] that a K3 surface of infinite height has
Néron–Severi group of rank 22, the maximum possible, and so this conjecture
together with [46] would imply the Tate conjecture for $K3$ surfaces over fields
of characteristic ≥ 5. Just as we are finishing these notes, Maulik and Madapusi
Pera have announced (independently) results that would lead to a proof of the
Tate conjecture for $K3$s with a polarization of low degree or of degree prime to
$p = \text{Char}(k)$.

  Finally, we note the "direct" approach: We have *a priori* inequalities

$$\text{Rank NS}\,(\mathcal{X}) \leq \dim_{\mathbb{Q}_\ell} H^2\big(\mathcal{X}, \mathbb{Q}_\ell(1)\big)^{G_k} \leq -\,\text{ord}_{s=1}\,\zeta\,(\mathcal{X}, s)$$

and if equality holds between the ends or between the first two terms, then we
have the full Tate conjectures. It is sometimes possible to find, say by a geometric
construction, enough cycles to force equality. We will give an example of this in
the context of elliptic surfaces in the next section.

9   Ranks of Jacobians

In this final section, we discuss how the preceding results on the BSD and Tate
conjectures can be used to find examples of Jacobians with large analytic rank and
algebraic rank over function fields over finite fields. In the first four subsections we
briefly review results from recent publications, and in the last three subsections
we state a few results that will appear in future publications.

9.1   Analytic ranks in the Kummer tower

In this subsection k is a finite field, $K = k(t)$, and $K_d = k(t^{1/d})$ for each positive
integer d prime to p.

It turns out that roughly half of all abelian varieties defined over K have unbounded analytic rank in the tower K_d. More precisely, those that satisfy a simple parity condition have unbounded rank.

Theorem 9.1. *Let A be an abelian variety over K with Artin conductor $\mathfrak{n}$ (an effective divisor on $\mathbb{P}^1_k$). Write $\mathfrak{n}'$ for the part of $\mathfrak{n}$ prime to the places 0 and ∞ of K. Let $\mathrm{Swan}_v(A)$ be the exponent of the Swan conductor of A at a place v of K. Suppose that the following integer is odd:*

$$\deg(\mathfrak{n}') + \mathrm{Swan}_0(A) + \mathrm{Swan}_\infty(A).$$

Then there is a constant c depending only on A such that if $d = p^f + 1$ then

$$\mathrm{ord}_{s=1} L\big(A/K_d, s\big) \geq \frac{p^f + 1}{o_d(q)} - c,$$

where $o_q(d)$ is the order of q in $(\mathbb{Z}/d\mathbb{Z})^\times$.

This is a slight variant of [68, Theorem 4.7] applied to the representation of G_K on $V_\ell A$. Note that $o_d(q) \leq 2f$, so the rank tends to infinity with f.

9.2 Jacobians of large rank

Theorem 9.1 gives an abundant supply of abelian varieties with large analytic rank. Using Corollary 8.3, we can use this to produce Jacobians of large algebraic rank.

Theorem 9.2. *Let p be an odd prime number, $K = \mathbb{F}_p(t)$, and $K_d = \mathbb{F}_p(t^{1/d})$. Choose a positive integer g such that $p \nmid (2g+2)(2g+1)$ and let X be the hyperelliptic curve over K defined by*

$$y^2 = x^{2g+2} + x^{2g+1} + t.$$

Let J_X be the Jacobian of X, an abelian variety of dimension g over K. Then for all d the BSD conjecture holds for J_X over K_d, and there is a constant depending only on p and g such that for all d of the form $d = p^f + 1$ we have

$$\mathrm{Rank}\, J_X\big(K_d\big) \geq \frac{p^f + 1}{2f} - c.$$

This is one of the main results of [68]. We sketch the key steps in the proof. The analytic rank of J_X is computed using Theorem 9.1. Because X is defined by an equation involving 4 monomials in 3 variables, by Shioda's construction it turns out that the surface $\mathcal{X} \to \mathbb{P}^1$ associated to X/K_d is dominated by a product of curves. Therefore T_2 holds for $\mathcal{X}$ and BSD holds for J_X. In [68] we also checked that J_X is absolutely simple and has trivial $K/\mathbb{F}_q$ trace for all d.

In [68] we gave other examples so that for every p and every $g > 0$ there is an absolutely simple Jacobian of dimension g over $\mathbb{F}_p(t)$ which satisfies BSD and has arbitrarily large rank in the tower of fields K_d.

9.3   Berger's construction

The results of the preceding subsections show that Shioda's 4-monomial construction is a powerful tool for deducing BSD for certain specific Jacobians. However, it has the defect that it is rigid – the property of being dominated by a Fermat surface, or more generally by a product of Fermat curves, does not deform.

In her thesis, Berger gave a more flexible construction which leads to *families* of Jacobians satisfying BSD in each layer of the Kummer tower. We quickly sketch the main idea; see [8] and [71] for more details.

Let k be a general field and fix two smooth, proper, geometrically irreducible curves $\mathcal{C}$ and $\mathcal{D}$ over k. Fix also two separable, non-constant rational functions $f \in k(\mathcal{C})^\times$ and $g \in k(\mathcal{D})^\times$ and form the rational map

$$\mathcal{C} \times \mathcal{D} \dashrightarrow \mathbb{P}^1, \qquad (x, y) \longmapsto f(x)/g(y).$$

Under mild hypotheses on f and g, the generic fiber of this map has a smooth, proper model $X \to \operatorname{Spec} K = k(t)$.

Berger's construction is designed so that the following is true.

Theorem 9.3 (Berger). *With notation as above, for all d prime to $p = \operatorname{Char}(k)$, the surface $\mathcal{X}_d \to \mathbb{P}^1$ associated to X over $K_d = k(t^{1/d}) \cong k(u)$ is dominated by a product of curves. Thus if k is finite, the BSD conjecture holds for J_X over K_d.*

It is convenient to think of the data in Berger's construction as consisting of a discrete part, namely the genera of $\mathcal{C}$ and $\mathcal{D}$ and the combinatorial type of the divisors of f and g, and a continuous part, namely the moduli of the curves and the locations of the zeroes and poles of the functions. From this point of view it is clear that there is enormous flexibility in the choice of data. In [8], Berger used this flexibility to produce families of elliptic curves over $\mathbb{F}_p(t)$ such that for almost all specializations of the parameters to values in $\mathbb{F}_q$, the resulting elliptic curve over $\mathbb{F}_q(t)$ satisfies BSD and has unbounded rank in the tower $K_d = \mathbb{F}_q(t^{1/d})$.

Note that the values of d which give high ranks (via Theorem 9.1) are those of the form $p^f + 1$ and there is a well-known connection between such values and supersingularity of abelian varieties over extensions of $\mathbb{F}_p$. Using Berger's construction and ideas related to those in the next subsection, Occhipinti produced elliptic curves over $\mathbb{F}_p(t)$ which have high rank at *every* layer of the tower $\overline{\mathbb{F}}_p(t^{1/d})$. More precisely, he found curves $E/\mathbb{F}_p(t)$ such that $\operatorname{Rank} E(\overline{\mathbb{F}}_p(t^{1/d})) \geq d$ for all d prime to p. This seems to be a completely new phenomenon. See [47] and the paper based on it for details.

9.4   Geometry of Berger's construction

In [71], we made a study of the geometry of Berger's construction with a view toward more explicit results on ranks and Mordell–Weil groups.

To state one of the main results, let $\mathcal{C}$, $\mathcal{D}$, f, and g be as in Berger's construction. Define covers $\mathcal{C}_d \to \mathcal{C}$ and $\mathcal{D}_d \to \mathcal{D}$ by the equations

$$z^d = f(x), \qquad w^d = g(x).$$

We have an action of μ_d on $\mathcal{C}_d$ and $\mathcal{D}_d$ and their Jacobians. It turns out that the surface $\mathcal{X}_d \to \mathbb{P}^1$ associated to X over $K_d = k(t^{1/d})$ is birational to the quotient of $\mathcal{C}_d \times \mathcal{D}_d$ by the diagonal action of μ_d. This explicit domination by a product of curves makes it possible to compute the rank of J_X over K_d in terms of invariants of $\mathcal{C}_d$ and $\mathcal{D}_d$. More precisely,

Theorem 9.4. *Suppose that k is algebraically closed. Then there exist non-negative integers c_1, c_2, and N such that for all d relatively prime to N we have*

$$\operatorname{Rank} \operatorname{MW}\left(J_X/K_d\right) = \operatorname{Rank} \operatorname{Hom}_{k-av}\left(J_{\mathcal{C}_d}, J_{\mathcal{D}_d}\right)^{\mu_d} - dc_1 + c_2.$$

Here the superscript indicates those homomorphisms which commute with the action of μ_d on the two Jacobians.

The constants c_1 and c_2 are given explicitly in terms of the input data, and the integer N is there just to make the statement simple – there is a formula for the rank for any value of d. See [71, Theorem 6.4] for the details.

The theorem relates ranks of abelian varieties over a function field K_d to homomorphisms of abelian varieties over the constant field k. The latter is sometimes more tractable. For example, when $k = \overline{\mathbb{F}}_p$, the homomorphism groups can be made explicit via Honda–Tate theory.

Here is another example where the endomorphism side of the formula in Theorem 9.4 is tractable: Using it and Zarhin's results on endomorphism rings of superelliptic Jacobians, we gave examples in [72] of Jacobians over function fields of characteristic zero with bounded ranks in certain towers. More precisely:

Theorem 9.5. *Let g_X be an integer ≥ 2 and let X be the smooth, proper curve of genus g_X over $\mathbb{Q}(t)$ with affine plane model*

$$ty^2 = x^{2g_X+1} - x + t - 1.$$

Then for every prime number p and every integer $n \geq 0$ we have

$$\operatorname{Rank} J_X\left(\overline{\mathbb{Q}}(t^{1/p^n})\right) = 2g_X.$$

Another dividend of the geometric analysis is that it can sometimes be used to find explicit points in high rank situations. (In principle this is also true in the context of Shioda's 4-monomial construction, but I know of few cases where it has been successfully carried out, a notable exception being [57], which relates to an isotrivial elliptic curve.)

The following example is [71, Theorem 8.1]. The points which appear here are related to the graphs of Frobenius endomorphisms of the curves $\mathcal{C}_d$ and $\mathcal{D}_d$ for a suitable choice of data in Berger's construction.

Theorem 9.6. *Fix an odd prime number p, let $k = \overline{\mathbb{F}}_p$, and let $K = k(t)$. Let X be the elliptic curve over K defined by*

$$y^2 + xy + ty = x^3 + tx^2.$$

For each d prime to p, let $K_d = k(t^{1/d}) \cong k(u)$. We write ζ_d for a fixed primitive dth root of unity in k. Let $d = p^n + 1$, let $q = p^n$, and let

$$P(u) = \left(\frac{u^q(u^q - u)}{(1 + 4u)^q}, \; \frac{u^{2q}(1 + 2u + 2u^q)}{2(1 + 4u)^{(3q-1)/2}} - \frac{u^{2q}}{2(1 + 4u)^{q-1}} \right).$$

Then the points $P_i = P(\zeta_d^i u)$ for $i = 0, \ldots, d-1$ lie in $X(K_d)$ and they generate a finite index subgroup of $X(K_d)$, which has rank $d - 2$. The relations among them are that $\sum_{i=0}^{d-1} P_i$ and $\sum_{i=0}^{d-1}(-1)^i P_i$ are torsion.

The main result of [71] also has implications for elliptic curves over $\mathbb{C}(t)$. We refer the reader to the last section of that paper for details.

9.5   Artin–Schreier analogues

The results of the preceding four subsections are all related to the arithmetic of Jacobians in the Kummer tower $K_d = k(t^{1/d})$. It turns out that all the results – high analytic ranks, a Berger-style construction of Jacobians satisfying BSD, a rank formula, and explicit points – have analogues for the Artin–Schreier tower, that is, for extensions $k(u)/k(t)$ where $u^q - u = f(t)$. See a forthcoming paper with Rachel Pries [49] for more details.

9.6   Explicit points on the Legendre curve

As we mentioned before, if one can write down enough points to fill out a finite index subgroup of the Mordell–Weil group of an abelian variety, then the full BSD conjecture follows. This plays out in a very satisfying way for the Legendre curve.

More precisely, let p be an odd prime, let $K = \mathbb{F}_p(t)$, and let

$$K_d = \mathbb{F}_p(\mu_d)(t^{1/d}) \cong \mathbb{F}_p(\mu_d)(u).$$

Consider the Legendre curve

$$E: \qquad y^2 = x(x + 1)(x + t)$$

over K. (The signs are not the traditional ones, and E is a twist of the usual Legendre curve. It turns out that the $+$ signs are more convenient, somewhat analogously to the situation with signs of Gauss sums.)

If we take d of the form $d = p^f + 1$, then there is an obvious point on $E(K_d)$, namely $P(u) = (u, u(u+1)^{d/2})$. Translating this point by $\mathrm{Gal}(K_d/K)$ yields more points: $P_i = P(\zeta_d^i u)$ where $\zeta_d \in \mathbb{F}_p(\mu_d)$ is a primitive dth root of unity and $i = 0, \ldots, d-1$.

In [69] we give an elementary proof of the following.

Theorem 9.7. *The points P_i generate a subgroup of $E(K_d)$ of finite index and rank $d-2$. The relations among them are that $\sum_{i=0}^{d-1} P_i$ and $\sum_{i=0}^{d-1}(-1)^i P_i$ are torsion.*

It is easy to see that the L-function of E/K_d has order of zero at $s=1$ a priori bounded by $d-2$, so we have the equality $\operatorname{ord}_{s=1} L(E/K_d, s) = \operatorname{Rank} E(K_d)$, i.e., the *BSD* conjecture. Thus we also have the refined *BSD* conjecture. Examining this leads to a beautiful "analytic class number formula": Let V_d be the subgroup of $E(K_d)$ generated by the P_i and $E(K_d)_{\text{tor}}$. (The latter is easily seen to be of order 8.) Then we have

$$\left[E(K_d) : V_d\right]^2 = \left|\mathrm{Ш}(E/K_d)\right|$$

and these numbers are powers of p.

It then becomes a very interesting question to make the displayed quantity more explicit. It turns out that E is closely related to the X of the previous section, and this brings the geometry of "domination by a product of curves" into play. Using that, we are able to give a complete description of $\mathrm{Ш}(E/K_d)$ and $E(K_d)/V_d$ as modules over the group ring $\mathbb{Z}_p[\operatorname{Gal}(K_d/K)]$. These results will appear in a paper in preparation.

It also turns out that we can control the rank of $E(K_d)$ for general d, not just those of the form $p^f + 1$. Surprisingly, we find that there is high rank for many other values of d, and in a precise quantitative sense, among the values of d where $E(K_d)$ has large rank, those prime to all $p^f + 1$ are more numerous than those not prime to $p^f + 1$. The results of this paragraph will appear in publications of the author and collaborators, including [48] and works cited there.

9.7 Characteristic 0

Despite significant effort, I have not been able to exploit Berger's construction to produce elliptic curves (or Jacobians of any fixed dimension) over $\mathbb{C}(t)$ with unbounded rank. Roughly speaking, efforts to increase the $\operatorname{Hom}(\ldots)^{\mu_d}$ term, say by forcing symmetry, seem also to increase the value of c_1. (Although one benefit of this effort came from examination of an example that led to the explicit points in Theorem 9.6 above.)

At the workshop I explained a heuristic which suggests that ranks might be bounded in a Kummer tower in characteristic zero. Theorem 9.5 above is an example in this direction. I do not know how to prove anything that strong, but we do have the following negative result.

Recall that an elliptic surface $\pi \colon \mathcal{E} \to \mathcal{C}$ has a height, which can be defined as the degree of the invertible sheaf $(R^1\pi_*\mathcal{O}_\mathcal{E})^{-1}$. When $\mathcal{C} = \mathbb{P}^1$, the height is ≥ 3 if and only if the Kodaira dimension of $\mathcal{E}$ is 1. Recall also that there is a reasonable moduli space for elliptic surfaces of height d over $\mathbb{C}$ and it has dimension $10d - 2$.

In what follows, "very general" means "belongs to the complement of countably many divisors in the moduli space".

Theorem 9.8. *Suppose that $E/\mathbb{C}(t)$ is the generic fiber of $\mathcal{E} \to \mathbb{P}^1$ over $\mathbb{C}$ where $\mathcal{E}$ is a very general elliptic surface of height $d \geq 3$. Then for every finite extension $L/\mathbb{C}(t)$ where L is a rational field (i.e., $L \cong \mathbb{C}(u)$) we have $E(L) = 0$.*

The theorem is proven by controlling the collection of rational curves on $\mathcal{E}$ and is a significant strengthening of a Noether–Lefschetz type result (cf. [12]) according to which the Néron–Severi group of a very general elliptic surface is generated by the zero section and a fiber. More details on Theorem 9.8 and the heuristic that suggested it will appear in a paper currently in preparation.

Bibliography

[1] M. Artin. Supersingular $K3$ surfaces. *Ann. Sci. École Norm. Sup.* (4), 7:543–567 (1975), 1974.

[2] M. Artin. Lipman's proof of resolution of singularities for surfaces. In *Arithmetic Geometry (Storrs, Conn., 1984)*, pages 267–287. Springer, New York, 1986.

[3] M. Artin. Néron models. In *Arithmetic Geometry (Storrs, Conn., 1984)*, pages 213–230. Springer, New York, 1986.

[4] M. Artin and H.P.F. Swinnerton-Dyer. The Shafarevich–Tate conjecture for pencils of elliptic curves on $K3$ surfaces. *Invent. Math.*, 20:249–266, 1973.

[5] M. Artin and G. Winters. Degenerate fibres and stable reduction of curves. *Topology*, 10:373–383, 1971.

[6] L. Bădescu. *Algebraic Surfaces*. Universitext. Springer-Verlag, New York, 2001. Translated from the 1981 Romanian original by Vladimir Maşek and revised by the author.

[7] W. Bauer. On the conjecture of Birch and Swinnerton-Dyer for abelian varieties over function fields in characteristic $p > 0$. *Invent. Math.*, 108:263–287, 1992.

[8] L. Berger. Towers of surfaces dominated by products of curves and elliptic curves of large rank over function fields. *J. Number Theory*, 128:3013–3030, 2008.

[9] S. Bosch, W. Lütkebohmert, and M. Raynaud. *Néron Models*, vol. 21 of *Ergebnisse der Mathematik und ihrer Grenzgebiete (3) [Results in Mathematics and Related Areas (3)]*. Springer-Verlag, Berlin, 1990.

[10] P. Colmez and J.-P. Serre, eds. *Correspondance Grothendieck–Serre*. Documents Mathématiques (Paris) [Mathematical Documents (Paris)], 2. Société Mathématique de France, Paris, 2001.

[11] B. Conrad. Chow's K/k-image and K/k-trace, and the Lang–Néron theorem. *Enseign. Math.* (2), 52:37–108, 2006.

[12] D.A. Cox. The Noether–Lefschetz locus of regular elliptic surfaces with section and $p_g \geq 2$. *Amer. J. Math.*, 112:289–329, 1990.

[13] P. Deligne. La conjecture de Weil, II. *Inst. Hautes Études Sci. Publ. Math.*, (52):137–252, 1980.

[14] G. Faltings. Finiteness theorems for abelian varieties over number fields. In *Arithmetic Geometry (Storrs, Conn., 1984)*, pages 9–27. Springer, New York, 1986. Translated from the German original [Invent. Math. **73** (1983), no. 3, 349–366; ibid. **75** (1984), no. 2, 381; MR 85g:11026ab] by Edward Shipz.

[15] M. Fontana. Sur le plongement projectif des surfaces définies sur un corps de base fini. *Compositio Math.*, 31:1–22, 1975.

[16] W.J. Gordon. Linking the conjectures of Artin–Tate and Birch–Swinnerton-Dyer. *Compositio Math.*, 38:163–199, 1979.

[17] B.H. Gross. Local heights on curves. In *Arithmetic Geometry (Storrs, Conn., 1984)*, pages 327–339. Springer, New York, 1986.

[18] A. Grothendieck. *Fondements de la géométrie algébrique. [Extraits du Séminaire Bourbaki, 1957–1962.]*. Secrétariat mathématique, Paris, 1962.

[19] A. Grothendieck. Le groupe de Brauer, III. Exemples et compléments. In *Dix exposés sur la cohomologie des schémas*, pages 88–188. North-Holland, Amsterdam, 1968.

[20] R. Hartshorne. *Algebraic Geometry*. Graduate Texts in Mathematics, 52. Springer-Verlag, New York, 1977.

[21] M. Hindry and A. Pacheco. Sur le rang des jacobiennes sur un corps de fonctions. *Bull. Soc. Math. France*, 133:275–295, 2005.

[22] L. Illusie. Complexe de de Rham–Witt. In *Journées de Géométrie Algébrique de Rennes (Rennes, 1978), Vol. I*, volume 63 of *Astérisque*, pages 83–112. Soc. Math. France, Paris, 1979.

[23] L. Illusie. Finiteness, duality, and Künneth theorems in the cohomology of the de Rham–Witt complex. In *Algebraic Geometry (Tokyo/Kyoto, 1982)*, volume 1016 of *Lecture Notes in Math.*, pages 20–72. Springer, Berlin, 1983.

[24] K. Kato and F. Trihan. On the conjectures of Birch and Swinnerton-Dyer in characteristic $p > 0$. *Invent. Math.*, 153:537–592, 2003.

[25] N.M. Katz. *Moments, Monodromy, and Perversity: a Diophantine Perspective*, volume 159 of *Annals of Mathematics Studies*. Princeton University Press, Princeton, NJ, 2005.

[26] S.L. Kleiman. The Picard scheme. In *Fundamental Algebraic Geometry*, volume 123 of *Math. Surveys Monogr.*, pages 235–321. Amer. Math. Soc., Providence, RI, 2005.

[27] K. Kramer. Two-descent for elliptic curves in characteristic two. *Trans. Amer. Math. Soc.*, 232:279–295, 1977.

[28] S. Lichtenbaum. Curves over discrete valuation rings. *Amer. J. Math.*, 90:380–405, 1968.

[29] S. Lichtenbaum. Duality theorems for curves over p-adic fields. *Invent. Math.*, 7:120–136, 1969.

[30] S. Lichtenbaum. Zeta functions of varieties over finite fields at $s = 1$. In *Arithmetic and Geometry, Vol. I*, volume 35 of *Progr. Math.*, pages 173–194. Birkhäuser Boston, Boston, MA, 1983.

[31] C. Liedtke. A note on non-reduced Picard schemes. *J. Pure Appl. Algebra*, 213:737–741, 2009.

[32] J. Lipman. Desingularization of two-dimensional schemes. *Ann. of Math. (2)*, 107:151–207, 1978.

[33] Q. Liu. *Algebraic Geometry and Arithmetic Curves*, volume 6 of *Oxford Graduate Texts in Mathematics*. Oxford University Press, Oxford, 2002. Translated from the French by Reinie Erné, Oxford Science Publications.

[34] Q. Liu, D. Lorenzini, and M. Raynaud. Néron models, Lie algebras, and reduction of curves of genus one. *Invent. Math.*, 157:455–518, 2004.

[35] Q. Liu, D. Lorenzini, and M. Raynaud. On the Brauer group of a surface. *Invent. Math.*, 159:673–676, 2005.

[36] Ju.I. Manin. Cyclotomic fields and modular curves. *Uspekhi Mat. Nauk*, 26:7–71, 1971.

[37] J.S. Milne. Elements of order p in the Tate–Šafarevič group. *Bull. London Math. Soc.*, 2:293–296, 1970.

[38] J.S. Milne. On a conjecture of Artin and Tate. *Ann. of Math. (2)*, 102:517–533, 1975.

[39] J.S. Milne. *Étale Cohomology*, volume 33 of *Princeton Mathematical Series*. Princeton University Press, Princeton, NJ, 1980.

[40] J.S. Milne. *Arithmetic Duality Theorems*, volume 1 of *Perspectives in Mathematics*. Academic Press Inc., Boston, MA, 1986.

[41] J.S. Milne. Jacobian varieties. In *Arithmetic Geometry (Storrs, Conn., 1984)*, pages 167–212. Springer, New York, 1986.

[42] J.S. Milne. Values of zeta functions of varieties over finite fields. *Amer. J. Math.*, 108:297–360, 1986.

[43] L. Moret-Bailly. Pinceaux de variétés abéliennes. *Astérisque*, 129:266, 1985.

[44] D. Mumford. *Lectures on Curves on an Algebraic Surface*. With a section by G.M. Bergman. Annals of Mathematics Studies, No. 59. Princeton University Press, Princeton, N.J., 1966.

[45] D. Mumford and T. Oda. *Algebraic Geometry II*. In preparation.

[46] N. Nygaard and A. Ogus. Tate's conjecture for $K3$ surfaces of finite height. *Ann. of Math. (2)*, 122:461–507, 1985.

[47] T. Occhipinti. *Mordell–Weil groups of large rank in towers*. ProQuest LLC, Ann Arbor, MI, 2010. Ph.D. Thesis, The University of Arizona.

[48] C. Pomerance and D. Ulmer. *On balanced subgroups of the multiplicative group*, in *Number theory and related fields*, volume 43 of Springer Proc. Math. Stat., pages 253–270, Springer, New York, 2013.

[49] R. Pries and D. Ulmer. Arithmetic of abelian varieties in Artin–Schreier extensions, preprint, 2013, arxiv:1305.5247

[50] M. Raynaud. Spécialisation du foncteur de Picard. *Inst. Hautes Études Sci. Publ. Math.*, 38:27–76, 1970.

[51] P. Schneider. Zur Vermutung von Birch und Swinnerton-Dyer über globalen Funktionenkörpern. *Math. Ann.*, 260:495–510, 1982.

[52] C. Schoen. Varieties dominated by product varieties. *Internat. J. Math.*, 7:541–571, 1996.

[53] J.-P. Serre. Sur la topologie des variétés algébriques en caractéristique *p*. In *Symposium internacional de topología algebraica, International Symposium on Algebraic Topology*, pages 24–53. Universidad Nacional Autónoma de México and UNESCO, Mexico City, 1958.

[54] J.-P. Serre. *Local Fields*, volume 67 of *Graduate Texts in Mathematics*. Springer-Verlag, New York, 1979. Translated from the French by Marvin Jay Greenberg.

[55] *Théorie des intersections et théorème de Riemann-Roch*. Lecture Notes in Mathematics, Vol. 225. Springer-Verlag, Berlin, 1971. Séminaire de Géométrie Algébrique du Bois-Marie 1966–1967 (SGA 6). Dirigé par P. Berthelot, A. Grothendieck et L. Illusie. Avec la collaboration de D. Ferrand, J.P. Jouanolou, O. Jussila, S. Kleiman, M. Raynaud et J.-P. Serre.

[56] T. Shioda. An explicit algorithm for computing the Picard number of certain algebraic surfaces. *Amer. J. Math.*, 108:415–432, 1986.

[57] T. Shioda. Mordell–Weil lattices and sphere packings. *Amer. J. Math.*, 113:931–948, 1991.

[58] T. Shioda. Mordell--Weil lattices for higher genus fibration over a curve. In *New Trends in Algebraic Geometry (Warwick, 1996)*, volume 264 of *London Math. Soc. Lecture Note Ser.*, pages 359–373. Cambridge Univ. Press, Cambridge, 1999.

[59] M. Szydlo. Elliptic fibers over non-perfect residue fields. *J. Number Theory*, 104:75–99, 2004.

[60] J.T. Tate. Algebraic cycles and poles of zeta functions. In *Arithmetical Algebraic Geometry (Proc. Conf. Purdue Univ., 1963)*, pages 93–110. Harper & Row, New York, 1965.

[61] J.T. Tate. Endomorphisms of abelian varieties over finite fields. *Invent. Math.*, 2:134–144, 1966.

[62] J.T. Tate. On the conjectures of Birch and Swinnerton-Dyer and a geometric analog. In *Séminaire Bourbaki, Vol. 9*, Exp. No. 306, pages 415–440. Soc. Math. France, Paris, 1966.

[63] J.T. Tate. Conjectures on algebraic cycles in ℓ-adic cohomology. In *Motives (Seattle, WA, 1991)*, volume 55 of *Proc. Sympos. Pure Math.*, pages 71–83. Amer. Math. Soc., Providence, RI, 1994.

[64] D. Ulmer. p-descent in characteristic p. *Duke Math. J.*, 62:237–265, 1991.

[65] D. Ulmer. On the Fourier coefficients of modular forms, II. *Math. Ann.*, 304:363–422, 1996.

[66] D. Ulmer. Elliptic curves with large rank over function fields. *Ann. of Math.* (2), 155:295–315, 2002.

[67] D. Ulmer. Geometric non-vanishing. *Invent. Math.*, 159:133–186, 2005.

[68] D. Ulmer. L-functions with large analytic rank and abelian varieties with large algebraic rank over function fields. *Invent. Math.*, 167:379–408, 2007.

[69] D. Ulmer. Explicit points on the Legendre curve, *J. Number Theory*, 136:165–194, 2014.

[70] D. Ulmer. Elliptic curves over function fields. In *Arithmetic of L-functions (Park City, UT,* 2009), volume 18 of *IAS/Park City Math. Ser.*, pages 211–280. Amer. Math. Soc., Providence, RI, 2011.

[71] D. Ulmer. On Mordell–Weil groups of Jacobians over function fields, *J. Inst. Math. Jussieu*, 12:1–29, 2013.

[72] D. Ulmer and Y.G. Zarhin. Ranks of Jacobians in towers of function fields. *Math. Res. Lett.*, 17:637–645, 2010.

[73] Ju.G. Zarhin. Abelian varieties in characteristic p. *Mat. Zametki*, 19:393–400, 1976.

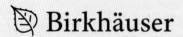

birkhauser-science.com

Advanced Courses in Mathematics – CRM Barcelona (ACM)

Edited by

Carles Casacuberta, Universitat de Barcelona, Spain

Since 1995 the Centre de Recerca Matemàtica (CRM) has organised a number of Advanced Courses at the post-doctoral or advanced graduate level on forefront research topics in Barcelona. The books in this series contain revised and expanded versions of the material presented by the authors in their lectures.

■ Asaoka, M. / El Kacimi Alaoui, A. / Hurder, S. / Richardson, K., Foliations: Dynamics, Geometry and Topology (2014).
ISBN 978-3-0348-0870-5

This book is an introduction to several active research topics in foliation theory and its connections with other areas. It includes expository lectures showing the diversity of ideas and methods arising and used in the study of foliations.

The lectures by A. El Kacimi Alaoui offer an introduction to foliation theory, with emphasis on examples and transverse structures. S. Hurder's lectures apply ideas from smooth dynamical systems to develop useful concepts in the study of foliations, like limit sets and cycles for leaves, leafwise geodesic flow, transverse exponents, stable manifolds, Pesin theory, and hyperbolic, parabolic, and elliptic types of foliations, all of them illustrated with examples. The lectures by M. Asaoka are devoted to the computation of the leafwise cohomology of orbit foliations given by locally free actions of certain Lie groups, and its application to the description of the deformation of those actions. In the lectures by K. Richardson, he studies the geometric and analytic properties of transverse Dirac operators for Riemannian foliations and compact Lie group actions, and explains a recently proved index formula.

Besides students and researchers of foliation theory, this book will appeal to mathematicians interested in the applications to foliations of subjects like topology of manifolds, dynamics, cohomology or global analysis.

■ Alesker, S. / Fu, J.H.G., Integral Geometry and Valuations (2014).
ISBN 978-3-0348-0873-6

This book contains a revised and expanded version of the lecture notes of the Advanced Course on Integral Geometry and Valuation Theory held at the CRM in September 2010. The first part, by Semyon Alesker, provides an introduction to the theory of convex valuations with emphasis on recent developments. In particular, it presents the new structures

on the space of valuations discovered after Alesker's irreducibility theorem. The newly developed theory of valuations on manifolds is also described.

In the second part, Joseph H. G. Fu gives a modern introduction to integral geometry in the sense of Blaschke and Santaló. The approach is new and based on the notions and tools presented in the first part. This original viewpoint not only enlightens the classical integral geometry of euclidean space, but it also allows the computation of kinematic formulas in other geometries, such as hermitian spaces. The book will appeal to graduate students and interested researchers from related fields including convex, stochastic, and differential geometry.

■ Cruz-Uribe, D. / Fiorenza, A. / Ruzhansky, M. / Wirth, J., Variable Lebesgue Spaces and Hyperbolic Systems (2014).
ISBN 978-3-0348-0839-2

This book targets graduate students and researchers who want to learn about Lebesgue spaces and solutions to hyperbolic equations. It is divided into two parts.

Part 1 provides an introduction to the theory of variable Lebesgue spaces: Banach function spaces like the classical Lebesgue spaces but with the constant exponent replaced by an exponent function. Part 2 gives an overview of the asymptotic properties of solutions to hyperbolic equations and systems with time-dependent coefficients. A number of examples is considered and the sharpness of results is discussed. An exemplary treatment of dissipative terms shows how effective lower order terms can change asymptotic properties and thus complements the exposition.

■ Berger, L. / Böckle, G. / Dembélé, L. / Dimitrov, M. / Dokchitser, T. / Voight, J., Elliptic Curves, Hilbert Modular Forms and Galois Deformations (2013).
ISBN 978-3-0348-0617-6

■ Cominetti, R. / Facchinei, F. / Lasserre, J.B., Modern Optimization Modelling Techniques (2012).
ISBN 978-3-0348-0290-1